Optimization
of Solid-Phase
Combinatorial Synthesis

Optimization of Solid-Phase Combinatorial Synthesis

edited by

Bing Yan
ChemRx Advanced Technologies
South San Francisco, California

Anthony W. Czarnik
Sensors for Medicine and Science, Inc.
Germantown, Maryland

CRC Press
Taylor & Francis Group
Boca Raton London New York

CRC Press is an imprint of the
Taylor & Francis Group, an **informa** business

CRC Press
Taylor & Francis Group
6000 Broken Sound Parkway NW, Suite 300
Boca Raton, FL 33487-2742

First issued in paperback 2019

© 2002 by Taylor & Francis Group, LLC
CRC Press is an imprint of Taylor & Francis Group, an Informa business

No claim to original U.S. Government works

ISBN-13: 978-0-8247-0654-8 (hbk)
ISBN-13: 978-0-367-39653-4 (pbk)

**Visit the Taylor & Francis Web site at
http://www.taylorandfrancis.com**

**and the CRC Press Web site at
http://www.crcpress.com**

Preface

In one sense, combinatorial libraries are like extracts from natural sources. Each provides valuable sources of compound diversity from which useful substances may be discovered. However, in several important ways they are not alike. The theoretical diversity accessible via combinatorial synthesis is far wider than that accessible from natural products. The components of a combinatorial library can be of high purity without fractionation. Also, critically, the chemical structures of combinatorial library members do not need to be elucidated. The enormous benefits of synthesized libraries are realized only when synthetic steps actually occur as expected and in high yield. The bad news is that this requires substantial development effort prior to library synthesis. The good news is that this is a controllable variable, one that chemists are equipped to solve.

The need for quality in compound libraries—known structures of high purity and yield—is more urgent today than ever. Even 5–10 years ago, popular sentiment had it that only compounds made for lead optimization and structure–activity relationship (SAR) studies required higher purity; i.e., the purity of lead discovery libraries could be compromised. Now it has become evident that the purity and quantity of compounds made for lead discovery screening have become increasingly important because of the false-positive issue and the enormous amount of time and resources needed for resynthesis and reconfirmation.

It is not very difficult to make high-quality libraries using easy chemistry. The real challenge is to make high-quality libraries using sophisticated chemistry. Although methods for monitoring and optimizing solution-phase reactions are well developed, similar methods for solid-phase organic reactions are still in the early stages of development. There are two approaches to higher-quality lead discovery libraries: massive-scale purification and intensive chemistry optimization. Current purification methods are generally of low throughput and expensive. Therefore, optimization of combinatorial synthesis has become indispensable.

A recent survey shows that solid-phase synthesis continues to hold a dominant position (80% during 1992–1995, 50% in 1996, 67% in 1997 and 1998, and 80% in 1999) in combinatorial synthesis as more and more areas of chemistry are redeveloped in this medium (1). Although solid-phase chemistry continues to play an important role, the distinction between solid-phase and solution-phase combinatorial synthesis is blurred. Some solid-phase syntheses are better performed when one or more solution-phase reaction steps are incorporated. On the other hand, solution-phase synthesis often requires polymer-bound reagents or scavenger resins for effective reaction and product cleanup. This book focuses mainly on solid-phase synthesis optimization issues with solution-phase issues in mind and does not make a clear-cut distinction.

Chapters 1–3 present three approaches to combinatorial synthesis. In Chapter 1, Boldi and Johnson summarize the massively parallel synthesis using a 96-well plate to make ~5000 compounds per library and 20–30 mg per well. They describe a library development system for optimizing synthesis protocols.

In Chapter 2, Dolle describes split-and-pool synthesis to make 0.5–1 million compound libraries with discrete compounds on the individual bead level.

In Chapter 3, Li reports on split-and-pool synthesis to make hundreds to thousands of discrete compounds in multimilligram quantities using IRORI MicroKan™ technology.

In combination, all these authors demonstrate that syntheses using different methods often face common optimization issues. Their systematic approaches to these issues are presented.

In Chapter 4, Hermkens, Hamersma, and de Man recount some key experiences in the translation of solution-phase chemistry to solid phase. It is important to use these as guidelines, not rules. A rough optimization stage is introduced before a full-blown optimization effort in their four-step approach. Their experience is documented by an unsuccessful oxazole synthesis, a successful Stille reaction on a pyrimidine core, and a successful diketopiperazine ring formation.

Solid-phase organic synthesis has benefited immensely from solid-phase peptide synthesis. By providing an overview of solid-phase peptide synthesis in Chapter 5, Van Den Nest and Albericio outline some resin selection criteria and review a variety of resin supports and resin characterization methods.

In Chapter 6, Hochlowski reviews the applications of fluorine-19 nuclear magnetic resonance (NMR) in combinatorial synthesis, including the quantitation of standards (internal, external, and mixed-phase) for calibration, kinetic measurements of compound cleavage from resin, and bead encoding.

Through the use of many real-life examples in Chapter 7, Yan, Dunn, Tang,

Coppola, and Vlattas document the use of single bead infrared (IR) as a versatile reaction monitoring tool for synthesis optimization.

In Chapter 8, Sefler and Minick describe the combined application of resin NMR and attenuated total reflectance Fourier transform infrared (ATR FTIR) to improve the quality of libraries.

In Chapter 9, Yan, Kshiragar, Fang, Fantauzzi, and Pan describe an analytical tool kit consisting of a range of analytical methods for conducting feasibility, validation, and rehearsal of library synthesis.

In Chapter 10, Congreve, Kay, and Scicinski cite many examples of reaction monitoring by product derivatization. Mass spectrometry has been used to control the quality of solid-bound compounds by using dural linker analytical constructs.

Shah, Detre, Raillard, and Fitch report their recent work on compound-independent quantitation in Chapter 11. Analyzing the identity and purity of the compound released from a single bead can be done by liquid chromatography/mass spectrometry/ultraviolet spectrometry (LC/MS/UV). The quantitation of compounds from a single bead is accomplished by quantifying the bulk-released compounds using chemiluminescent nitrogen detection (CLND) and estimating the average concentration and synthesis yield.

For molecules undergoing reaction while linked to a solid support, the support literally serves as an unusual kind of solvent. In Chapter 12, Gerritz discusses topics from the application of the classic Hammett relationship to predictions of solid-state reactivity.

In Chapter 13, Fenniri and McFadden summarize the state of the art in multispectral imaging of resin-supported libraries and describe the development of new screening methods based on multispectral imaging.

Yang, Chen, and Chu discuss applying microwave technology to increase the reaction rate without altering the synthesis yield in Chapter 14. They also describe the analysis of stereoisomer formation during combinatorial synthesis using capillary electrophoresis.

Scavenger resins, polymer-bound reagents, catch-and-release methods, and resin capture have become increasingly important in the production of combinatorial libraries using solution-phase chemistry. Such strategies allow organic synthesis using simple, parallel product purification by filtration, avoiding silica-gel chromatography and/or extraction. Organ and Mallik review a wide range of solid-supported reagents and their application in synthesis in Chapter 15. (By including this chapter, we emphasize that, with the sole purpose of improving library quality, we have not limited the book to only solid-phase or solution-phase synthesis methods).

Combinatorial chemistry is a new and valuable discipline. Today, no one involved in discovery can argue (with a straight face) that large screening libraries are not important. In this book, we have tried to make the case that because chemists *can* control the quality of synthetic libraries, we should. This is important for discovery efforts, but equally important as a step toward recognition of the pivotal role combinatorial chemists play in the discovery process today.

Bing Yan
Anthony W. Czarnik

REFERENCE

1. RE Dolle. Comprehensive survey of combinatorial library synthesis: 1999. J Comb Chem 2:383–433, 2000.

Contents

Contributors

Fernando Albericio Department of Organic Chemistry, University of Barcelona, Barcelona, Spain

Armen M. Boldi Department of Development, ChemRx Advanced Technologies, Inc., South San Francisco, California

Shui-Tein Chen Institute of Biological Chemistry, Academia Sinica, Taipei, Taiwan, Republic of China

Yen-Ho Chu Department of Chemistry, National Chung Cheng University, Chia-Yi, Taiwan, Republic of China

Miles S. Congreve Cambridge University Chemical Laboratory, GlaxoSmithKline, Cambridge, England

Gary M. Coppola Novartis Pharmaceuticals Corporation, Summit, New Jersey

Adrianus P. A. de Man Lead Discovery Unit, N.V. Organon, Oss, The Netherlands

George Detre Department of Analytical Chemistry, Affymax Research Institute, Palo Alto, California

Roland E. Dolle Department of Chemistry, Pharmacopeia, Inc., Cranbury, New Jersey

Robert F. Dunn   Novartis Pharmaceuticals Corporation, Summit, New Jersey

Liling Fang   ChemRx Advanced Technologies, Inc., South San Francisco, California

Pascal Fantauzzi   ChemRx Advanced Technologies, Inc., South San Francisco, California

Hicham Fenniri   H. C. Brown Laboratory of Chemistry, Purdue University, West Lafayette, Indiana

William L. Fitch   Department of Analytical Chemistry, Affymax Research Institute, Palo Alto, California

Samuel W. Gerritz   Department of Lead Synthesis, Bristol-Myers Squibb, Wallingford, Connecticut

Hans Hamersma   Department of Molecular Design and Informatics, N.V. Organon, Oss, The Netherlands

Pedro H. H. Hermkens   Lead Discovery Unit, N.V. Organon, Oss, The Netherlands

Jill Hochlowski   Medicinal Chemistry Technologies, Abbott Laboratories, North Chicago, Illinois

Charles R. Johnson   Department of Development, ChemRx Advanced Technologies, Inc., South San Francisco, California

Corinne Kay   Cambridge University Chemical Laboratory, GlaxoSmithKline, Cambridge, England

Tushar Kshiragar   ChemRx Advanced Technologies, Inc., South San Francisco, California

Rongshi Li   Department of Chemistry, ChemBridge Research Laboratories, San Diego, California

Debasis Mallik   Department of Chemistry, York University, Toronto, Ontario, Canada

Ryan M. McFadden H. C. Brown Laboratory of Chemistry, Purdue University, West Lafayette, Indiana

Douglas J. Minick Department of Computational, Analytical, and Structural Sciences, GlaxoSmithKline, Research Triangle Park, North Carolina

Michael G. Organ Department of Chemistry, York University, Toronto, Ontario, Canada

Jianmin Pan ChemRx Advanced Technologies, Inc., South San Francisco, California

Stephen Raillard Department of Process Chemistry, Affymax Research Institute, Palo Alto, California

Jan Scicinski Cambridge University Chemical Laboratory, GlaxoSmithKline, Cambridge, England

Andrea M. Sefler Department of Computational, Analytical, and Structural Sciences, GlaxoSmithKline, Research Triangle Park, North Carolina

Nikhil Shah Department of Analytical Chemistry, Affymax Research Institute, Palo Alto, California

Qing Tang Novartis Pharmaceuticals Corporation, Summit, New Jersey

Wim Van Den Nest Organic Chemistry Department, DiverDrugs SL, L'Hospitalet de Llobregat, Barcelona, Spain

Isidoros Vlattas Novartis Pharmaceuticals Corporation, Summit, New Jersey

Bing Yan Department of Analytical Sciences, ChemRx Advanced Technologies, Inc., South San Francisco, California

Joy Jing-Yi Yang Department of Chemistry, National Chung Cheng University, Chia-Yi, Taiwan, Republic of China

1

A System for the Development and Production of Small-Molecule Libraries

Armen M. Boldi and Charles R. Johnson
ChemRx Advanced Technologies, Inc., South San Francisco, California

I. INTRODUCTION

New combinatorial synthetic methods, both solid-phase and solution-phase, now allow the expeditious preparation of organic compound arrays (1–11). In the context of an ongoing development program, ChemRx Advanced Technologies has devised a number of solution-phase and solid-phase spatially separated small-molecule libraries for screening against a broad range of biological targets. Adapting carbon–carbon and carbon–heteroatom bond-forming reactions for array synthesis has been our primary focus.*

The use of solid phase has been a central feature of combinatorial chemistry. Peptide synthesis on polystyrene supports produced the field of combinatorial chemistry. For peptides and DNA synthesis, a large variety of solid supports from controlled pore glass to cellulose in cotton have been used (12,13). Small-molecule combinatorial chemistry has focused almost entirely on polystyrene-based resins (14,15). We have used polystyrene almost exclusively to generate spatially separated diversity libraries of 5000 compounds with greater than 50 µmol per well.

---

* Our current library development and production systems were initially established in 1996 in Arris Pharmaceutical, Inc. in a program to synthesize diverse screening libraries. In 1998, Axys Pharmaceuticals (formerly Arris Pharmaceutical, Inc.) separated the combinatorial chemistry group as a subsidiary, Axys Advanced Technologies, Inc. In April 2000, Axys Advanced Technologies, Inc. merged with ChemRx to become ChemRx Advanced Technologies as a partner company in the Discovery Partners International (DPI) family of companies, which also includes IRORI, Discovery Technologies, Ltd., and Structural Proteomics. In late 2000, SIDDCO joined DPI as part of ChemRx Advanced Technologies.

Furthermore, methods of solution-phase combinatorial synthesis that use various purification methods have been developed (99–105). For solution-phase synthesis, resin-based chemistry has played an important role. Solution-phase synthesis generally relies on solid-phase materials as reagents, scavengers, and supports for parallel cleanup and purification. Both scavenger resins and solid-supported liquid extraction (SLE) have played a major role in our solution-phase efforts. Other groups have reported extensively on the use of solid-phase methods for high throughput purification with ion-exchange resins, solid-phase extraction (SPE), and preparative high performance liquid chromatography (HPLC) (16–19).

Medicinal chemistry, process chemistry, and manufacturing practiced in the traditional pharmaceutical company are all integral to the ChemRx approach to small-molecule library generation (Fig. 1) (19). Library design uses medicinal chemistry guidelines, chemistry development is related to process research, and production is a manufacturing operation. First of all, compounds should possess good pharmacological properties such as low $c \log P$ (<5)* and low molecular weight (<500) (20). Both the scaffolds and the diversity elements used to construct libraries should be druglike. Second, the chemistry used in making these molecules must be robust and reliable. Process chemistry principles are applied to library development by thorough reaction optimization and evaluation of all synthesis precursors. However, unlike the development of a process to produce a drug substance, one set of conditions must be optimized for a range of substrates. Variable reactivity and solubility of diverse building blocks complicate the identification of a single set of optimal reaction conditions.

Third, the type of organization in which libraries are produced is also important. We have found that one group for chemistry development and another group for library manufacture adhering to the principles of total quality management leads to high quality libraries. The separation between chemistry development and production has many benefits including job specialization for consistent results and improved quality, expert knowledge and maintenance of equipment, increased efficiency, and higher productivity. An important factor is that chemists can focus on chemistry, and the production group can focus on manufacturing. As a consequence, synthetic methods must be designed for production in a manufacturing environment. These methods must be tested in the manufacturing setting on a pilot scale prior to full-scale production. Automation, purification, and analytical methods are some issues that are addressed prior to library manufacture. All of these elements make library development a multivariable problem that must be solved prior to library generation. In this chapter we describe and illustrate the processes essential to library generation.

---

* $c \log P$ = calculated log P where P is the partition coefficient between l-octanol and water. For a further discussion, see Ref. 19a.

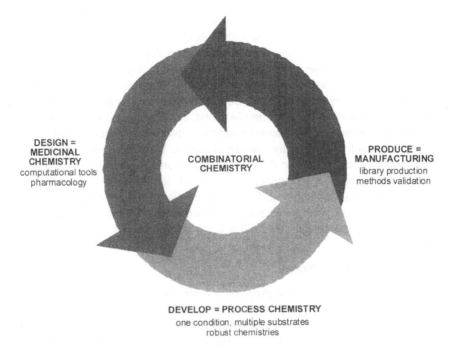

DESIGN =
MEDICINAL
CHEMISTRY
computational tools
pharmacology

COMBINATORIAL
CHEMISTRY

PRODUCE =
MANUFACTURING
library production
methods validation

DEVELOP = PROCESS CHEMISTRY
one condition, multiple substrates
robust chemistries

Figure 1   An integrated approach to library generation.

II. LIBRARY DESIGN

A. Starting Principles

What type of library is desirable for screening? Because the primary goal of any lead discovery library is the identification of new target pharmacophores, one must first determine whether the compound array will be used in a variety of assays or for one specific target. Discovery libraries are designed for covering a large diversity space. Consequently, one library will ideally provide several active compounds across a broad range of biological targets. Alternatively, targeted libraries are designed for a specific gene family or a particular biological target. Many of the library compounds will therefore be active. In both cases, libraries are designed to give a low rate of false positive activity and generate verifiable hits in biological assays.

B. Pharmacological Guidelines

The most useful libraries should possess some known druglike properties. Lipinski's "rule of five" is a convenient aid in designing libraries that look like drugs to medicinal chemists (20). Molecules with desirable ADMET properties enhance the utility of small-molecule libraries in accelerated lead discovery (21). Clearly, library structures must encompass a pharmacophore-like structure in order to have a reasonable chance of interacting with biological targets.

The pharmacological richness of natural products also provides a useful guideline for library design (22,23). Several groups have reported the synthesis of "natural product–like" libraries (24). First, compound arrays can possess the structural characteristics of natural products. Alternatively, libraries can be derived from naturally occurring starting materials. We have found amino acids and carbohydrates useful for the construction of small-molecule libraries (Fig. 2). In this regard, representative libraries of tetrahydro-β-carbolines (25), benzopyrans (26), guanidines (27), pyrrolidines, alkoxyprolines (28), and various sugar derivatives (22) have been prepared. We envision that natural products will continue to be a fertile source of biologically active libraries.

C. Computational Tools

We have used commercial and proprietary computational tools in the library design process. The focus of our early software efforts was in two areas. The first was the creation of structures that could populate a database for the library. Database structures could then be used to visualize the library members and calculate

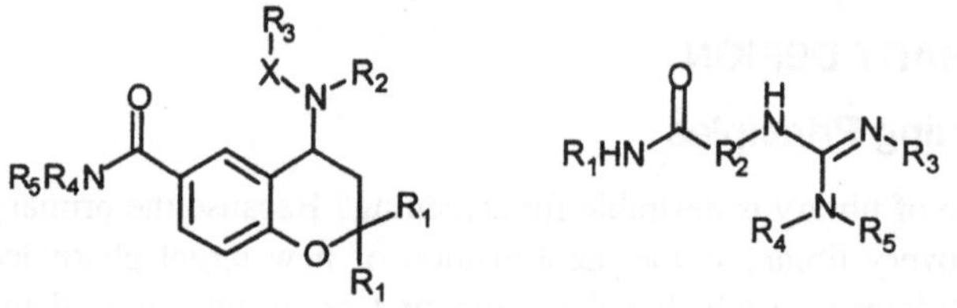

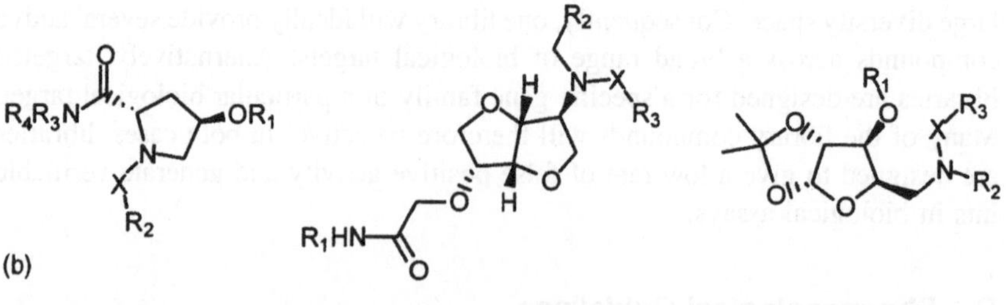

Figure 2 Representative libraries with natural product core structures. (a) Natural product motif libraries. (b) Natural product derived libraries.

properties such as $c \log P$ and molecular weight. The molecular weight database was then transferred to the mass spectrometer for comparison to the molecular weight data. This direct comparison of the database to the mass spectral data is an important quality check. This cross-check shows errors in database preparation and mistakes in the synthesis. The database software effort eventually led to software that could be trained to do virtual chemistry (29).

The second software initiative was to calculate molecular diversity. The method is based on the concept of the comparison of three-dimensional shapes called Icepick (30). The diversity substituents are treated as individual molecules. Low energy conformations are calculated, and the resulting shapes are compared with other members of the set. The diversity score is calculated on the basis of how the shapes overlapped. This allows the computational group to select a small set of side chains (e.g., 20 amines) that represent a large set of commercially available compounds (e.g., 2000 amines). Diversity is selected for chemistry compatibility, removal of pharmaceutically undesired functionality (alkylators, long alkyl chains), and materials that would give undesired physical properties such as high molecular weight and a high $c \log P$ values. From these initial needs, the role of software and databases continues as a critical element in the design and analysis of libraries.

D. Synthetic and Manufacturing Considerations

As part of the design process, a viable synthetic route must be identified. Incorporation of new synthetic methodology in the library process is highly desirable. The availability of library building blocks and the performance of various precursors in the library chemistry will determine what chemotypes are synthetically accessible. Because the quality of the final compounds affects screening results, purification methods need to be developed. Criteria for library quality must be prospectively established. Potential screening liabilities (i.e., presence of reactive impurities such as alkylators and aldehydes) must be avoided because these impurities will adversely affect screening.

III. LIBRARY DEVELOPMENT

A. Library Proposals

Ideas are peer-reviewed and evaluated for pharmaceutically acceptable properties, synthetic methods, and library manufacturability. This arrangement encourages input from all our chemists and ensures successful manufacture and use of each library.

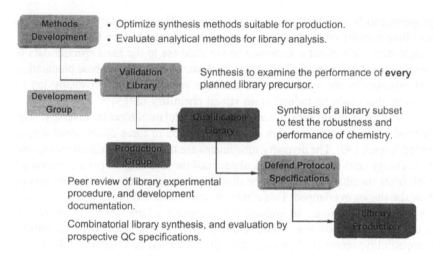

Methods Development
- Optimize synthesis methods suitable for production.
- Evaluate analytical methods for library analysis.

Development Group

Validation Library
Synthesis to examine the performance of **every** planned library precursor.

Qualification Library
Synthesis of a library subset to test the robustness and performance of chemistry.

Production Group

Defend Protocol, Specifications

Peer review of library experimental procedure, and development documentation.

Combinatorial library synthesis, and evaluation by prospective QC specifications.

Library Production

Figure 3   Key steps in library development.

B.   Methods Development and Validation

After initial design, feasibility of the library chemistry must be demonstrated through the synthesis of compounds using a method that can be further developed into a process for library manufacture (Fig. 3). From the synthesis of these compounds, the chemist gains a better idea of the stability and physical properties of the molecules. Optimization of synthesis methods and of the scope of the chemistry begin. Analytical methods for library analysis are also evaluated.

Central to the development process is optimization of the chemistry. The objective is to enhance the quality of the final product by improving each individual step. General optimization principles apply to both solution-phase and solid-phase synthetic routes. There are, however, some specific issues to each reaction medium. Suitable solvent for resin swelling (31), resin agitation, reaction concentration, reagent stoichiometry, and reaction time must all be optimized. Quantitative conversion of reactants to products and minimization of side reactions ensure high purity of final products. Reaction conditions must also be robust. Finally, suitable postsynthesis purification steps ensure high quality libraries. The following questions are generally addressed during library development:

What are the key reaction parameters?
What are the scope and range of reaction conditions that can be used to generate the final products in high purity and high yield?
What reactants can be employed?
Have suitable library analytical methods been developed?

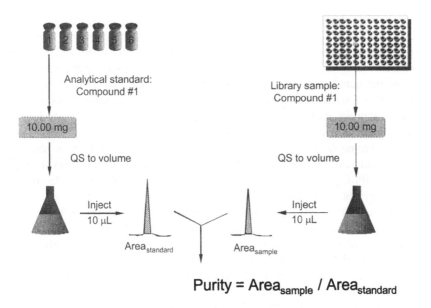

Figure 4 Weight percent method of purity determination.

Is the synthetic method suitable for production of the library in a production group?
Is the chemistry robust and reproducible?
Have hazardous steps been minimized, or eliminated wherever possible?
Does the chemistry work in a manufacturing environment?
Are all precursors and final products soluble in suitable solvents?

Optimization of solid-phase reactions is accomplished by numerous different methods. One approach to converting normal classical syntheses to high throughput parallel syntheses is to define unit operations. Operations such as reagent additions are easily converted to highly parallel operations by using multichannel pipetters or multichannel liquid-handling robotics such as the Robbins Hydra 96-channel solvent dispenser. However, many synthetic operations for workup of reactions are more difficult to convert to a high throughput format. For instance, how does one remove excess reagents from the desired product by parallel methods?

One tool we use in reaction optimization is external standards. The concept of using standards to measure purity is shown in Figure 4.* We typically synthe-

---

* We thank Douglas Livingston for the drawings in Figures 4 and 6. For a complete discussion of the analytical methods used in library development and production, see Chapter 9.

size, purify, and characterize several representative compounds that will be found
in the final library. We then prepare solutions of known concentration of these
compounds and generate HPLC curves. During the course of development, we
measure the purity of product wells by weighing the mass found in the well and
analyzing the samples by HPLC. The ratio of the area under the curve of the two
samples gives a weight-based true purity. This analysis method has been useful
to detect impurities that are not detected by HPLC. Impurities include insoluble
materials, excess solvents such as trifluoroacetic acid (TFA), inorganics, materials
with poor chromophores, materials that elute over a wide range such as plastic
oligomers, and materials that would typically be retained by a guard column during
preparative HPLC purification. Analytical methods used for library development
and chemistry optimization are further discussed in Chapter 9.

Once a suitable library production method has been developed, all the
building blocks that will be used to prepare the library must be tested in what
we refer to as validation libraries. There are two formats for validation libraries
(Fig. 5). In parallel validation libraries, one site of diversity is varied and the
remaining sites of diversity are held constant. In matrix validation libraries, all
sites of diversity are varied. Whereas a parallel library helps determine how one
set of precursors performs with a specific substrate, a matrix library provides
information of cross-compatibility of all sites of diversity. Validation must be
performed in a manner as close as possible to the eventual production method.
Ultimately, all precursors are validated by the methods used to produce the library. Changes in reaction conditions require revalidation in order to ensure that
all precursors will still meet the purity and yield criteria set during validation.

C. Triazolopyridazine Case Study for Methods Development

An overview of the chemistry used to prepare triazolopyridazine libraries helps
illustrate some of the typical issues we face in library development. The develop-

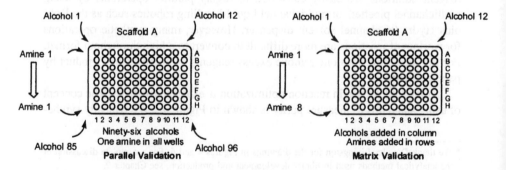

Figure 5   Typical plate layouts for validation libraries.

Table 1 Horner–Wadsworth–Emmons Olefination on PEG-PAL resin

| | | Conversion[a] |
Entry	R_2CHO	(%)
1	PhCHO	>95
2	i-PrCHO	>95
3	n-HexCHO	85
4	(structure) PhCH=C(CH₃)CHO	75

[a] Determined by ^1H NMR integration of the ratio of **4** to **5**.

ment of a solid-phase synthesis of triazolopyridazines (32,33) had two synthetic steps that required optimization: the Horner–Wadsworth–Emmons reaction and the Diels–Alder reaction. Through a series of optimization studies, suitable conditions for library production were developed.

The Horner–Wadsworth–Emmons (HWE) reaction is a reliable method for generating olefins.* Masamune and coworkers reported mild conditions for this olefination procedure using triethylamine as base and LiBr as an additive (36). Recently, several groups reported variants of the solid-supported HWE reaction and the Wittig reaction.† We reported a solid-supported HWE reaction using polyethylene glycol–linked PEG-PAL resin (Table 1) (44). Initially, diethylphosphonoacetic acid was coupled to the N-terminus of a solid-phase peptide support with PyBOP and N-methylmorpholine in DMF. Solid-supported **1** was treated with excess aldehyde **2**, LiBr, and triethylamine. Upon washing away excess reagents from the resin-bound alkene **3**, the reaction was stopped. The α,β-unsaturated amide **4** was cleaved from the solid support with trifluoroacetic acid. Shown in Table 1 are the conversions, after 24 h, of diethylphosphonoacetamide linked to PEG-PAL resin in the reaction with a tenfold excess of various aldehydes. Conversion to alkene **4** was poorest when α,β-unsaturated aldehydes were used. Conversion was calculated by comparing the ratio of phosphonoacetamide **5** and alkene **4** by integration of the ^1H NMR spectra after TFA cleavage. Only the E-alkene was observed, as judged by the coupling constants of the alkene protons. In optimization of the HWE reaction, reaction progress was monitored by gel-phase ^{31}P NMR of the solid support suspended in acetonitrile. The ^{31}P NMR showed the resonance of the starting resin-bound diethylphosphonoace-

---

* For reviews on the HWE reactions, see Refs. 34 and 35.
† For other examples of olefinations on support, see Refs. 37–43.

tamide **1** as a narrow multiplet at 22 ppm. When the reaction was complete, the 22 ppm peaks disappeared and a new broad resonance for the diethylphosphate in solution appeared near 0 ppm.

For the efficient synthesis of solid-supported dienes in a library format, we chose to use the higher loading Rink amide resin (0.6–0.7 mmol/g) instead of PEG-PAL resin (0.2–0.3 mmol/g) (Table 2) (32). However, with the Rink amide

Table 2 Horner–Wadsworth–Emmons Olefination on Rink Amide Resin

Entry	R$_2$CHO	Base[a]	Solvent	Conversion[b] (%)
1	OHC (Ph)	Et$_3$N	CH$_3$CN	<1
2	OHC (Ph)	DBU	CH$_3$CN	<1
3	OHC (Ph)	Et$_3$N	THF	50
4	OHC (Ph)	DBU	THF	>99
5	OHC (Ph)	DBU	THF	>99
6	OHC	DBU	THF	>99

[a] DBU = 1,8-diazabicyclo[4.3.0]non-5-ene.
[b] Determined by area under the curve at 214 nm by analytical HPLC.

resin, we modified the reaction conditions to ensure complete olefination. Rink amide resin does not swell in polar solvents such as acetonitrile (45). Therefore, the reaction did not work even with a stronger base such as DBU (Table 2, entries 1 and 2). In solution studies, Rathke and Nowak (46) demonstrated that both acetonitrile and THF are suitable solvents for the HWE reaction. They also showed that DBU accelerated the HWE reaction. Using DBU instead of triethylamine as base and anhydrous THF instead of acetonitrile as solvent ensured complete olefination within 4 h (Table 2, entries 4–6). These modifications extended the scope and utility of the HWE reaction on solid support.

With the HWE reaction optimized, cycloaddition with 4-substituted urazine dienophiles yielded the triazolopyridazine structure.* The solid-supported [4 + 2] cycloaddition has attracted considerable attention.† 4-Substituted urazoles synthesized from isocyanates and ethyl carbazate (78,79) were oxidized to 4-substituted urazines **11** with iodobenzene diacetate (80) and added to the solid-supported dienes **10** to give triazolopyridazines **12** after cleavage off solid support (Table 3, entries 1 and 2). We found these urazoles to be soluble in dimethylformamide (DMF), a practical solvent for library production. All reactions proceeded smoothly to give on average 30% overall yield (seven linear steps) and 70% HPLC purity (area under the curve at 214 nm) with a range of amino acids, α,β-unsaturated aldehydes, and urazoles. Unreacted diene from the HWE olefination was occasionally observed as a minor impurity ($\sim$ <5%).

The cycloaddition methodology was extended to the synthesis of imides (Table 3, entries 7–9). These compounds could then be further functionalized through a Mitsunobu reaction. Because 4-substituted urazoles performed well in cycloadditions with solid-supported dienes in DMF, we were surprised that urazine **11** (R_4 = H), generated by oxidation of urazole with PhI(OAc)$_2$, decomposed in DMF, DMA, and DMSO (Table 3, entries 3–5). 1,4-Dioxane proved to be the solvent of choice. To ensure complete cycloaddition, we adapted a reported in situ urazole oxidation procedure (81) whereby urazine was generated by portionwise addition of oxidant to a premixed solution of urazole and diene, yielding resin-bound imide (Table 3, entry 9). Mixing was also very important for the success of this reaction.

D. Control of Intermolecular Versus Intramolecular Cleavage off Solid Support

Process optimization to direct cleavage off solid support by either an intermolecular or an intramolecular pathway has also been demonstrated. We developed

---

* For a review on 1,2,4-triazoline-3,5-diones (urazoles), see Ref. 47. For reviews on azo dienophiles, see Refs. 48 and 49. For an intramolecular example, see Refs. 50 and 51.

† For examples of solid-supported [4 + 2] cycloadditions, see Refs. 52–77.

Table 3 Optimization of [4 + 2] Cycloaddition with Urazines

Entry	R$_4$	Stoichiometry of PhI(OAc)$_2$	Solvent[a]	Conversion[b] (%)
1	Ph	1 equiv	DMF	>99
2	Me	1 equiv	DMF	>99
3	H	1 equiv	DMF	0[c]
4	H	1 equiv	DMA	0[c]
5	H	1 equiv	DMSO	0[c]
6	H	1 equiv	CH$_2$Cl$_2$	0[d]
7	H	1 equiv	THF	65
8	H	1 equiv	1,4-Dioxane	70
9	H	6 equiv	1,4-Dioxane	>99

DMF = dimethylformamide; DMA = n,n-dimethylacetamide; DMSO = dimethyl sulfoxide; THF = tetrahydrofuran.
[b] Determined by area under the curve at 214 nm by analytical HPLC.
[c] Decomposition of urazole.
[d] No urazine generated.

a [3 + 2] cycloaddition of trimethylsilyldiazomethane with solid-supported acrylates on the Marshall linker (25,26,28,82–84) as a method for generating Δ^2-pyrazolines (19). After coupling various 2-substituted acrylic acids **14** to the Marshall linker **13** attached to polystyrene resin, cycloaddition of olefin **15** with trimethylsilyldiazomethane **16** gave silyl diazo **17** (Scheme 1). The cycloaddition consistently worked well except with 2,3-disubstituted acrylic acids. Upon treatment of the silyl intermediate **17** with trifluoroacetic acid (85), S_N2'-type protodesilylation generated solid-supported Δ^2-pyrazoline **18**.*

Subsequent acylation of pyrazoline **18** was problematic. Acid chlorides gave a mixture of two acylated regioisomers; these reactions were generally not clean (86). Sulfonyl chlorides and isothiocyanates did not give the desired products. Acylation with carboxylic acids under various conditions (DIC, PyBOP,

---

* The mechanism of this transformation presumably involves protonation of the 2-nitrogen of the 2,3-diazo species with loss of the silane analogous to the known reaction of allylic silanes with electrophiles.

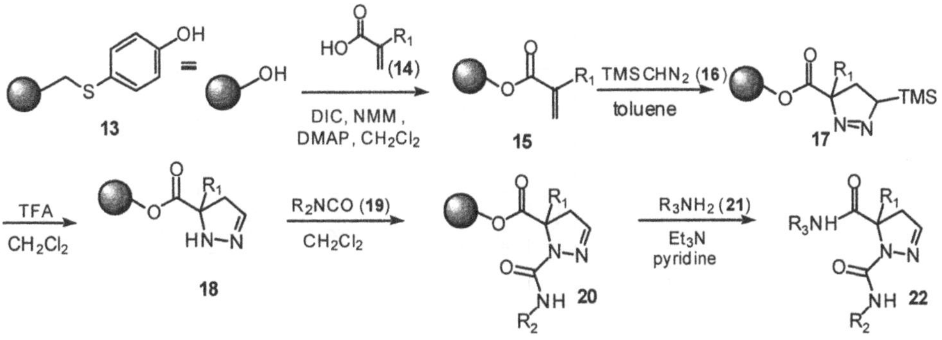

Scheme 1 Solid-phase synthesis of Δ^2-pyrazolines **22**.

PyBrOP, or HATU) slowly gave the desired product. However, acylation with aliphatic isocyanates (CH_2Cl_2, ambient temperature, 4 days) gave the desired product **20**. Upon heating to 50°C in toluene, acylation with aliphatic isocyanates was complete within 16 h. Treatment of solid-supported intermediate **20** with various primary amines **21** gave Δ^2-pyrazolines **22**. A variety of 2-substituted acrylic acids, aliphatic isocyanates, and primary amines were used to generate spatially separated arrays of Δ^2-pyrazolines.

We noted that after acylation of the intermediate pyrazoline **18** with aryl isocyanates to yield **20**, cleavage off solid support did not yield the expected Δ^2-pyrazolines **22** (Scheme 2). Instead, intramolecular cyclization occurred to yield

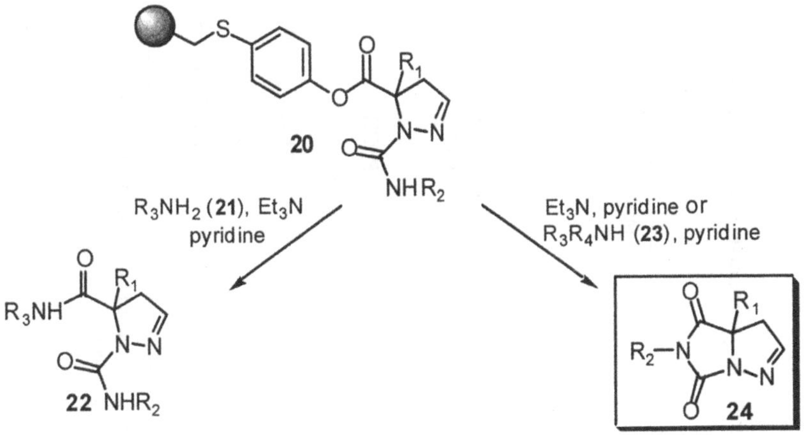

Scheme 2 Solid-phase synthesis of triazabicyclooctanes **24**.

Table 4 Representative Triazabicyclooctanes 24 Prepared via Solid-Phase Synthesis

Entry	Compound	R₂	Pyridine	Et₃N	i-Pr₂NEt	i-PrNH₂	Piperidine
1	24a	Ph—	94 (>99)[a]	90 (>99) <5%	85 (>99)	82 (94)	49 (85)
2	24b	PhCH₂—	<5 (NA)[b]	(NA) <5%	<5 (NA)	17 (>99)	>95 (89)
3	24c	CH₂=CH—CH₂—	<5 (NA)	(NA) <5%	<5 (14)	28 (94)	>95 (83)
4	24d	CH₃CH₂CH₂—	<5 (NA)	(NA)	<5 (NA)	<5 (NA)	>95 (52)

[a] % Purity determined by analytical HPLC area under the curve at 214 nm (% crude yield).
[b] NA = not available.

an unprecedented class of structures, triazabicyclooctanes **24**. To evaluate the scope of this chemistry, we prepared a small library with various isocyanates and different bases (Table 4). Pyrazoline **18** was acylated with a range of isocyanates (R_2 = Ph, PhCH$_2$, CH$_2$=CHCH$_2$, n-Pr) to give urea **20**. We then treated each row of ureas **20** with a different base (pyridine, triethylamine, diisopropylethylamine, isopropylamine, or piperidine) in pyridine solutions. Aryl ureas generally cyclized with pyridine, tertiary amines, and the secondary amine piperidine (Table 4, entry 1). Piperidine generally promoted cyclization of various ureas (Table 4, entries 1–4). α-Branched primary amines such as isopropylamine gave some of the desired product **24**, but pyrazoline **22** was the major product. When primary amines were used as previously described, the triazabicyclooctane **24** was generally not observed. We observed that aryl ureas treated with tertiary amines gave the purest products (entry 1). In all four cases, piperidine gave the best conversion to triazabicyclooctanes (Table 4, entries 1–4).

IV. QUALIFICATION

Once the production method has been optimized and all precursors meet specific validation criteria, the library method is evaluated in what we call a qualification library (Fig. 3). Unlike the validation library, the qualification library is produced by the production method in the production group. A detailed protocol is written that describes the library experimental procedure and includes the bill of materials and the plate layout for the qualification library. After the library is produced, the quality control methods used to analyze the final production library are used to evaluate the qualification library. The qualification library serves as a predictor of performance of the chemistry in full-scale library production.

V. PRODUCTION

One advantage of technology transfer from a development group to a separate production group is the requirement for documentation. In this transition, the quality assurance group plays a critical role in document and data control. Once this information is transferred to the production group, the qualification library helps our evaluation of the library chemistry. If the write-up is not sufficiently detailed, the development group further refines the written documentation. Prior to library production, a final review of the library chemistry, results of the qualification library, and a discussion of the proposed production library synthesis are presented to the scientific staff by the lead chemist in a protocol defense. The protocol defense is designed to review the library chemistry and the underlying production technology. Upon successful defense, the production group assumes responsibility for production of the library, and the analytical group evaluates the quality of the library.

A critical aspect of production is the translation of the library experimental method into a series of unit operations. A detailed description of each unit process is contained in a document called the production protocol. These are standard methods that the production group uses to generate libraries.

VI. POSTSYNTHESIS CLEANUP

Traditionally, solid-phase cleavage reagents were removed by evaporation and were limited to TFA, HF, or NH_3. For solid-phase chemistry, the variety and quantity of cleavage reagents have greatly expanded with the addition of post-cleavage workup procedures. We have used resin capture and liquid–liquid extraction for postsynthesis cleanup for both solid-phase and solution-phase libraries. Many solutions to converting extractive workups to high throughput have been proposed and examined.* The most straightforward is the use of robotics and measured liquid transfers. Because one cannot achieve good layer separation, this method can result in material losses. Shortcomings to this approach in a high throughput format included poor mixing of layers, poor phase separation, and, at small scale, droplet adherence to containers. In addition, emulsions complicated automation of this procedure. Finally, measured volume liquid–liquid extraction is difficult to run in a highly parallel fashion. A single-needle robot can rapidly distribute an extraction solvent to each well of a 96-well plate; however, removal of the separated layers requires the robot to wash the needle between wells.

Our solution to automated extraction workups was to use partition chromatography to achieve separations similar to liquid–liquid extraction (Fig. 6) (26,28,83,92,93). In partition chromatography, a liquid layer is immobilized on a solid support and a second, nonmiscible liquid is passed over the immobilized layer. Separation is obtained when compounds in the mixture have different partition coefficients in the two layers. The method was invented about 60 years ago and was widely used until it was replaced by bonded phase chromatography supports (94). Fast separations with selectivity similar to that of liquid–liquid extraction could be achieved with gravity elution, large particle sizes, and short column lengths. We developed a high throughput postsynthesis cleanup in which the immobilized liquid is an aqueous buffer. Silica gel, diatomaceous earth, cellulose powder, starch, and cotton were good candidates for a solid support to immobilize aqueous buffers (95,96). We chose to use diatomaceous earth for its wide range of particle sizes and because it is essentially chemically inert. The aqueous buffer coats the surface of the particles and exposes a large surface area of water for

---

* For measured volume high throughput extraction, see Refs. 87–89. For the lollipops extractor, see Ref. 90. For a discussion of the hydrophobic membrane, see Ref. 91.

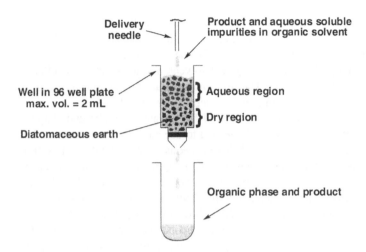

Figure 6 Supported liquid extraction (SLE).

contact with the organic solvent layer. We have referred to this liquid–liquid extraction method as supported liquid extraction (SLE).

A. Parallel Supported Liquid Extraction

Removal of amines by SLE with acidic buffers has been achieved in a number of applications both for solution-phase library synthesis and for cleanup following cleavage from solid supports. (26,28,83,92,93,98). In either case, a high through-put parallel system is required for library synthesis. As shown in Figure 7, each well of a filter plate is filled with about 2 mL of coarse diatomaceous earth. The support is treated with 300 μL aqueous buffer and allowed to equilibrate for 5–10 min. When aqueous buffer is added to the diatomaceous earth, the support

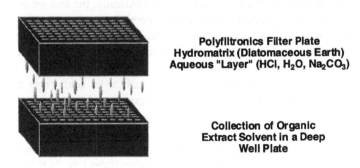

Polyfiltronics Filter Plate
Hydromatrix (Diatomaceous Earth)
Aqueous "Layer" (HCl, H₂O, Na₂CO₃)

Collection of Organic
Extract Solvent in a Deep
Well Plate

Figure 7 Stacked plates for extract collection.

appears wet. However, once the water has absorbed onto the surface of the particles, the appearance is identical to that of the dry diatomaceous earth. The reaction mixture (0.1 mmol) in less than 1 mL of a water-immiscible solvent is added to a well and gravity eluted with additional solvent, and the solution is collected in another deep-well filter plate. The resulting separation is similar to using the same solvent and buffer combination in a separatory funnel.

Since we chose to work with 96-well filter plates, the size and shape of the column was that of one well. We needed to use particle size to control flow rates and separations. Our preferred diatomaceous earth is Hydromatrix from Varian.* It is described as a flux calcined filter aid derived from plankton marine diatomite that is 95% larger than 100 mesh. For the filter plate, the ideal frit material is polypropylene, because it is resistant to passing droplets of water. The amount of buffer can vary from about 300 to 600 μL. We use 300 μL for each 2 mL of support when stripping of the aqueous phase must be avoided. For this volume, the bottom of the column of diatomaceous earth remains dry and can trap any droplets of aqueous buffer that are released from the top half of the column. Nonvolatile buffers such as citric acid or sulfuric acid should be used in the smaller volumes. Also, some absorption of the water from the immobilized layer can occur when eluting with solvents that can absorb water, and a dry layer of support may help to remove that water from the product solution. Larger amounts of buffer (600 μL) can be used when minor stripping of the aqueous phase and thus contamination of the product would not pose a problem, for example, if neat water or a volatile buffer such as hydrochloric acid is used.

The elution solvent can be any of a wide range of compositions, including solvent mixtures containing up to 20% water-miscible solvent (e.g., THF, *i*-PrOH). Higher amounts of water-miscible solvent may achieve good separation but will absorb water from the column. However, stripping water from the column does not necessarily mean that the buffer component will be stripped. As for any other step, optimization and validation of the extractive step are necessary. In addition, large particle size allows the use of carbonate buffers in the extraction of acids as the sodium salts. The column bed generally allows the carbon dioxide produced to escape without disturbing the support bed. Generally, the amount of material per well was 0.05–0.2 mmol. Because the collection vessel used was a deep-well plate, the elution volume was limited to 2 mL. Better recovery of product resulted when a more concentrated solution was added to the column and a larger portion of the 2 mL was used for elution.

In Table 5, we show data on the separation of carboxylic acids from a neutral standard, 1-naphthalenemethanol (98). These separation data were pro-

---

* Chem Elut and Tox Elut extraction columns and Hydromatrix are listed on pp 65–67 of the 1996 catalog of Varian Sample Preparation Products, 24201 Frampton Avenue, Harbor City, CA 90710.

Table 5 SLE Separation of Acids from 1-Naphthalenemethanol

Entry	Acid	Acid before extraction[a]	Acid after extraction[a]	% Acid remaining	% 1-Naphthalene-methanol recovered[b]
1	(structure)	2708	0	<1%	98%
2	(structure)	3032	0	<1%	97%
3	(structure)	6231	0	<1%	98%
4	(structure)	4341	0	<1%	89%
5	(structure)	4485	0	<1%	90%
6	(structure)	5057	0	<1%	98%

[a] Determined by area under the curve at 214 nm by analytical HPLC.
[b] Determined by area under the curve at 254 nm by analytical HPLC.

duced from experiments that closely simulated the volumes and concentrations that we use in library synthesis. A mixture of the test carboxylic acid (0.10 mmol) and 1-naphthalenemethanol (0.025 mmol) was dissolved in 1 mL of dichloromethane. A 3 mL Chem Elut Extube (Varian) was treated with 0.4 mL of 2 N sodium carbonate and allowed to equilibrate for 5 min. The test acid mixture was added to the column and eluted with 2 mL of dichloromethane. The eluent was evaporated, and the residue was dissolved in 10 mL of methanol. The compound concentrations were estimated from peak areas in chromatographs with UV detection at 214 nm or 254 nm and adjusted for the differences between starting and final volumes. These data demonstrate that liquid–liquid extraction can be done in a rapid, parallel format that is suitable for high throughput synthesis. Some favorable aspects of the method are the use of readily available 96-well filter plates, inexpensive solid support, and normal liquid-handling robotics. In addition, the method is reliable, works on very small volumes, can be scaled to large quantities, and gives predictable separations.

In addition to SLE, resin capture can be used to effect similar separations (99–105). For instance, amines can be removed from neutral compounds either by capture on acidic ion-exchange resin or by extraction into aqueous acid. In many cases both methods work equally well. We have found that the polarity of the material to be removed is a good guide to which method will work best. As shown in Figure 8, a water-soluble impurity is predicted to be easily removed by SLE whereas a hydrophobic impurity is likely to be best removed by resin capture. In addition, SLE can be used to remove water-soluble impurities, such as mixed inorganics, that are difficult to remove by resin capture (106). These two separation technologies are complementary and have been used sequentially to optimize product purity. The case study shown in Sec. VI.C demonstrates chemistry that was difficult to optimize using resin capture.

B. Purification of Products Cleaved Off the Marshall Linker by SLE

In several reports describing the solid-phase synthesis of tetrahydro-β-carbolines (25), piperazines (83), benzopyrans (26), and alkoxyprolines (28) we have re-

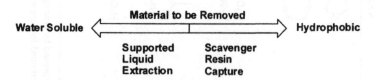

Figure 8 Purification method selection based on physical properties.

ported the use of SLE to remove excess amine after cleavage off the Marshall linker. In the synthesis of Δ^2-pyrazolines described in Sec. III.D, excess amine was also removed from compounds **22** by supported liquid extraction. The SLE conditions used 2 N HCl or water as the aqueous priming buffer and 4:1 CHCl$_3$/THF as the elution solvent.

C. Pyridazine Case Study for SLE

A solution-phase spatially separated library that illustrates some of the important aspects of the SLE technology is presented in Scheme 3 (98,107). This example illustrates several methods that are applicable to postcleavage cleanup of libraries made on solid-supports. For the library represented in Scheme 3, a set of 3,6-dichloropyridazine-4-carboxamides **25** was prepared in bulk (108). Each amide **25** was distributed into a 96-well plate. The most active chloride was displaced with 1.05 equiv of secondary amines **26** with *N,N,N',N'*-tetraethylethylenediamine (TEED) as a HCl scavenger to produce **27**. TEED was chosen as a nonvola-

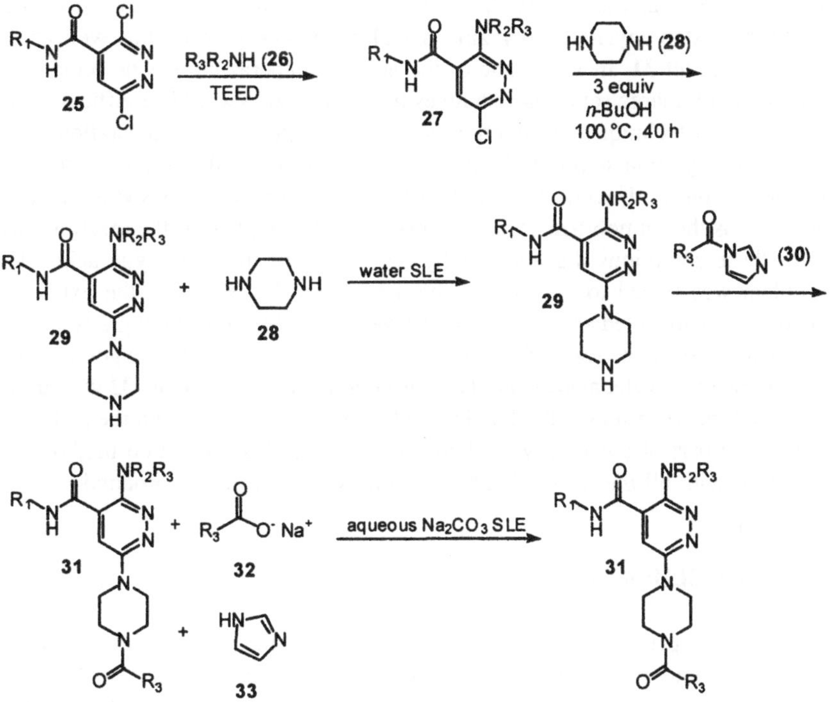

Scheme 3   Solution-phase synthesis and purification of 3,4,6-trisubstituted pyridazines.

tile tertiary amine base that could be easily removed by extraction. The plates were sealed, then heated to give essentially complete reaction. Each well was treated with piperazine **28** in *n*-butanol. Development studies had shown that at least 3 equiv of piperazine was required to reduce or eliminate dimer formation by bisaddition to piperazine **28**. After heating at 100°C for 40 h the reaction was essentially complete for a variety of substrates to give intermediate **29**. Prior to acylation of the monoalkylated piperazine, excess piperazine was removed and the monoalkylated piperazine was free-based. Supported liquid extraction using unbuffered water and chloroform achieved both goals. Each well was treated with chloroform. The resulting solution was transferred to the top of a column of diatomaceous earth pretreated with water in the corresponding well of an SLE plate. The chloroform solution passed through the immobilized water on the diatomaceous earth, and water-soluble components were extracted. Owing to its high water solubility, piperazine was extracted and retained on the column by the immobilized water. Equilibrium and favorable partition coefficients ensured that piperazine hydrochloride was formed, extracted into the aqueous layer, and retained on the column. The basic but hydrophobic amine **29** remained in the organic layer. The separation was accomplished because the basic amines have very different water solubility. If any piperazine remained after separation, then in the next step acylated piperazine would have been produced as well as the desired product **31**. In fact, none of the acylated piperazine was detected. This separation of a large set of basic amines from piperazine would be difficult using resin capture techniques. Extraction is uniquely suited for this separation.

The acylation step used 2 equiv of acylimidazoles **30** to drive the reaction to completion. Carboxylic acids were removed by SLE with 2 N sodium carbonate used as the immobilized phase. However, when applied to the reaction mixtures these conditions did not initially remove all of the carboxylic acids. The problem was traced to stability of the acylimidazoles **30** under the extraction conditions. Unreacted acylimidazoles **30** were hydrolyzed by adding 2 N sodium carbonate to the plate wells. SLE following this hydrolysis step gave good separation of the carboxylic acids from the neutral product **31**. Imidazole **33** was highly water-soluble regardless of pH and was also removed by SLE with the carbonate buffer. As part of the library method optimization, this separation method was validated with all the materials and conditions used in library production.

VII. CONCLUSION

With the emergence of parallel synthesis techniques, the pharmaceutical industry has rapidly integrated these methods into the lead discovery process. Intensive optimization is required to convert a classical solution-phase synthesis into a series of rapid, highly parallel synthetic methods suitable to produce libraries.

The choice between solution phase and solid phase is directed by the nature of the library chemistry. Even solution-phase libraries require development of suitable conditions to use solid-phase reagents for synthesis and purification.

We have successfully applied the system described above for the generation of both solution-phase and solid-phase libraries. Although the generation of libraries is a very complex process, this system ensures a successful outcome. More synthetic transformations will continue to be developed on solid phase and for high throughput organic synthesis. As library syntheses become more sophisticated and new methods are developed, our approach will continue to accommodate such advances.

ACKNOWLEDGMENTS

We are grateful to Doug Livingston for advocating the Total Quality Management (TQM) philosophy that is the underlying basis of our current systems. The core of the ChemRx Advanced Technologies systems approach to library development was established by many other members of the combinatorial chemistry group, including Guy Breitenbucher, Barry Bunin, Nathan Collins, Jeff Dener, Pascal Fantauzzi, Liling Fang, Mike Hocker, Thutam Hopkins, John Mount, Chris Phelan, Matt Plunkett, David Robinette, Kraig Yager, and Birong Zhang. We thank all the members of the team who have contributed to the development of high quality libraries. We also thank Jeff Dener for his helpful comments and suggestions regarding this manuscript.

REFERENCES

1. P Seneci. Solid-Phase Synthesis and Combinatorial Technologies. New York: Wiley, 2000.
2. BA Bunin, JM Dener, DA Livingston. Application of combinatorial and parallel synthesis to medicinal chemistry. In: A Doherty, ed. Annual Reports in Medicinal Chemistry, Vol 34. San Diego: Academic Press, 1999, pp 267–286.
3. WA Loughin. Aust J Chem 51:875–893, 1998.
4. JW Corbett. Org Prep Proc Int 30:489–550, 1998.
5. BA Bunin. The Combinatorial Index. San Diego: Academic Press, 1998.
6. LA Thompson, JA Ellman. Chem Rev 96:555–600, 1996.
7. JS Fruchtel, G Jung. Angew Chem Int Ed Engl 35:17–42, 1996.
8. PHH Hermkens, HCJ Ottenheijm, D Rees. Tetrahedron 52:4527–4554, 1996.
9. NK Terrett, M Gardner, DW Gordon, RJ Kobylecki, J Steele. Tetrahedron 51: 8135–8173, 1995.
10. MA Gallop, RW Barrett, WJ Dower, SPA Fodor, EM Gordon. J Med Chem 37: 1233–1251, 1994.

11. EM Gordon, RW Barrett, WJ Dower, SPA Fodor, MA Gallop. J Med Chem 37: 1385–1401, 1994.
12. E Atherton, RC Sheppard. Solid Phase Peptide Synthesis: A Practical Approach. Oxford, UK: IRL Press, 1989.
13. F Eckstein, ed. Oligonucleotides and Analogues: A Practical Approach, Oxford, UK: IRL Press, 1991.
14. F Guillier, D Orain, M Bradley. Chem Rev 100:2091–2157, 2000.
15. I Sucholeiki. Selection of supports for solid-phase organic synthesis. In: WH Moos, MR Pavia, BK Kay, AD Ellington, eds. Annual Reports in Combinatorial Chemistry and Molecular Diversity; Vol 1. Leiden: ESCOM, 1997, pp 41–49.
16. MJ Suto, LH Gayo-Fung, MSS Palanki, R Sullivan. Tetrahedron 54:4141–4150, 1998.
17. RM Lawrence, OM Fryszman, MA Poss, JA Biller, HN Weller. Am Biotechnol Lab, October 1996.
18. L Zeng, DB Kassel. Anal Chem 70:4380–4388, 1998.
19. AM Boldi. Emerging trends in solid-phase combinatorial library methodology. 4th Lake Tahoe Symp on Molecular Diversity. Granlibakken at Lake Tahoe, CA, Mar 21–25, 2000.
19a. RB Silverman. The Organic Chemistry of Drug Design and Drug Action. San Diego, CA: Academic Press, 1992, pp 26–34.
20. CA Lipinski, F Lombardo, BW Dominy, PJ Freeney. Adv Drug Delivery Rev 23: 3–25, 1997.
21. RA Fecik, KE Frank, EJ Gentry, JR Menon, LA Mitscher, H Telikepalli. Med Res Rev 18:149–185, 1998.
22. AM Boldi. Libraries containing natural product structural motifs. CHI Sixth Annual Conf on High-Throughput Organic Synthesis, San Diego, CA, Feb 14–16, 2001.
23. M-L Lee, G Schneider. J Comb Chem 3:284–289, 2001.
24. DG Hall, S Manku, F Wang. J Comb Chem 3:125–150, 2001.
25. PP Fantauzzi, KM Yager. Tetrahedron Lett 39:1291–1294, 1998.
26. JG Breitenbucher, HC Hui. Tetrahedron Lett 39:8207–8210, 1998.
27. AM Boldi, JM Dener, TP Hopkins. Solid-phase library synthesis of trisubstituted guanidines. Book of Abstracts, 220th Natl Meeting Am Chem Soc, Washington, DC, Aug 20–24, 2000. Washington, DC: Am Chem Soc, 2000, ORGN 333.
28. AM Boldi, JM Dener, TP Hopkins. J Comb Chem 3: in press, 2001.
29. D Chapman. Reaction-centered informatics for combinatorial chemistry. Book of Abstracts, 213th Natl Meeting Am Chem Soc, San Francisco, CA, Apr 13–17, 1997. Washington, DC: Am Chem Soc, 1997, CINF 64.
30. J Mount, J Ruppert, W Welch, AN Jain. J Med Chem 42:60–66, 1999.
31. R Santini, MC Griffith, M Qi. Tetrahedron Lett 39:8951–8954, 1998.
32. AM Boldi, CR Johnson, HO Eissa. Tetrahedron Lett 40:619–622, 1999.
33. CO Ogbu, MN Qabar, PD Boatman, J Urban, JP Meara, MD Ferguson, J Tulinsky, C Lum, S Babu, MA Blaskovich, H Nakanishi, F Ruan, B Cao, R Minarik, T Little, S Nelson, M Nguyen, A Gall, M Kahn. Bioorg Med Chem Lett 8:2321–2326, 1998.
34. SE Kelly. In: BM Trost, ed. Comprehensive Organic Synthesis. Vol. 1. New York: Pergamon; 1991, pp 729–817.
35. BE Maryanoff, AB Reitz. Chem Rev 89:863–927, 1989.

36. MA Blanchette, W Choy, JT Davis, AP Essenfeld, S Masamune, WR Roush, T Sakai. Tetrahedron Lett 25:2183–2186, 1984.
37. T Groth, M Meldal. J Comb Chem 3:45–63, 2001.
38. JM Salvino, TJ Kiesow, S Darnbrough, R Labaudiniere. J Comb Chem 1:134–139, 1999.
39. MA Blaskovich, M Kahn. J Org Chem 63:1119–1125, 1998.
40. NW Hird, K Irie, K Nagai. Tetrahedron Lett 38:7111–7114, 1997.
41. P Wipf, TC Henninger. J Org Chem 62:1586–1587, 1997.
42. R Williard, V Jammalamadaka, D Zava, CC Benz, CA Hunt, PJ Kushner, TS Scanlan. Chem Biol 2:45–51, 1995.
43. C Chen, LA Ahlberg Randall, RB Miller, AD Jones, MJ Kurth. J Am Chem Soc 116:2661–2662, 1994.
44. CR Johnson, B Zhang. Tetrahedron Lett 36:9253–9256, 1995.
45. W Li, B Yan. J Org Chem 63:4092–4097, 1998.
46. MW Rathke, M Nowak. J Org Chem 50:2624–2626, 1985.
47. S Rádl. Adv Hetrocycl Chem 67:119–205, 1997.
48. A Adam, O De Lucchi. Angew Chem Int Ed Engl 19:762–779, 1980.
49. SM Weinreb. In: BM Trost, ed. Comprehensive Organic Synthesis, Vol 5. New York: Pergamon, 1991, pp 401–449.
50. M Kahn, S Wilke, B Chen, K Fujita. J Am Chem Soc 110:1638–1639, 1988.
51. M Kahn, RA Chrusciel, T Su, T Xuan. Synlett 1991:31–32, 1991.
52. V Yedidia, CC Leznoff. Can J Chem 58:1144–1150, 1980.
53. LF Tietze, T Hippe, A Steinmetz. Synlett 1996:1043–1044, 1996.
54. JS Panek, B Zhu. Tetrahedron Lett 37:8151–8154, 1996.
55. NW Hird, M Crawshaw. Tetrahedron Lett 38:7115–7118, 1997.
56. BR Stranix, GD Darling. J Org Chem 62:9001–9004, 1997.
57. D Craig, MJ Robson, SJ Shaw. Synlett 1998:1381–1383, 1998.
58. JD Winkler, Y-S Kwak. J Org Chem 63:8634–8635, 1998.
59. DA Heerding, DT Takata, C Kwon, WF Huffman, J Samanen. Tetrahedron Lett 39:6815–6818, 1998.
60. S Wendeborn, A De Mesmaeker, WK-D Brill. Synlett 1998:865–868, 1998.
61. JD Winkler, W McCoull. Tetrahedron Lett 39:4935–4936, 1998.
62. BA Burkett, CLL Chai. Tetrahedron Lett 40:7035–7038, 1999.
63. S Sun, WV Murray. J Org Chem 64:5941–5945, 1999.
64. EM Smith. Tetrahedron Lett 40:3285–3288, 1999.
65. C Chen, B Munoz. Tetrahedron Lett 40:3491–3494, 1999.
66. K Paulvannan. Tetrahedron Lett 40:1851–1854, 1999.
67. SC Schurer, S Blechert. Synlett 1999:1879–1882, 1999.
68. W Zhang, W Xie, J Fang, PG Wang. Tetrahedron Lett 40:7929–7933, 1999.
69. K Paulvannan, T Chen, JW Jacobs. Synlett 1999:1609–1611, 1999.
70. CJ Creighton, CW Zapf, JH Bu, M Goodman. Org Lett 1:1407–1409, 1999.
71. L Blanco, R Bloch, E Bugnet, S Deloisy. Tetrahedron Lett 41:7875–7878, 2000.
72. S Kobayashi, K Kusakabe, H Ishitani. Org Lett 2:1225–1227, 2000.
73. S Sun, IJ Turchi, D Xu, WV Murray. J Org Chem 65:2555–2559, 2000.
74. GJ Kuster, HW Scheeren. Tetrahedron Lett 41:515–519, 2000.

75. CW Lindsley, LK Chan, BC Goess, R Joseph, MD Shair. J Am Chem Soc 122: 422–423, 2000.
76. LF Tietze, H Evers, E Topken. Angew Chem Int Ed 40:903–905, 2001.
77. U Reiser, J Jauch. Synlett 2001:90–92, 2001.
78. RC Cookson, SS Gupte, IDR Stevens, CT Watts. Org Synth 51:121–125, 1971.
79. W Adam, N Carballeira. J Am Chem Soc 106:2874–2882, 1984.
80. RM Moriarity, I Prakash, R Penmasta. Synth Commun 17:409–413, 1987.
81. DHR Barton, X Lusinchi, JS Ramirez. Bull Soc Chim Fr 1985:849–858, 1985.
82. DL Marshall, IE Liener. J Org Chem 35:867–868, 1970.
83. JG Breitenbucher, CR Johnson, M Haight, JC Phelan. Tetrahedron Lett 39:1295–1298, 1998.
84. BA Dressman, U Singh, SW Kaldor. Tetrahedron Lett 39:3631–3634, 1998.
85. AA Ouyahia, G Leroy, J Weiler, R Touillaux. Bull Soc Chim Belg 85:545–555, 1976.
86. I Grandbert, D Vei-pi, AN Kost, VI Kozlova. J Gen Chem USSR 1961:501–504, 1961.
87. A Studer, S Hadida, R Ferrito, S Kim, P Jeger, P Wipf, DP Curran. Science 275: 823–826, 1997.
88. S Cheng, DD Comer, JP Williams, PL Myers, DL Boger, J Am Chem Soc 118: 2567–2573, 1996.
89. AR Frisbee, MH Nantz, GW Kramer, PL Fuchs. J Am Chem Soc 106:7143–7145, 1984.
90. N Bailey, AWJ Cooper, MJ Deal, AW Dean, AL Gore, MC Hawes, DB Judd, AT Merritt, R Storer, S Travers, ST Watson. Chimia 51:832–837, 1997.
91. TJ Hamilton, M Bush, JC Robbins, SH DeWitt. Pfizer Global Research-Ann Arbor, 2800 Plymouth Road Ann Arbor, MI 48105. Phase Separation Devices for High-Throughput Synthesis, Isolation, & Purification. Poster Day at Wayne State University, Detroit, MI, May 22, 1997.
92. CR Johnson, B Zhang, P Fantauzzi, M Hocker, K Yager. Tetrahedron 54:4097–4106, 1998.
93. CR Johnson, B Zhang, P Fantauzzi, M Hocker, K Yager. Application of supported liquid extraction (SLE) to solution and solid phase library synthesis. Fifth Int Symp, Solid Phase Synthesis and Combinatorial Libraries, London, UK, September 1997.
94. AJP Martin, RLM Synge. Biochem J 35:1358–1366, 1941.
95. AJ Gordon, RA Ford. The Chemist's Companion. New York: Wiley, 1972, pp 371, 382.
96. AS Kertes, M Zangen, G Schmuckler. Analytical and other applications of the principle of solvent extraction. In: J Rydberg, C Musikas, GR Choppin, eds. Principles and Practice of Solvent Extraction. New York: Marcel Dekker, 1992, pp 516–524.
97. SX Peng, C Henson, MJ Strojnowski, A Golebiowski, SR Klopfenstein. Anal Chem 72:261–266, 2000.
98. B Zhang, CR Johnson. Application of solid supported liquid-liquid extraction (SLE): Purification of a library of substituted 3,6-diamino-pyridazine-4-carboxamides from acidic, basic, and water soluble impurities. Pacifichem 2000, Honolulu, HI, December 2000.

99. LA Thompson. Curr Opin Chem Biol 4:324–337, 2000.
100. RJ Booth, JC Hodges. Acc Chem Res 32:18–26, 1999.
101. JC Hodges. Synlett 1999:152–158, 1999.
102. DL Flynn, RV Devraj, W Naing, JJ Parlow, JJ Weidner, S Yang. Med Chem Res 8:219–243, 1998.
103. RJ Booth, JC Hodges. J Am Chem Soc 119:4882–4886, 1997.
104. SW Kaldor, MG Siegel. Curr Opin Chem Biol 1:101–106, 1997.
105. A Cheminat, C Benezra, MJ Farrall, JMJ Frechet. Can J Chem 59:1405–1414, 1981.
106. MG Organ, SW Kaldor, CE Dixon, DJ Parks, U Singh, DJ Lavorato, PK Isbester, MG Siegel. Tetrahedron Lett 41:8407–8411, 2000.
107. CR Johnson, B Zhang. Application of supported liquid extraction to solution and solid phase library synthesis. Sixth Int Symp on Solid Phase Synthesis and Combinatorial Chemical Libraries, York, UK, September 1999.
108. M Winn, D Biswanath, TM Zydowsky, RJ Altenbach, FZ Basha, SA Boyd, ME Brune, SA Buckner, D Crowell, I Drizin, AA Hancock, H-S Jae, JA Kester, JY Lee, RA Mantei, KC Marsh, EI Novosad, KW Oheim, SH Rosenberg, K Shiosaki, BK Sorensen, K Spina, GM Sullivan, AS Tasker, TW von Geldern, RB Warner, TJ Opgenorth, DJ Kerkman, JF DeBernardis. J Med Chem 36:2676–2688, 1993.

2

Quality Control of Combinatorial Libraries Encoded with Electrophoric Tags

A Case Study

Roland E. Dolle
Pharmacopeia, Inc., Cranbury, New Jersey

I. INTRODUCTION

The analysis of parallel libraries produced by solid-phase or solution-phase chemistries is a relatively straightforward problem. Typically, milligram quantities of spatially separated discrete compounds are synthesized, optionally purified by high throughput parallel high performance liquid chromatography (HPLC) and characterized by a variety of classical analytical methods including HPLC, ultraviolet spectrometry (UV), evaporative light scattering detection (ELSD), mass spectrometry (MS), and nuclear magnetic resonance (NMR). The ongoing concern in parallel library compound analysis is the speed (throughput) at which this can be accomplished and associated informatics. It is also highly desirable to have a robust method of direct compound quantitation that does not require the use of an authentic standard. A number of commercial enterprises are working diligently to solve these issues.

The characterization of large encoded combinatorial libraries created through split-pool synthesis is a more challenging venture (1,2). In addition to the shared concerns of analysis speed, data capture, and universal quantitation, the library size and per-bead yield bring another dimension of complexity. A 50,000-member library encoded with electrophoric molecular tags prepared in 200-fold redundancy contains 10 million beads. Depending on the resin (bead) loading and the overall yield of the synthetic sequence, each bead in the library possesses ≤1 nmole of compound and 1–5 pmole of associated tags. The enormity of the library's bead count and the subnanomole quantity of compounds and tags present on the beads preclude characterization of every compound. However, in the context of using large encoded libraries as a lead discovery tool, it can be argued that the rigorous analysis of each library compound is unnecessary. A more appropriate and pragmatic approach is to develop, validate, and implement quality control protocols that ensure a high degree of library fidelity.

In addition to the extensive synthon profiling that is performed by the combinatorial chemist during solid-phase reaction optimization, there are six quality control (QC) measures routinely employed at Pharmacopeia for library creation and compound elution (Fig. 1). These include (a) QC compound synthesis, (b) tag QC, (c) construction-in-process QC, (d) intermediate quality assurance

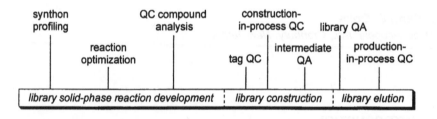

Synthon profiling/reaction optimization: Confirmation of reagent viability, establishing suitability of monomer combinations in the proposed synthesis. >50% profiling of one monomer set against representative members of a second set. Gravimetric yield and purity analysis obtained for >30 compounds.

QC compound analysis: 5-10 library compounds synthesized as on-bead and purified off-bead standards rigorously analyzed to estimate overall library yield and purity.

Tag QC: Tag fidelity confirmed after each tagging step prior to pooling or direct divide.

Construction-in-process QC: Resynthesis of 2-3 QC compounds during actual library construction using bulk reagents and solvents.

Intermediate/library QA: Decode-assisted single-bead LC/MS to confirm physical existence of cleaved library members. Provides information regarding composite library fidelity and the performance of individual synthons.

Production-in-process QC: Production safeguard to ensure optimal elution of library compounds into assay plates.

Figure 1 Quality control protocols along the library pipeline.

(QA), (e) final library QA, and (f) production-in-process QC. Quality control procedures a–c occur during the library synthesis cycle. Intermediate and final library QA are protocols that provide confirmation of the overall success of the synthesis and the performance of individual synthons. Production-in-process QC is a protocol deployed in the production department to ensure the proper elution of compounds from beads into wells of assay plates on a mass scale. The description and application of these QC protocols for library generation are presented in the following case study of a 25,200-member library of statine amides (2,3).

II. SYNTHESIS OVERVIEW OF THE STATINE AMIDE LIBRARY 1

The synthesis of the statine amide library **1** is illustrated in Schemes 1 and 2 (3). Tentagel resin **2** is prepared for the first round of molecular encoding by first derivatizing the resin with bis-Fmoc lysine to furnish resin **3** (Scheme 1). Approximately 17 g of resin is derivatized, corresponding to the desire to prepare the library in 200-fold redundancy (25,200 members × 200 = 5,040,000 beads; each gram of resin contains approximately 300,000 beads, therefore ∼17 g total resin). The derivatization step increases the resin loading capacity by a factor of ∼2 and eliminates primary amine residues that are not compatible with tagging. Resin **3** is split into 40 reaction vessels and tagged with six electrophoric tags in binary fashion. Tagged resins **4** are deprotected and coupled to 4-bromomethyl-3-nitrobenzoic acid. These transformations generate resin batches **5** containing the photolabile linker required for compound attachment and synthesis. Each of the 40 resin batches **5** is reacted with one of 40 primary amines (synthons **A1–A40**, Table 1) as 0.5 M solutions in tetrahydrofuran (THF). Resins **6** so obtained are pooled and split into 63 batches. Each resin batch is coupled to one of 63 Fmoc-amino acids (synthons **B1–B63**) and then submitted to a second binary encoding step, again using six tags. Resin batches **7** are pooled to give the advanced tagged intermediate resin **8**. The library synthesis is completed by subjecting fully tagged intermediate resin **8** to a two-step derivatization sequence as shown in Scheme 2. This is accomplished by first dividing the resin into two batches and removing the Fmoc-protecting group to give **9a, 9b**, then coupling to either Fmoc-statine or Fmoc-phenylalanine statine (synthons **C1, C2**; **9a, 9b** to **10a, 10b**). Resins **10a**, and **10b** are further subdivided into five portions each and subjected to the following sequence: (a) Fmoc deprotection, (b) acylation with one of 10 carboxylic acids (synthons **D1–D10**), (c) amino acid side-chain deprotection. In the case of capping element **D8**, the resin is additionally washed with a 1 M solution of hydrazine in methanol to remove the acetoxy protecting group. The two-step

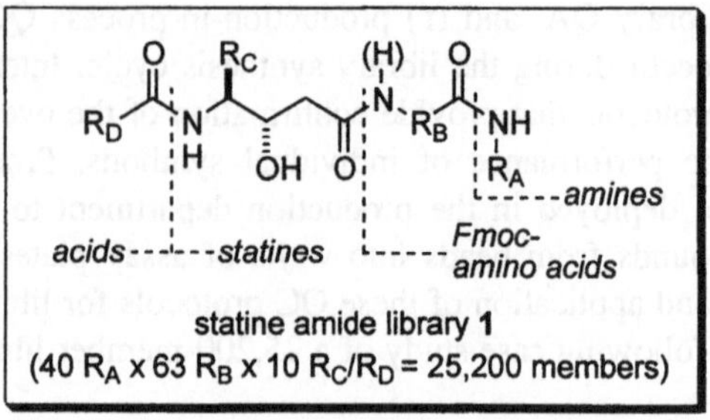

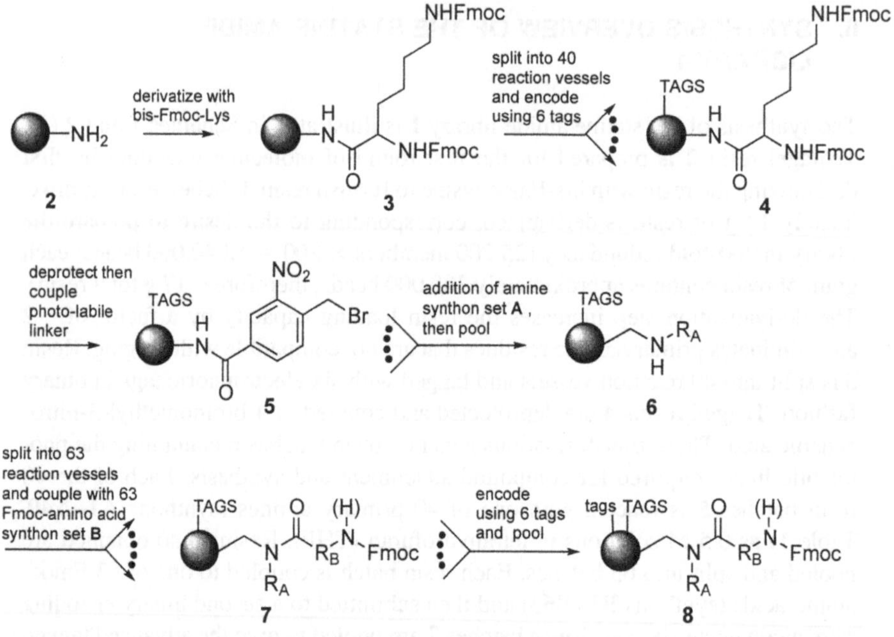

Scheme 1 Encoded synthesis of library **1**: Preparation of advanced intermediate **8**. (From Ref. 3.)

derivatization sequence gave rise to 10 sublibraries **11a–11j**, defined by synthons **C/D**. Each sublibrary contains 2520 unique compounds, or 504,000 compounds considering the 200-fold redundancy (~1.7 g per sublibrary). Compounds are released from the beads (**11a–11j** to **1a–1j**) by exposure to UV light (365 nm) in 80:20 EtOH-water (see Sec. III.B). The statine amide library **1** is thus composed of $40 \times 63 \times 10 = 25,200$ unique compounds.

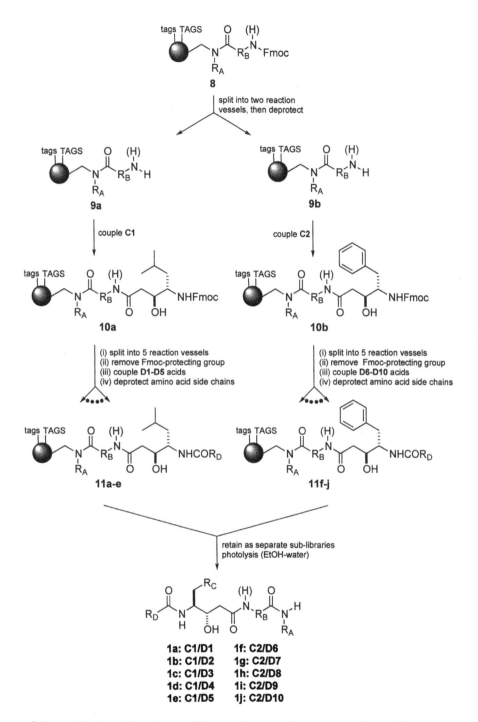

Scheme 2 Encoded synthesis of library **1**: Two-step derivatization of tagged intermediate **8** and completion of synthesis.

Table 1 Basis Set for Library 1

A synthon set

B synthon set

C/D synthon set

III. LIBRARY PRECONSTRUCTION QUALITY CONTROLS

A. Synthon Profiling

Approximately two-thirds of the time invested in a library synthesis is dedicated to defining and optimizing solid-phase reaction conditions. As part of the solid-phase development process, synthon profiling is conducted to establish the suitability of the monomers and combinations thereof in the proposed synthesis. Successful profiling confirms the viability of a reagent and its inclusion in the library, whereas unsuccessful profiling usually leads to its exclusion from the library. Profiling includes testing and analyzing combinatorial reagents in their respective steps, and in many instances each profiled reagent is subsequently carried through to a final library product to verify its compatibility with the whole synthesis. Analyses of profiled reactions are performed by LC/MS. Although it is impractical to examine every synthon combination in the library, 50–100% profiling of monomers in one synthon set is coupled with at least one other monomer from a second monomer set on solid phase. For library 1, batches of resin were derivatized with the photolabile linker and were sequentially (a) reacted with 20 or more of the 40 amines in synthon set **A**, (b) coupled to one or more of the Fmoc-amino acids in synthon set **B**, (c) cleaved from resin, and (d) characterized by LC/MS. Photocleavage is typically carried out by irradiating a suspension of beads in methanol in a glass vial at 365 nm for 30–60 min. The criteria used to decide whether to keep a synthon or a combination of synthons in a library is >85% purity and, in cases where gravimetric yields are obtained, a yield of >200 pM per bead.

Analogously, a small number of **A** synthons attached through the photolabile linker would be used to profile >50% of the Fmoc-amino acids. In the case of profiling synthon sets **C/D**, resin containing the advance intermediate **A1-B3-C1-NH$_2$** was prepared and the 10 carboxylic acid monomers **D1–D10** were individually coupled, cleaved, and subjected to purity analysis. Profiling led to the final selection of synthons in Table 1 and optimized reaction conditions (3).

B. QC Compound Selection and Analysis

During solid-phase reaction development, 5–10 quality control compounds are synthesized in bulk and submitted for more rigorous analysis (Fig. 2). The function of the QC compounds is twofold. First, they are used to establish the optimal photolysis conditions to elute compounds from the whole library. Second, they are used to estimate the overall library yield and purity. These yield data are utilized by biology when reconstituting dried bead eluent to obtain a desired library screening concentration.

Guidelines for QC compound selection include the following.

1. Quality control compounds are representative of the chemistry and cLogP distribution of the library, the distributed range of the cLogP values to equal the central 80% section of the library.

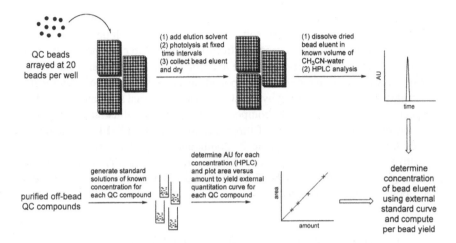

Figure 2 Analysis of QC compounds.

2. A QC compound is synthesized for all significantly different chemistries and library spurs.

3. When criteria permit, a good chromophore should be incorporated into each QC compound.

4. A minimum of five QC compounds per library are synthesized using the same resin and reaction conditions that will be used for library production.

5. Quality control compounds are supplied both as the resin-bound (on-bead) sample (~200 mg beads) and as the corresponding purified off-bead standards (~10 mg compound). Crude and purified ^{1}H NMR, HPLC, and MS together with accurate gravimetric yield and purity estimate (LC/MS) are included with each QC submission. QC standard purity is ≥85%.

A series of internally generated standard operating procedures (SOPs) are used by the analysts to evaluate QC compounds. The SOPs capture reagent and elution solvent selection, apparatus and supplies (e.g., HPLC, detector, master plate and derivative plate type and vendor, centrifuge, heated custom-made UV light box), preparation of QC compound standard solutions, detailed elution procedures for both photochemical and acid-cleavable libraries, HPLC analysis, and data processing and reporting.

Operationally, approximately 20 beads per QC are arrayed as a suspension in isopropanol (IPA) into 96-well filter bottom plates and dried using a speed-vac. The number of beads is accurately recorded, and then they are resuspended in one of two elution solvents for photocleavage: EtOH–water (80:20) or IPA–

water–TFA (80:20:3).* The plates are irradiated at 365 nm in an in-house UV oven, taking four time points per solvent, and subjected to 2 h post-soak at 50°C to ensure diffusion of the compound from the bead. After the post-soak, bead eluent is collected in a derivative plate and dried. The average per-bead yield is obtained by redissolving the dried compounds in a precise volume of acetonitrile–water (80:20) and quantitating by HPLC against an external standard curve of the respective purified off-bead QC compound. This process is illustrated in Figure 2, and the representative analysis of three QC compounds, **12–14**, for the statine amide library **1** is shown in Table 2. Compound **14** contains an indole and is sensitive to photolysis in the presence of trifluoroacetic acid (TFA). On the basis of the QC data, EtOH–water 80:20, and a 2 h irradiation time were selected as the library elution conditions.

IV. LIBRARY CONSTRUCTION QUALITY CONTROLS

A. Construction-in-Process QC

After reagent profiling and reaction optimization are complete and the library's synthons are defined, the next step is to construct the library. Depending on the complexity of the chemistry and the number of library spurs to be generated, the construction process takes 3–4 weeks to complete. This is a relatively short period of time in comparison to the months needed for solid-phase reaction development. The library construction process is initiated by securing a set of reaction vessels in which the solid-phase chemistry will be carried out. The reaction vessels are positioned on a wrist-action shaker and charged with an appropriate amount of resin. Both chemistry and tagging are conducted in the reaction vessels. Set up alongside these library reaction vessels are a second set of companion reaction vessels for construction-in-process QCs. In this QC protocol, two or three of the library's QC compounds are resynthesized in the companion vessels—in real time—using the bulk reagents and solvents employed during library synthesis. The yield and purity of both the resynthesized QC intermediates and final compounds are determined. A robust library construction is inferred upon favorable comparison of the construction-in-process QC analytical data to the original set of data obtained for the same QC compounds.

B. Tag QC

Binary encoding of the library with electrophoric molecular tags also takes place during the construction phase. The molecular tags are diazoketones derived from

---

* Alternatively, 60:40 and 80:20 TFA–acetonitrile are preferred elution solvents for acid cleavage.

Table 2 Representative QC Compounds from Library 1. Average Bead Yield in Two Elution Solvents at Three UV Exposure Intervals.

Compound	UV exposure (min)	80/20 EtOH–water (pmol/bead)				80/20/3 i-PrOH–water–TFA (pmol/bead)			
		Assay 1	Assay 2	Assay 3	Average	Assay 1	Assay 2	Assay 3	Average
12	30	272	332	313	306	432	434	440	435
	60	364	336	369	356	424	469	448	447
	120	444	398	331	391	407	462	421	430
13	30	183	208	233	208	0	0	0	0
	60	245	294	303	281	0	0	0	0
	120	309	354	357	340	0	0	0	0
14	30	624	712	531	622	527	562	554	558
	60	572	419	576	575	587	552	556	572
	120	620	621	675	639	611	692	545	616

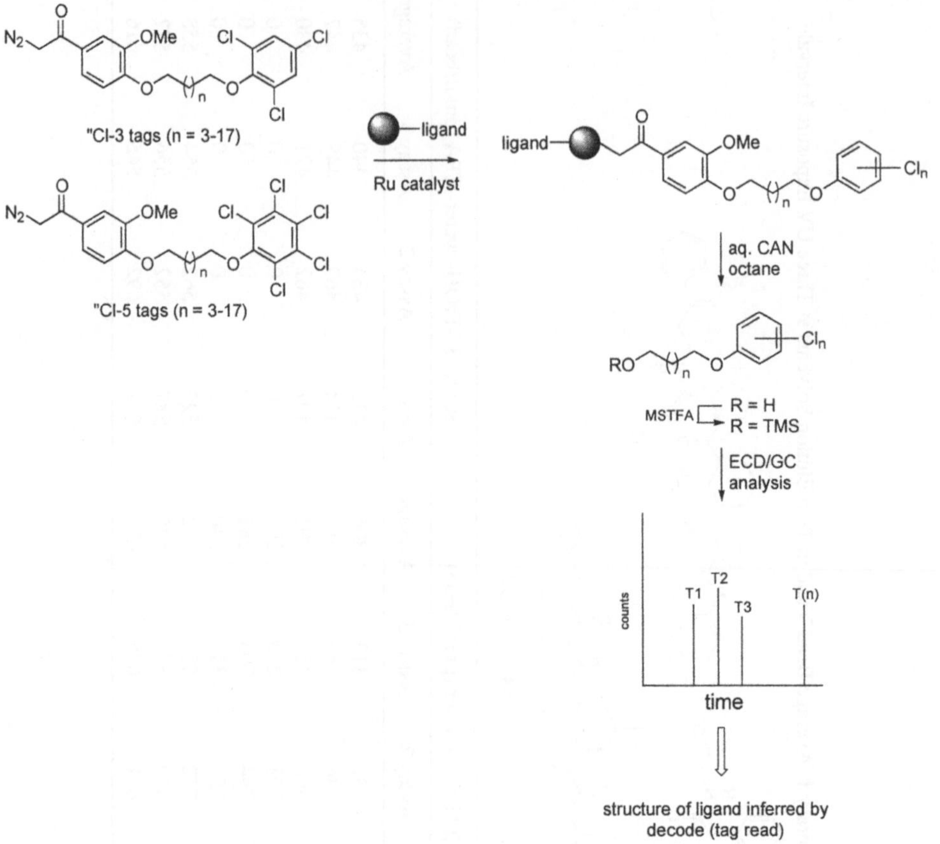

Figure 3 Electrophoric molecular tags and the encoding process.

halophenoxyethers of vanillic acid and are incorporated into the beads (encoded) by a ruthenium-catalyzed carbene insertion reaction (Fig. 3). In a process known as tag decoding, tags are detached from the beads directly in 96-well filter plates upon treatment with aqueous solution of cerium ammonium nitrate. After oxidation, octane is added to extract the tag alcohols, which in turn are converted to their trimethylsilane (TMS) derivatives and analyzed by electron capture detection gas chromatography (ECD/GC). A database maintains a record of each monomer and its associated set of tags, and thus the structure of any given library compound may be inferred by decoding the bead. The number of tags required to encode a set of synthons is given by the formula 2^n. In library **1**, 12 tags were used to encode the entire library of 25,200 compounds. Six tags sufficed to encode the 40 amines of synthon set **A** (2^6 = 64 bits), and a second set of six tags encoded the 63 Fmoc-amino acid synthon set **B**. The **C/D** synthon set was not tagged but was kept spatially separate as 10 sublibraries.

Because library deconvolution is made possible only by ECD/GC tag analysis, it is essential that the tags be incorporated accurately and in sufficient quantity during library synthesis prior to any pooling step. Technicians in a dedicated tagging laboratory encode all library reaction vessels. A tagging SOP defines the amount and tag used per gram of resin, appropriate solvent, and reaction time (the carbene insertion reaction is monitored for the disappearance of diazoketone by HPLC) for reliable and consistent encoding. As a final safeguard, tag QC is performed by retrieving 5–10 beads per reaction vessel, arraying them as single beads either in glass vials or in a 96-well assay plate, and detaching the tags (Fig. 4). Once again a careful record is kept regarding which tags are used to encode which lot of beads. If the QC tag analysis matches the record for that vessel and the tags are present at 1–5 pmoles/bead (peak intensity in the GC

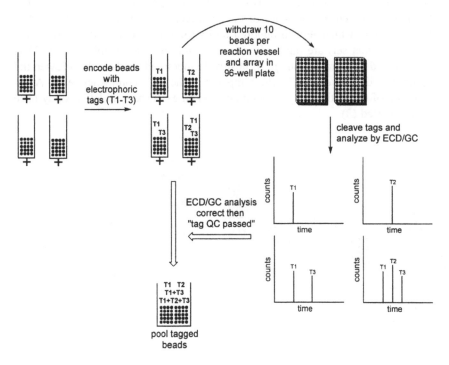

Figure 4 Tag QC protocol.

chromatogram versus standard tag set), a ''pass'' status is acknowledged and bead pooling may proceed. In very rare instances tag QC may fail and the reaction vessel is simply resubmitted for a second round of encoding and tag QC.

C. Intermediate Library QA

The desire for empirical data to confirm compound synthesis led to the development of quality assurance (QA) protocols. QA is based on the combined application of tag decode analysis and single-bead LC/MS and may be performed on tagged intermediates and on the final library. QA provides information on the overall success of an intermediate or final library synthesis and as a measure of the performance of individual synthons.

Figure 5 illustrates the single-bead LC/MS analysis. A tagged library bead is eluted to remove its attached compound. The corresponding bead is then decoded, the putative structure is revealed, and a predicted molecular weight is obtained. The bead eluent is then analyzed by LC/MS to provide the molecular weight of the compound. A comparison of the predicted versus experimentally determined molecular weights gives a ''found (F)/not found (NF)'' answer, confirming the presence or absence of the compound inferred from its tags. QA is the process in which several hundred beads (minimal sample size = $10 \times$ the largest synthon set) are randomly retrieved and analyzed in this fashion. Data are reported as a proportion (p) of positively identified beads with its lower and upper boundary interval (lb–ub) computed at the 95% confidence level. A description of the statistical considerations and an extensive validation study have been reported (3).

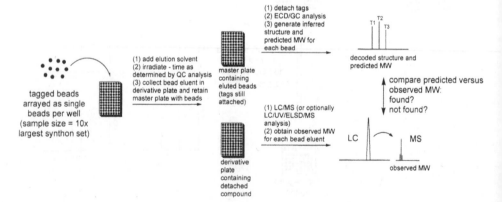

Figure 5 Library QA. Combined application of tag decode analysis and single-bead LC/MS.

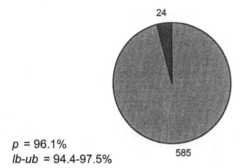

p = 96.1%
lb–ub = 94.4-97.5%

Figure 6 Composite QA for intermediate **8** (F = gray; NF = black).

The advance library intermediate **8** (Scheme 1) was subjected to QA testing. The intermediate represents the fully encoded pooled mixture of the first two synthon sets: 40 **A** × 63 **B** = 2520 compounds. The QA sample size is defined as 10 times the largest synthon set and in this case translates to 630 beads. Experimentally, 609 beads were analyzed, of which 585 compounds were confirmed by LC/MS: p = 96.1%, *lb–ub* = 94.4–97.5% (Fig. 6, Table 3). This high value of p for the composite intermediate underscores the robustness of the synthetic chemistry for the first two steps in the library. One question that may be asked is whether the 4% unconfirmed compounds is due to a single poorly performing synthon. Further details of the performance of the individual synthons are shown in Figures 7a and 7b and in Table 3 (synthon sets **A** and **B**, respectively). Inspection of these histograms reveals that there were no gross synthon failures, many of the synthons showing 100% found rates (Table 3). The only possible exception is the moderate success of **B19**, L-arginine: p = 63.6%, *lb–ub* = 30.1–94.5%.

In addition to compound identification by LC/MS, the semiquantitation of library intermediates **8** is also possible using HPLC/UV/ELSD/MS instrumentation. To accomplish this task, it is first necessary to generate "universal" UV and ELSD quantitation curves with representative standards from the library. In this instance, solutions of four Fmoc-protected amino acids were prepared at five concentrations and UV and ELSD data were collected. The concentration curves are averaged to yield separate general UV and ELSD curves for single-bead quantitation (data not shown). Some 100 beads from resin batch **8** were retrieved, with compounds cleaved and tags decoded as described above. The results of 45 random analyses are shown in Figure 8 (ascending order of UV yield data). The average bead yield obtained was 565 pM/bead by UV and 854 pM/bead by ELSD. Traditionally, UV is not useful for universal quantitation.

Table 3 Proportion (*p*) and Boundaries (*lb–ub*) for the Composite Intermediate **8** and Selected **A** and **B** Synthons

QA	*n*	F	*p* (%)	*lb–ub* (%)
Composite	609	585	96.1	94.4–97.5
Synthon **A**				
1	14	14	100	80.7–100
3	7	7	100	65.2–100
6	11	10	90.9	70.6–100
10	22	22	100	87.3–100
11	10	9	90.9	67.8–100
15	15	15	100	81.9–100
18	14	12	85.7	65.7–100
19	23	21	91.3	78.7–100
25	16	16	100	82.9–100
32	15	15	100	81.9–100
33	19	15	78.9	56.9–97.8
40	16	16	100	82.9–100
Synthon **B**				
1	9	9	100	71.7–100
3	12	11	91.7	73.0–100
9	18	18	100	84.7–100
19	11	7	63.6	30.1–94.5
20	6	5	83.3	47.9–100
30	13	10	76.9	53.0–100
31	11	9	81.8	56.9–100
32	8	8	100	68.8–100
46	5	5	100	54.9–100
58	10	9	90	67.8–100
59	12	12	100	77.9–100
60	5	5	100	54.9–100
61	8	8	100	68.8–100
63	7	7	100	65.2–100

Because each compound from pooled intermediate **8** contains an Fmoc group and it would be expected to be the dominant chromophore, UV can be used here for semiquantitation with a reasonable level of confidence. As reflected in Figure 8, evaporative light scattering consistently overestimates single-bead yields relative to UV and in several instances exceeds theoretical yield. It is our general experience that ELSD is not well suited for single-bead quantitation. This is due to the fact that data are acquired near the instrument's limit of sensitivity and

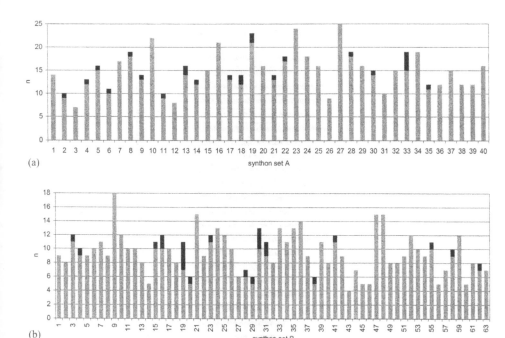

(a)

(b)

Figure 7 Quality assurance for intermediate **8**. (a) Synthon set **A**; (b) Synthon set **B**.

the detector's response is sensitive to changes in HPLC gradient and bead matrix, especially at single-bead compound concentration (J. Guo, unpublished observation).

V. LIBRARY POSTCONSTRUCTION AND ELUTION QUALITY CONTROLS

A. Library QA

As a final assessment of library fidelity and the performance of individual synthons, single-bead LC/MS is performed on the fully tagged library (see Figure 5 and QA discussion in Section IV) (3). The QA data obtained from library **1** are shown in Figures 9 and 10 and Table 4. A total of 1902 beads (30 × sampling of largest synthon set) were analyzed, and 1598 were positively confirmed: $p = 84\%$, $lb-ub = 82.3-85.6\%$. Examining the analysis at the sublibrary level reveals that the two-step derivatization sequence to convert intermediate **8** into sublibra-

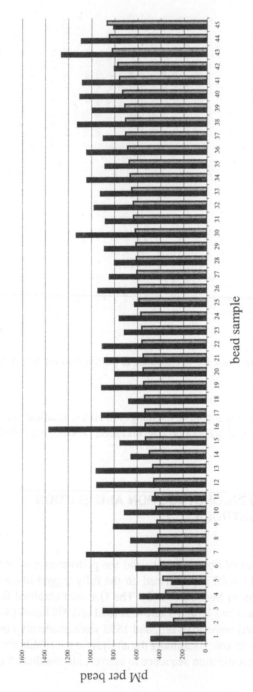

Figure 8 Ultraviolet (gray) and ELSD (black) results from pooled intermediate **8** computed using calibration curves.

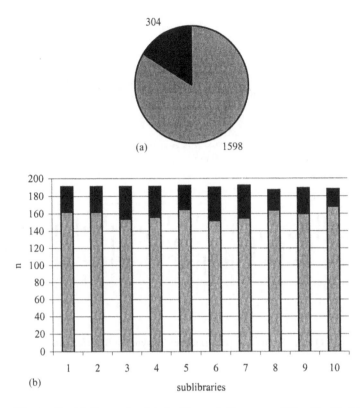

Figure 9 (a) Composite QA for library **1** (F = gray; NF = black). (b) Library **1** QA: synthons **C/D**.

ries **1a–1j** was successful; i.e., overt synthetic failures were not detected. Values for p for the sublibraries ranged from about 80% to 90%. Individual synthon performances (p_i) also generated from the data reveal some interesting trends (Fig. 10). The majority of the 40 R_A and 63 R_B units were found in cleaved library compounds ($p_i > 75\%$), including hydrophilic synthons possessing negatively charged moieties (**A29, A30, B21, B22**, and **B26, B27**). Noted exceptions include the basic charge amines (**A17–A19** and **A32**), potentially photosensitive synthons (sulfur-containing synthons, **A6, B32, B33**, and furfurylamine, **A11**), a limited set of polar **B** amino acid synthons, e.g., L- and D-asparagine and glutamine (**B20, B21** and **B28, B29**), and peptidomimetic synthon **B46**. The poor to mediocre performance of these synthons arose after the first two combinatorial steps because problems were not detected during the QA analysis of intermediate **8**. The 10–12% decrease in positive ID observed in final versus intermediate QA (Fig. 7,

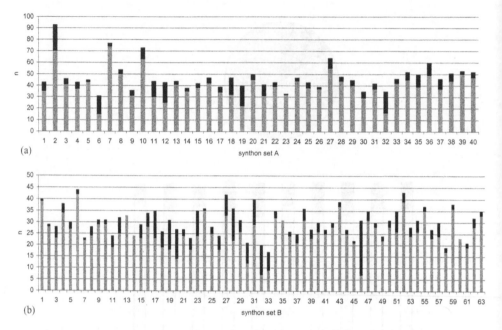

Figure 10   Library **1** QA. (a) Synthon **A** performance; (b) synthon **B** performance. (F = gray; NF = black.)

Table 3) may be attributed to the poor showing of library compounds containing photosensitive and basic charged synthons. The data suggest that compounds containing such synthons would be underrepresented during biological screening.

B.   Production-in-Process QC

Once QA is complete, the library is turned over to the production department for bead arraying and compound elution into assay plates. Prior to bead arraying, each sublibrary is wet filtered to remove broken beads and bead fragments that may have formed during construction. An isobuoyant suspension of beads is prepared, and the beads are arrayed (multiple or single beads per well) into 96-well filter-bottom master plates. During the arraying process, beads from three of the library's QC compounds are also arrayed and designated as production-in-process QCs (PIP-QCs). The suspension solvent is removed, and the beads are washed with methanol. The optimal elution solvent (EtOH–water 80:20 as determined for the QC compounds **12–14**) is added to each well. The plates are individually sealed in a plastic bag (to minimize solvent evaporation during cleavage) and

Table 4 Proportion (p) and Boundaries (lb–ub) for Composite Library **1** and Selected **A**, **B**, and **C/D** Synthons

QA	n	F	p (%)	lb–ub (%)
Composite	1902	1598	84.0	82.3–85.6
Sublibrary **C/D**				
1b	191	162	84.2	79.3–90.0
1d	191	156	81.7	75.8–87.3
1f	190	152	80.0	73.9–85.8
1h	187	164	87.7	82.6–92.5
1j	188	168	89.4	84.5–93.8
Synthon **A**				
1	50	42	84.0	72.3–94.4
4	50	46	92.0	82.8–99.5
6	34	14	41.2	23.5–59.3
9	44	41	93.2	85.3–100
11	44	13	29.5	15.5–44.5
18	55	31	56.4	42.3–70.2
19	39	12	30.8	15.6–46.9
32	33	10	30.3	13.8–47.9
33	50	46	92.0	82.8–99.5
39	39	38	97.4	91.9–100
Synthon **B**				
3	28	23	82.1	65.3–96.7
9	32	30	93.8	84.6–100
19	32	19	59.4	40.6–77.5
20	31	19	61.3	42.3–79.6
32	22	14	63.6	40.9–85.2
36	26	24	92.3	81.1–100
37	29	26	89.7	78.0–100
45	23	22	95.7	85.7–100
46	39	7	17.9	5.6–31.9
58	19	16	84.2	67.0–100
59	41	39	95.1	87.9–100
60	24	22	91.7	79.5–100
61	21	20	95.2	84.3–100

laid out in a 10 × 10 array on a plate tray. In the case of library **1**, the arrayed plates are placed in a custom high intensity UV light box and irradiated for 2 h followed by a 2 h post-soak at 50°C. The bags are opened, and each master plate is coupled with a derivative plate. The plate pairs are centrifuged. The master plates containing the beads are stored for subsequent decoding, and the derivative plates containing solutions of compounds and PIP-QCs are dried (GeneVac). The

dried PIP-QC plates are reconstituted in acetonitrile–water and analyzed by LC/
UV. If the yield of the PIP-QCs is >70% of the original QC yields, library elution
is deemed successful and the dried derivative plates are made available for
screening.

VI. BIOLOGICAL EVALUATION OF LIBRARY 1
AND CORROBORATION WITH QC/QA DATA

Statine, a known transition state mimetic for aspartic acid proteases (4) is the
central pharmacophore of library 1. The library was evaluated against two proteases of this class, human cathepsin D (cat D) and malarial plasmepsin II (plm
II) (3,5,6). Screening was initially carried out in "survey mode" at 30 beads per
well (0.5 library equivalent, 1260 beads per sublibrary) to establish the most
active sublibrary. Sublibraries 1a (isovalerylamido-Val-statine, C1/D1) and 1j
(Z-Val-phenylalanine statine, C2/D10) showed significant activity, with 1j dis-
playing greatest potency. This latter sublibrary was further evaluated at the single-
bead level (2 library equivalents, ~5000 beads), and only those wells (beads)
showing ≤30% activity remaining at the estimated screening concentration of
25 nM were decoded. A total of 131 decodes were obtained (55 for plm II and
76 for cat D), and the corresponding **A** and **B** synthon frequencies are shown in
Figure 11. Five compounds, **15–19** (replicate structures), which appeared as the
most potent inhibitors in the single-bead screen, were resynthesized to confirm
their activity (Table 5).

Nascent structure–activity relationships (SARs) emerging from the screen
suggested that the enzymes tolerate a rather broad structural range of neutral,
hydrophobic A synthons. Examples of preferred A synthons include the 2-OMe-,
3-OMe-, and 4-OMe-benzylamines (**A14–16**, cat D and plm II), phenethylamine
(**A23**, cat D), its 4-Cl analog (**A25**, cat D and plm II), and neopentylamine (**A33**,
plm II). In contrast, both enzymes possess a strong bias toward specific, stereo-
chemically defined, hydrophobic amino acids. Of the 63 amino acids (synthon
set **B**) available, only seven were found in the plm II decodes (L-thienylalanine,
B9; L-cyclohexylalanine, **B24**; and L-valine, **B42**, present in 38 of 55 total decodes). For cat D, eight L-amino acids were observed, including **B9** and **B42** and
a unique set of non-natural amino acids **B3**, **B45**, **B53**, and **B58** that imparted a
high degree of selectivity for this enzyme. These structural features are represented in the resynthesized compounds **15–19** (Table 5). In virtually every instance, the observed synthons correlated with high positive identification ($p_i >$
75%) in the library QA analysis (Figs. 10a and 10b and Table 4).

Conspicuously absent from the decoded structures were the neutral, lipo-
philic synthons methylthioethylamine (**A6**) and furfurylamine (**A11**) and those
synthons containing a basic nitrogen, e.g., the pyridinylmethylamines **A17–A19**,

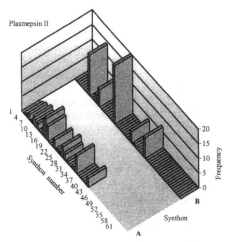

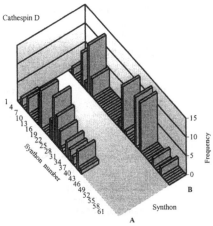

A synthon/frequency: 3/1, 4/1, 5/1, 7/1, 8/3, 9/2, 10/2, 13/2, 14/6, 15/7, 16/5, 20/3, 22/1, 23/1, 24/1, 25/4, 26/2, 28/1, 31/1, 33/6, 40/3
B synthon/frequency: 9/14, 13/3, 15/6, 24/20, 34/4, 36/1, 42/7

A synthon/frequency: 3/2, 7/4, 14/8, 15/10, 21/5, 22/5, 23/12, 25/8, 27/5, 28/1, 31/4, 33/3, 35/2, 38/5, 39/1, 40/1
B synthon/frequency: 3/2, 9/9, 11/2, 13/13, 15/5, 32/2, 34/12, 36/1, 42/13, 45/13, 53/2, 58/2

Figure 11 Frequency of the **A** and **B** synthons obtained from decoded structures (sublibrary **1j**).

A36, A37, and imidazoylpropylamine **A32**. Library QA data, however, indicate that the performance of these synthons, especially **A6, A11, A19**, and **A32**, cannot be estimated with reasonable certainty ($p \sim 50\%$; $lb–ub \sim 14–90\%$). Such synthons are mediocre performers at best, and compounds containing such elements will invariably be underrepresented during screening. The screening SAR cannot be relied upon as a predictive guide to the activity of compounds containing poorly performing synthons; their activity can be known only through resynthesis.

With this in mind, compounds **20–23** were synthesized and evaluated against cat D and plm II. These putative library compounds, not found in the screen, are analogs of inhibitor **15** ($K_i = 29$ nM, plm II; $K_i = 44$ nM, cat D) where the 2-methoxybenzylamine synthon **A14** is exchanged with four mediocre performing synthons: **A6, A11, A19**, and **A32**. Analogs **20** and **21** containing the neutral, hydrophobic synthons **A6** (**20**: $K_i = 12$ nM, plm II; $K_i = 44$ nM, cat D) and **A11** (**21**: $K_i = 26$ nM, plm II; $K_i = 46$ nM, cat D), respectively, possess activity and selectivity comparable to **15**. Most significant is the discovery that positively charged, hydrophilic amines are well tolerated by plm II. Statines **22** ($K_i = 15$ nM, plm II; $K_i = 140$ nM, cat D) and **23** ($K_i = 7.0$ nM, plm II; $K_i = 530$ nM, cat D) containing **A19** and **A32** are the most potent and selective plm II inhibitors identified from the library (e.g., **23**: 75-fold selective). In the absence of QA data, it would have been erroneously concluded that these hydrophilic

Table 5 Inhibition Constants (K_i) for Resynthesized Compounds Against Plasmepsin II (plm II) and Cathepsin D (cat D)

Resynthesized replicate structures **15-19** decoded from the library

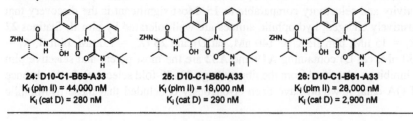

15: D10-C1-B24-A14
K_i (plm II) = 29 nM
K_i (cat D) = 44 nM

16: D10-C1-B24-A8
K_i (plm II) = 16 nM
K_i (cat D) = 32 nM

17: D10-C1-B58-A33
K_i (plm II) = 210 nM
K_i (cat D) = 3.0 nM

18: D10-C1-B45-A21
K_i (plm II) = 29 nM
K_i (cat D) = 4.0 nM

19: D10-C1-B53-A23
K_i (plm II) = 72 nM
K_i (cat D) = 18 nM

Analogs of **15** exchanging **A14** for weak performing synthons **A6, A11, A19,** and **A32** as determined by library QA.

20: D10-C1-B24-A6
K_i (plm II) = 12 nM
K_i (cat D) = 44 nM

21: D10-C1-B24-A11
K_i (plm II) = 26 nM
K_i (cat D) = 46 nM

22: D10-C1-B24-A19
K_i (plm II) = 15 nM
K_i (cat D) = 140 nM

23: D10-C1-B24-A32
K_i (plm II) = 7.0 nM
K_i (cat D) = 530 nM

Analogs of **17** exchanging **B58** for strong performing synthons **B59-61** as determined by library QA.

24: D10-C1-B59-A33
K_i (plm II) = 44,000 nM
K_i (cat D) = 280 nM

25: D10-C1-B60-A33
K_i (plm II) = 18,000 nM
K_i (cat D) = 290 nM

26: D10-C1-B61-A33
K_i (plm II) = 28,000 nM
K_i (cat D) = 2,900 nM

synthons are inferior to the hydrophobic amines frequently observed in the decoded structures, and valuable SARs leading to the synthesis of selective inhibitors **22** and **23** would have been lost.

As a concluding example, the D-Tiq (**B58**) residue was found to impart potent and selective cat D inhibition (**17**: $K_i = 3$ nM, 70-fold selective versus plm II). The compounds containing the isomeric tetrahydroisoquinoline synthons **B59–B61** are presumed less active, based on the screening data. Library QA substantiates this hypothesis by providing empirical evidence that synthons **B58–B61** are strong performers in the library ($p_i > 85\%$). Confirmation of the SAR supposition was secured by the synthesis and K_i determination for the series **24–26** ($K_i > 280$ nM) versus statine **17**.

VII. SUMMARY

Pharmacopoeia has developed and put into routine practice a series of quality control protocols to ensure the highest degree of library fidelity. Quality control steps occur throughout the entire library construction pipeline. Extensive synthon profiling, rigorous analysis of QC compounds, final confirmation of library chemistry (library QA), and PIP-QC analysis as described in this chapter are essential to ensure successful library synthesis and production elution. Final library QA provides empirical information regarding overall fidelity and an indication of the performance of individual synthons. Library QA is an especially powerful analytical tool that serves to corroborate and enhance the value of the SAR screening data. Library quality control is a dynamic process, and new and improved QC measures, e.g., statistically based single-bead quantitation, continue to be investigated.

REFERENCES

1. JJ Baldwin, RE Dolle. Annu Rep Comb Chem Mol Diversity 1:287–297, 1997.
2. RE Dolle, J Guo, W Li, N Zhao, JA Connelly. Mol Diversity 5:35–49, 2000.
3. RE Dolle, J Guo, L O'Brien, K Bowman, M Piznic, W Li, W Egan, CL Cavallaro, A Roughton, X Choa, J Reader, MW Orlowski, B Jacob-Samuel, C DiIanni Carroll. J Combin Chem 2:716–731, 2000.
4. AJ Barrett. In: AJ Barrett, ed. Proteinases in Mammalian Cells and Tissues. New York: Elsevier/North Holland Biomedical Press; 1977, pp 209–248.
5. CD Carroll, H Patel, TO Johnson, T Guo, M Orlowski, Z-M He, CL Cavallaro, J Guo, A Oksman, IY Gluzman, J Connelly, D Chelsky, DE Goldbert, RE Dolle. Bioorg Med Chem Lett 8:2315–2320, 1998.
6. CD Carroll, TO Johnson, S Tao, G Lauri, M Orlowski, IY Gluzman, DE Goldbert, RE Dolle. Bioorg Med Chem Lett 8:3203–3206, 1998.

3

How to Optimize Reactions for Solid-Phase Synthesis of Combinatorial Libraries Using RF-Tagged Microreactors

Rongshi Li
ChemBridge Research Laboratories, San Diego, California

I. INTRODUCTION

Combinatorial chemistry has expanded rapidly in the last decade, in both the development and the application of novel chemistry for the synthesis of compound libraries for lead discovery and in lead optimization in the fields of biotechnology, catalysts, new materials, agriculture, and pharmaceuticals. To achieve high productivity in combinatorial compound library synthesis, both parallel synthesis and the split-and-pool technique have evolved and gained significant acceptance. Parallel synthesis can produce discrete compounds in multimilligram quantities. However, only a limited number of compounds can be synthesized per unit time and per reaction. The split-and-pool technique can address this problem of parallel synthesis; however, it yields mixtures that need to be deconvoluted. A hybrid technique referred to as "directed sorting" addresses these disadvantages of parallel and split-and-pool synthesis.

In directed sorting, resin beads and a radiofrequency (RF) tag are enclosed in a microreactor for solid-phase reactions. RF tagging of microreactors provides convenient identification of compounds, because each microreactor is assigned

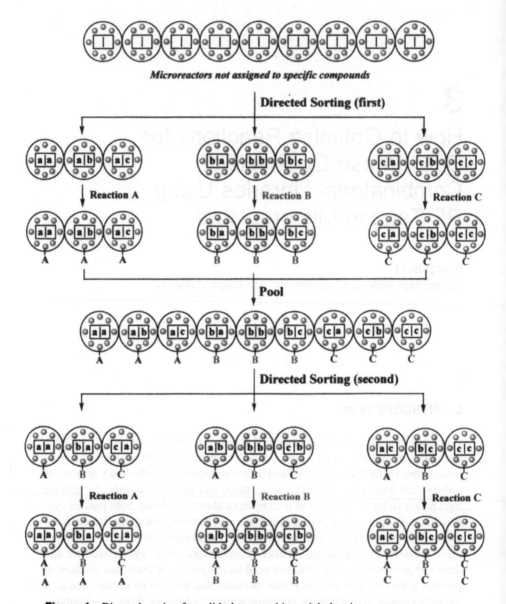

Microreactors not assigned to specific compounds

Directed Sorting (first)

Reaction A   Reaction B   Reaction C

Pool

Directed Sorting (second)

Reaction A   Reaction B   Reaction C

Figure 1   Directed sorting for solid-phase combinatorial chemistry.

to one specific compound at the first directed sorting step. It also enables the microreactors to be sorted between steps of the synthetic sequence. Synthetic history is tracked with the RF tagging method (1,2), and the final product is archived at the end of combinatorial library synthesis. Split-and-pool combinatorial synthesis using RF-tagged microreactors can generate libraries of a large number of discrete compounds (3–5). The directed sorting method is illustrated in Figure 1.

II. CHOICE OF SOLID SUPPORTS

Like choosing the right reagents for conventional solution-phase reactions, the choice of solid supports is critical to the successful synthesis of combinatorial libraries. The following discussions are focused on the most commonly used aldehyde resins as examples.

Aldehyde functionalized resins (Fig. 2; **1**, FMPB; **2**, FMPE; **3**, DFPEM; **4**, DFPE; **5**, BAL; **6**, indole resin) are versatile linker systems in solid-phase organic synthesis for the production of secondary amides, sulfonamides, ureas, carbamates, and guanidines (Fig. 3). These linkers are particularly attractive because a wide range of primary amines, acyl chlorides, sulfonyl chlorides, isocyanates, and chloroformates are commercially available, and carbodiimides (**6**) are readily accessible for combinatorial library construction. Most of these aldehyde functionalized resins are acid-labile linkers based on alkoxybenzaldehydes (**1–5**) except for the indole resin (**6**). Early linkers used for synthesis of carboxyamides and sulfonamides were based on trialkoxy systems related to the amide-releasing PAL resin (7–11), although such electron-rich systems were not always required to facilitate cleavage when the leaving group was a secondary amide. Therefore, dialkoxybenzaldehyde linkers have been developed (12–14). Indole resin has been used in the same manner as the alkoxybenzaldehyde-based supports (15).

Most reactions using these aldehyde functionalized resins involve reductive amination with excess of an amine and a reducing agent such as sodium triacetoxyborohydride in dimethylformamide (DMF) or sodium cyanoborohydride in trimethylorthoformate (TMOF), converting the formyl group to the appropriate secondary amines. The choice of aldehyde functionalized resins depends on the chemistry sequence. For instance, if reduction using lithium aluminum hydride (LAH) is part of the synthetic sequence, then aldehyde functionalized resins FMPB (**1**), BAL (**5**), and indole resin (**6**) are not suitable because of premature linker cleavage. In that case, FMPE (**2**), DFPEM (**3**), and DFPE (**4**) should be used. Another consideration in choosing the solid support is the cleavage rate, especially for compounds that are prone to decomposition under strong acidic

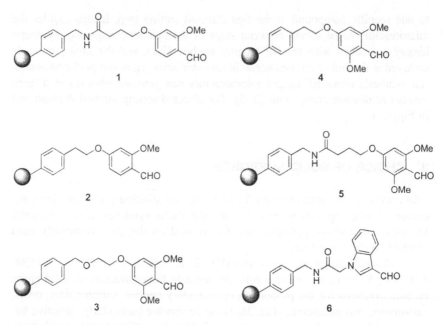

Figure 2 Structures of aldehyde functionalized resins. **FMPB (1)**, 4-(4-formyl-3-methoxyphenoxy)butyryl aminomethyl polystyrene; **FMPE (2)**, 2-(4-formyl-3-methoxyphenoxy)ethyl polystyrene; **DFPEM (3)**, 2-(3,5-dimethoxy-4-formylphenoxy)ethoxymethyl polystyrene; **DFPE (4)**, 2-(3,5-dimethoxy-4-formylphenoxy)ethyl polystyrene; **BAL (5)**, 4-(3,5-dimethoxy-4-formylphenoxy)butyryl aminomethyl polystyrene; **indole resin (6)**, polystyrene-indole-CHO.

conditions. Among these alkoxybenzaldehyde and indole resins, the indole resin was found to be the most acid-labile (16). Other factors such as interference of the linker with monitoring reactions should also be taken into account. FMPB (1) has a linker amide bond with a stretching frequency very similar to that of the aldehyde carbonyl group. This interferes with monitoring the reductive amination step using Fourier transform infrared analysis (FTIR). In that case, FMPE (2) resin can be used instead. A frequently observed limitation of FMPB (1) resin is that the release of the secondary amides containing multiple basic sites close to the amide backbone resin linkage is extremely sluggish.

Once solid supports are selected, an immediate action will be to try out the chemistry on them. Several examples have demonstrated that solid-phase supports are like solvents, and reaction optimization should be done on exactly the same support on which the library is produced (17,18).

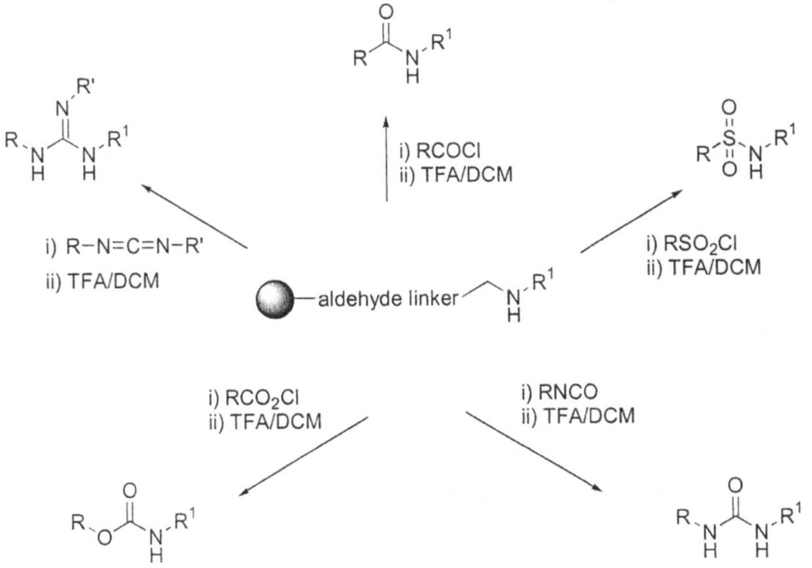

Figure 3 Applications of aldehyde functionalized resins.

The reactivities of 2-chlorotrityl chloride resins from two different commercial sources were compared. Synthesis of α,β-unsaturated ketones was performed on these resins using the Horner–Emmons reaction (19) (Scheme 1) and Claisen–Schmidt condensations (20) (Scheme 2). As shown in Table 1, the product yields ranged from 69% to 88% under Horner–Emmons reaction conditions, and the percentage of the unreacted starting material was about 10% or less regardless of the kind of resins used. This is a very clean reaction on the solid phase, because nothing but the desired product and the starting material was observed from the NMR spectra. For Claisen–Schmidt condensations, the conversion yield varies according to the resins used. For resin A (Table 1, entry 1), 88% of the desired product was obtained and less than 5% of the unreacted starting material was recovered. Claisen–Schmidt condensations do not work for resin B (Table 1, entry 2); all that was recovered after cleavage was the starting material, 4-hydroxyacetophenone. The conversion using resin C (entry 3) was only about one-half, whereas 81% of the desired product was observed for resin D (entry 4), with <5% of the unreacted starting material left. These results are quite convincing, because resin-bound 4-hydroxyacetophenone in all four microreactors was treated under identical reaction conditions with 4-bromobenzaldehyde and NaOH in the same reaction vessel.

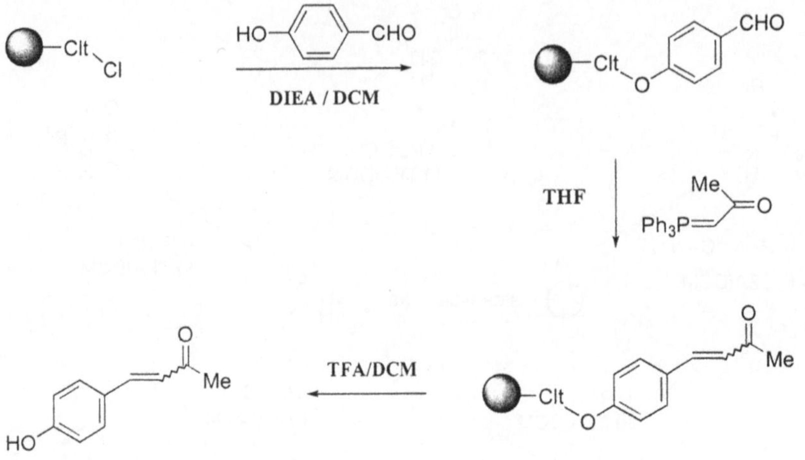

Scheme 1 Synthesis of α,β-unsaturated ketone via Horner–Emmons reaction.

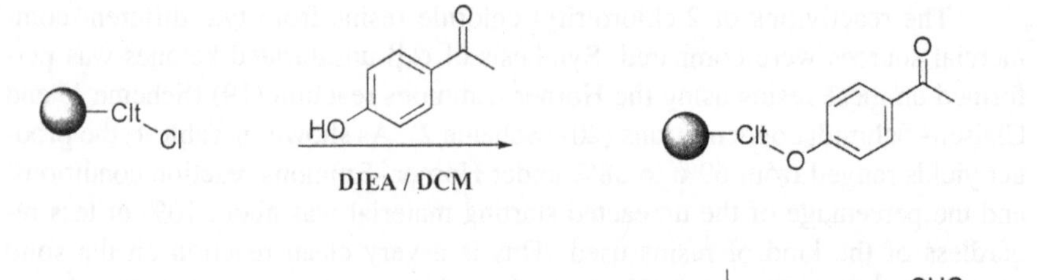

Scheme 2 Synthesis of α,β-unsaturated ketone via Claisen–Schmidt reaction.

Table 1 Comparison of Horner–Emmons and Claisen–Schmidt Reactions

			Yield (%)[a]			
			Horner–Emmons		Claisen–Schmidt	
Entry	Resin source	Loading (mmol/g)	Product	Starting material	Product	Starting material
1	A	1.5	69	7	88	<5
2	B	2.1	80	9	0	100
3	C	1.22	88	12	44	56
4	D	1.34	84	6	81	<5

[a] Yields were determined by ^{1}H NMR.

III. SYNTHETIC ROUTE DEVELOPMENT AND REACTION OPTIMIZATION

Development of synthetic routes normally takes time, especially for those solid-phase chemistries that do not have literature precedents. Synthetic route validation and optimization of reaction conditions are key to the successful synthesis of a combinatorial library. Optimization of reaction conditions on the solid phase is more convenient and more efficient than that of solution chemistry using RF-tagged microreactors such as the Kan reactors developed by IRORI.* Each microreactor can be "labeled" with an RF tag inside, and any combination of reaction conditions can be performed except for some conditions that are unsuitable for the polypropylene mesh of the microreactors. After the reactions, the microreactors from different reaction conditions can be pooled and worked up together. For instance, it will take no more than 50 reactions for the optimization of a five-step sequence, with 10 different reaction conditions each step (time, temperature, solvent, catalyst, and reagents, etc.). The workup of 50 reactions can be achieved by a simple washing of the pooled microreactors rather than working up 50 individual reactions of solution-phase chemistry.

Most examples for amide bond formation reported on solid supports use 5–10 equivalents of an amino acid, a coupling reagent, and Hunig's base. There are some cases in which fewer equivalents of both amino acids and coupling reagents have to be used owing to cost considerations. For example, non-natural amino acids and coupling reagents such as HATU are very expensive. Even for inexpensive natural amino acids and cheap coupling reagents, unnecessary waste can be avoided if reaction conditions are optimized to use fewer equivalents of reagents.

An example of the use of RF-tagged microreactors for the optimization of coupling conditions is shown in Table 2. The purpose of this optimization is to find conditions using minimum amounts of non-natural amino acids and coupling reagents. Resin-bound benzylamine in microreactors with RF tags was coupled with Fmoc-piperidine-4-carboxylic acid, using five different coupling reagents, two different solvents, and two or five (as a control) equivalents of coupling reagents. The yield was determined spectrophotometrically after removal of the Fmoc group. After a total of only 12 reactions under nine conditions, optimized coupling reaction conditions were established. As shown in Table 2, only 2 equivalents of the non-natural amino acid and coupling reagent, instead of the 5–10 equivalents commonly used, is required for complete coupling reactions. The yield is generally higher using DCM as solvent than when using DMF. For instance, DIC/HOBt in DCM gave 90% yield, whereas only 12% was observed

---

* Kan reactors are designed to be loaded with solid phase resin beads and an RF tag. Reactions on resin take place by allowing reagents to flow through the outer mesh walls of the Kan.

Table 2 Optimization of Coupling Conditions

Entry	Acid equiv	Coupling reagent	Solvent	Yield (%)[a]
1	2	DIC/HOBt	DCM	90
2	2	PyBOP	DCM	92
3	2	PyBrop	DCM	99
4	2	HATU	DCM	100
5	2	TFFH	DCM	94
6	2	DIC/HOBt	DMF	12
7	2	PyBOP	DMF	100
8	2	PyBrop	DMF	83
9	2	HATU	DMF	100
10	2	TFFH	DMF	67
11	5	HATU	DCM	100
12	5	HATU	DMF	100

[a] Determined spectrophotometrically at 301 nm.

using DMF as the solvent. TFFH also works better in DCM than in DMF, and the same is true for PyBrop. On the other hand, PyBOP works better in DMF than in DCM. Quantitative yields can be achieved for HATU-mediated coupling using only 2 equivalents of amino acid with either DCM or DMF as the solvent.

IV. SELECTION AND VALIDATION OF BUILDING BLOCKS

There are many ways to select building blocks for combinatorial library synthesis. The most common one is based on both chemical diversity calculation results and medicinal chemistry knowledge. Once building blocks are selected, one-dimensional reactions can be performed. For instance, for a 10,000-compound library from a four-step reaction sequence and 10 building blocks at each step [A (10) × B (10) × C (10) × D (10)], at least four sublibraries must be synthesized to test the scope of the chemistry and the behavior of the building blocks. To test each set of building blocks, only one building block from other steps is selected for testing. This selection could be either arbitrary or based on knowledge of the building blocks. Normally, representative building blocks of variable reactivity toward the resin are desirable. A total of 40 reactions for four libraries should be carried out. This includes library 1 [A (10) × B (1) × C (1) × D (1)],

Table 3 Diagonal Testing of Building Blocks

BB	A1	A2	A3	A4	A5	A6	A7	A8	A9	A0
B1	A1B1									
B2		A2B2								
B3			A3B3							
B4				A4B4						
B5					A5B5					
B6						A6B6				
B7							A7B7			
B8								A8B8		
B9									A9B9	
B0										A0B0

library 2 [A (1) × B (10) × C (1) × D (1)], library 3 [A (1) × B (1) × C (10) × D (1)], and library 4 [A (1) × B (1) × C (1) × D (10)]. These 40 reactions can cover only a very small set of combinations of the 10,000-compound library. Normally, some problematic building blocks can be identified after the first run. Further testing is normally required. The problematic building blocks can be tested by using different combinations with other sets of building blocks. This could exclude the possibility of the bad combinations done previously. If the problematic building blocks still gave unsatisfactory results, they should be excluded from the current library, or different route(s) of chemistry should be pursued.

Another way of testing building blocks is the diagonal synthesis of a matrix of two-dimensional building blocks as shown in Table 3. Only 10 reactions are needed to test 20 building blocks. Normally this one test will find some problematic building blocks, and further testing will be necessary.

V. COMBINATORIAL SYNTHESIS OF A DISCRETE COMPOUND LIBRARY

Even though one-dimensional building block testing or two-dimensional diagonal building block testing gave satisfactory results, such testing cannot guarantee that the final library will behave in the same way, because only limited combinations of the building blocks are tested. It is always a good idea to use extra microreactors as "control Kans" to monitor each step of the library production. Control microreactors can be either of the same type with different colors or different sizes so they can be easily picked up during the sorting process. They can be included either in the main library or as a separate library. This process has several advantages for monitoring the library production. First, control Kans can provide information on how good the reaction of each step is. This can be achieved, for instance, by a color test such as the chloranil test or the Kaiser test if applicable. If results from a specific step are not satisfactory, the library synthesis in that step could be fixed by repeating the chemistry—for instance, by double coupling. Single-bead Fourier transform infrared spectroscopy (FTIR) (21) is a very powerful technique to monitor reactions of functional group transformations. Control Kans can be effectively used for this purpose. Second, the yield information from each step can be obtained after further reaction in the control microreactor or cleavage of resin in the control Kans. Finally, the overall yield of the library can be estimated after the cleavage of the final set of control compounds.

How should one design and include the control microreactors for the library synthesis? This depends on the kind of library to be produced. In general, if it is desirable to know the yield of each building block from each step, then one set of building blocks for each step must be included and the yield can normally be obtained by cleavage of the intermediates from that step. A more efficient design of control microreactors employs diagonal synthesis of the small library, the strategy discussed previously and shown in Table 3. Building blocks used as controls could be the whole set or a small selection of building blocks based on the knowledge obtained from the building block tests. For instance, a small 4 × 4 sublibrary can be designed by picking up four representative building blocks from each set. Only four control Kans for compounds A1B1, A2B2, A3B3, and A4B4 along the diagonal of the matrix need to be included.

VI. CHARACTERIZATION OF THE COMPOUND LIBRARY

Characterization of the compound library is one of the most important parts of the library production process, especially because higher and higher quality is being demanded for the discrete compound library. There are many methods for characterizing the compound library. The two most common are ^{1}H NMR and LC/MS. ^{1}H NMR using an appropriate internal standard is the method of choice if only 10–20% of the compound library needs to be analyzed. This analysis provides valuable information on both yield and purity. For 100% analysis of the compound library, LC/MS is commonly used.

HPLC/NMR is an information-rich method for characterization of a combinatorial library (22). Quality control may be performed either on-line or off-line. The method is relatively insensitive and requires more material and time than HPLC/MS, but the structural information obtained can be very valuable. For instance, when isomeric compounds or compounds with the same molecular mass are present in the library, HPLC/NMR provides a unique characterization.

Table 4 Synthetic Method Comparison of 10,000 Compounds

Method	Compound form	Quantity	Number of reactions
Parallel	Discrete	Milligrams	11,110–40,000
Split-and-pool	Mixture	Nanograms	40
Directed sorting	Discrete	Milligrams	40

VII. CONCLUSION

Choosing the right solid supports and optimizing reaction conditions on the solid phase are critical to the successful synthesis of combinatorial libraries. Directed sorting is a powerful synthetic tool for the generation of large discrete compound libraries. This hybrid technique combines advantageous features from both parallel synthesis and split-and-pool strategy using RF-tagged microreactors and has been used as a flexible, efficient, and reliable combinatorial chemistry system. For instance, it requires only 40 reactions to synthesize 10,000 discrete compounds in a four-step sequence with 10 building blocks for each step. However, it will require at least 11,110 and up to 40,000 reactions in a parallel synthesis (Table 4).

REFERENCES

1. KC Nicolaou, XY Xiao, Z Parandoosh, A Senyei, MP Nova. Radiofrequency encoded combinatorial chemistry. Angew Chem Int Ed Engl 34:2289–2291, 1995.
2. EJ Moran, S Sarshar, JF Cargill, MJM Shahbaz, A Lio, AM Mjalli, RW Armstrong. Radio frequency tag encoded combinatorial library method for the discovery of tripeptide-substituted cinnamic acid inhibitors of the protein tyrosine phosphatase PTP1B. J Am Chem Soc 117:10787–10788, 1995.
3. XY Xiao, R Li, H Zhuang, B Ewing, K Karunaratne, J Lillig, R Brown, KC Nicolaou. Solid-phase combinatorial synthesis using MicroKan reactors, Rf tagging, and directed sorting. Biotechnol Bioeng (Comb Chem) 71:44–50, 2000.
4. R Li, XY Xiao, KC Nicolaou. High-output solid phase synthesis of discrete compounds using split-and-pool strategy. In: I Suchleiki, ed. High Throughput Organic Synthesis. New York: Marcel Dekker, 2000, pp 207–213.
5. R Brown. Content and discontent: A combinatorial chemistry status report. Mod Drug Discov 2:63–71, 1999.
6. J Chen, M Pattarawarapan, AJ Zhang, K Burgess. Solution- and solid-phase syntheses of substituted guanidinocarboxylic acids. J Comb Chem 2:276–281, 2000.
7. F Albericio, N Knerb-Cordonier, S Biancalana, L Gera, RI Masada, D Hudson, G Barany. Preparation and application of the 5-(4-(9-fluorenylmethyloxycarbonal)am-inomethyl-3,5-dimexoyphenoxy)valeric acid (PAL) handle for the solid-phase syn-

thesis of C-terminal peptide amides under mild conditions. J Org Chem 55:3730–3743, 1990.

8. CG Boojamra, KM Burow, JA Ellman. An expedient and high-yielding method for the solid phase synthesis of diverse 1,4-benzodiazepine-2,5-diones. J Org Chem 60: 5742–5743, 1995.

9. CG Boojamra, KM Burow, LA Thompson, JA Ellman. Solid-phase synthesis of 1,4-benzodiazepine-2,5-diones. Library preparation and demonstration of synthesis generality. J Org Chem 62:1240–1256, 1997.

10. CP Holmes. Abstract ORGN 383. 213th Natl Meeting Am Chem Soc, San Francisco, CA, April 1997.

11. KJ Jensen, J Alsina, MF Songster, J Vagner, F Albericio, G Barany. Backbone amide linker (BAL) strategy for solid-phase synthesis of C-terminal modified and cyclic peptides. J Am Chem Soc 120:5441–5452, 1998.

12. D Sarantakis, JJ Bicksler. Solid phase synthesis of sec-amides and removal from the polymeric support under mild conditions. Tetrahedron Lett 38:7325–7328, 1997.

13. AM Fivush, TM Willson. AMEBA: An acid sensitive aldehyde resin for solid phase synthesis. Tetrahedron Lett 38:7151–7154, 1997.

14. EE Swayze. Secondary amide-based linkers for solid phase organic synthesis. Tetrahedron Lett 38:8465–8468, 1997.

15. KG Estep, CE Neipp, LMS Stramiello, MD Adam, MP Allen, S Robinson, EJ Roskamp. Indole resin: A versatile new support for the solid phase of organic molecules. J Org Chem 63:5300–5301, 1998.

16. B Yan, N Nguyen, L Liu, G Holland, B Raju. Kinetic comparison of trifluoroacetic acid cleavage reactions of resin-bound carbamates, ureas, secondary amides, and sulfonamides from benzyl-, benzhydryl- and indole-based linkers. J Comb Chem 2: 66–74, 2000.

17. AW Czarnik. Solid-phase synthesis supports are like solvents. Biotechnol Bioeng (Comb Chem) 61:77–79, 1998.

18. R Li, XY Xiao, AW Czarnik. Reactivity comparison of 2-chlorotrityl chloride resins (to be submitted for publication).

19. DP Rotella. Solid-phase synthesis of olefin and hydroxyethylene peptidomimetics. J Am Chem Soc 118:12246–12247, 1996.

20. R Li, FE Cohen, GL Kenyon. A methodology for the solid phase synthesis of α,β-unsaturated ketone combinatorial libraries. Abstracts, 210th ACS Natl Meeting, Chicago, Aug 20–24, 1995, MEDI 227.

21. B Yan, G Kumaravel, H Anjaria, A Wu, RC Petter, CF Jewell Jr, JR Wareing. Infrared spectrum of single resin bead for real-time monitoring of solid-phase reactions. J Org Chem 60:5736–5738, 1995.

22. J Chin, JB Fell, M Jarosinski, MJ Shapiro, JR Wareing. HPLC/NMR in combinatorial chemistry. J Org Chem 63:386–390, 1998.

4

Key Factors in the Translation of Solution-Phase Chemistry to the Solid Phase

Pedro H. H. Hermkens, Hans Hamersma, and Adrianus P. A. de Man
N.V. Organon, Oss, The Netherlands

I. INTRODUCTION

In the process of identifying new compounds with interesting properties, combinatorial chemistry (CC) has proven its value. These properties may lie in the area of new pharmaceuticals, novel materials, artificial enzymes, catalysts, surfactants, etc. In the pharmaceutical field, CC started with the generation of huge libraries of non-druglike compounds such as peptide oligomers, generally considered to be poor as drug candidates or even as leads. The pharmaceutical industry quickly understood that they had to focus on "good quality" compounds—compounds that were amenable to optimization toward more druglike behavior. Therefore, in the last few years CC methodology has been increasingly aimed at the synthesis of small organic molecules. Although more and more classical organic reactions are being adapted for solid-phase reactions (1–6) it has to be accepted that even with today's available repertoire only a small portion of the diversity space can be addressed. What is even more frustrating is that most of the time we are fishing in the same pond. Therefore, the synthesis of "good quality" compounds is relying heavily on the identification of new robust solid-phase or solution-phase processes.

Although there is a general feeling that solution-phase approaches will have an enormous impact on the CC area, this chapter will deal with only solid-phase

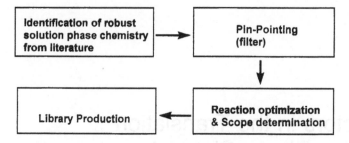

Figure 1 Four-step approach. (Reproduced with permission from Ref. 9.)

syntheses. The process of identifying suitable chemistry in the general literature and translating this to solid-phase conditions is a tedious one.

This process can be divided into four stages:

1. Identification of robust solution-phase chemistry from the literature
2. Pinpointing
3. Reaction optimization and scope determination
4. Library production

(see Fig. 1).

It is generally accepted that the rate-limiting step is the development time to identify the optimal reaction conditions for new chemistry on solid supports, rather than the actual production of the library (see Fig. 2).

In our laboratory the pinpointing phase has been introduced to filter out the promising ideas from the bad ones. The first step in this phase is to repeat the original solution-phase chemistry to test the robustness of the described conversion. If high yields are obtained, the same conditions (if compatible with resin) are tried on the solid support. Normally this is followed by a rough optimization program; a few variations are studied such as different resins and/or linkers, different equivalents of reagents (stoichiometry), and different temperatures.

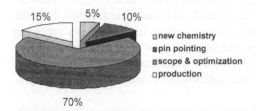

Figure 2 Estimation of percentage time used per stage. (Reproduced with permission from Ref. 9.)

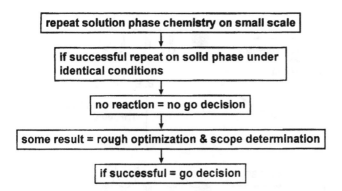

Figure 3 Pinpointing phase. (Reproduced with permission from Ref. 9.)

Only if some preliminary results are obtained is a full-blown optimization program started (see Fig. 3). The chemistry is stopped if no indications of successful reactions are found. This is a difficult decision, because no chemist likes to be beaten by chemistry. A frequently observed pitfall is that chemists go on much too long in their attempt to solve the problem.

Before the execution of such a full-blown reaction optimization program can start, some preparations and decisions have to be made (7). The most important ones are those related to the question of which resin, linker, and reaction-monitoring principle are the most ideal for the reaction under investigation. Some of these issues are dealt with in more detail in Chapters 5, 6, 7, and 8 and therefore need not be addressed here.

II. CASE STUDIES

In the following three subsections examples are described that follow the four-step approach as depicted in Figure 1. Because the highlight of this book is reaction optimization, we focus on the first three steps and do not address library production. The three examples chosen represent different starting points and demonstrate that the four-step approach should be used as a guideline and not as a rule.

A. Unsuccessful Oxazole Formation

Step 1. The conversion of glycine to oxazoles (8) was an attractive reaction sequence to be translated from the solution phase to the solid phase (see

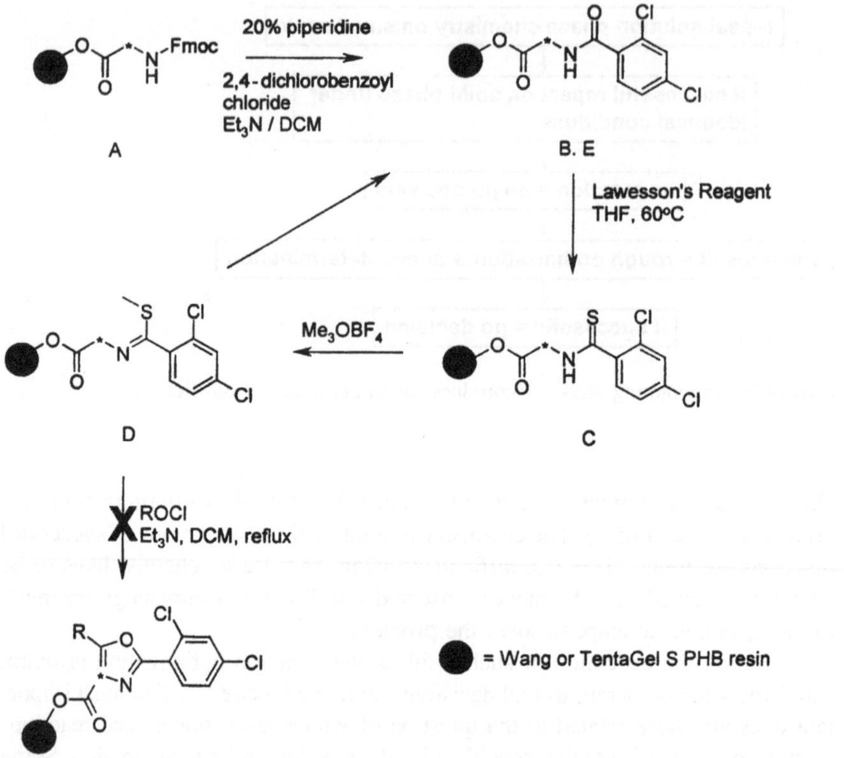

Figure 4   Oxazole formation on solid support. (Reproduced with permission from Ref. 7.)

Fig. 4). The starting materials are easily accessible, and oxazoles are well-known structural features in many drugs. There were no precedents for this cyclization on solid supports at the time this investigation started.

    Step 2.   Rehearsal of the solution-phase chemistry showed the reliability of the published conditions. It was therefore decided to use these exact conditions as a first attempt to study this reaction on the solid phase. The resins of choice were Wang and TentaGel S PHB.

    The reactions were monitored by ^{13}C gel NMR spectroscopy (Fig. 5). To this end, the reaction was carried out with a ^{13}C-enriched glycine derivative. The transformations proceeded smoothly until the final ring closure step. In this case the conditions described for solution-phase chemistry failed to give the anticipated ring closure product when the reaction was carried out on solid phase. Under these conditions the amide compound (E) was regenerated. Subsequently, this final cyclization step was studied in a rough optimization program in which

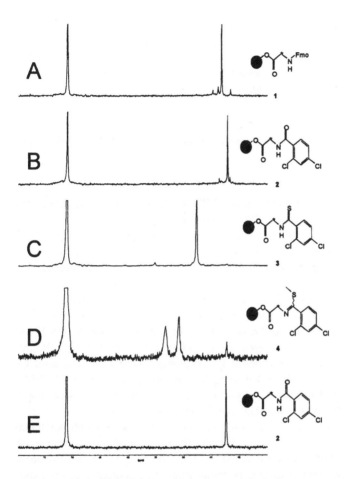

Figure 5 ^{13}C gel-NMR spectra of the reaction products at various stages of the syntheses (the glycine building block was ^{13}C-labeled at the C_α carbon). Resin (20–30 mg) was dispersed in benzene, and NMR spectra were recorded on a Bruker DRX600 spectrometer in approximately 15 min. (Reproduced with permission from Ref. 7.)

several conditions (different solvents, bases, and temperatures) were evaluated. None of these gave the required conversion, so further efforts on this reaction were abandoned.

B. Successful Stille Reaction on a Pyrimidine Core

Step 1. A second example is the Stille reaction on a 4-chloropyrimidine moiety (see Fig. 6), which has been studied extensively in solution-phase reac-

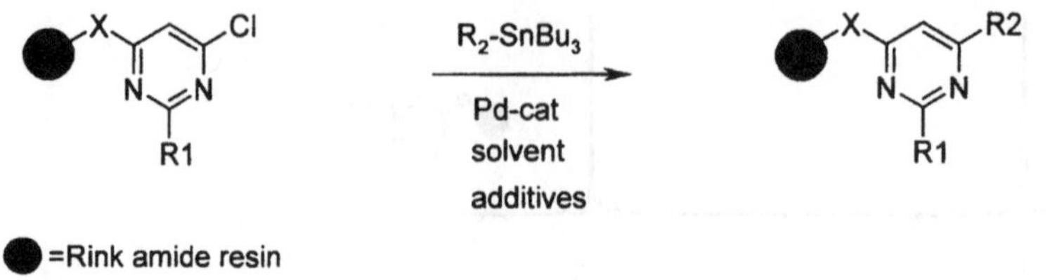

●=Rink amide resin

Figure 6   Solid-phase Stille reaction on 4-chloropyrimidine. (Reproduced with permission from Ref. 9.)

tions (10–15). Although accounts of the Stille reaction on solid phase have been published, there is no precedent for a reaction on a pyrimidine core.

Step 2.   Published solution-phase conditions vary in the following factors: palladium catalysts are $Pd(PPh_3)_4$ or $PdCl_2(PPh_3)_2$; solvents used are DMF, NMP, THF, or toluene; temperature range is 50–120°C; additives used are Et_4NCl, K_2CO_3, or LiCl (for triflates). Because of the wide range of conditions we skipped the rehearsal in the solution phase but immediately tried the solid-phase reaction using a rough cross section of the applied solution-phase conditions (see Table 1). We restricted ourselves to the nonvolatile high-boiling solvents DMF and NMP. However, we added two alternative palladium catalysts, $Pd_2(dba)_3$, [Pd_2(dibenzylideneacetone)$_3$] and $Pd(dppb)_2Cl_2$ [Pd(bisdiphenylphosphinebutane)Cl$_2$]. The only additives studied were LiCl and *N,N*-diisopropylethylamine (DIPEA). The reaction was performed with allyltributylstannane and with the 4-chloropyrimidine moiety attached to the Rink amide resin (see Fig. 6).

From the results, no conclusion could be drawn about which palladium catalyst is the optimal one. However, because the majority of the solution-phase methods used $PdCl_2(PPh_3)_2$, we decided to focus on this catalyst in subsequent studies. In addition, the results clearly demonstrate the importance of LiCl and/ or DIPEA (entries 4 and 6 in Table 1). This is a complete deviation from the

Table 1   Rough Study

Entry	Pd-cat	Temp (°C)	Solvent	Additive	Yield (%)
1	Pd_2dba_3	50	NMP		0
2	Pd_2dba_3	100	NMP		0
3	$PdCl_2(PPh_3)_2$	125	DMF		0
4	$PdCl_2(PPh_3)_2$	125	DMF	LiCl, DIPEA	76
5	$Pd(dppb)_2Cl_2$	125	DMF		0
6	$Pd(PPh_3)_4$	125	DMF	LiCl, DIPEA	81

solution-phase data. Usually, solution-phase conditions with 4-chloropyrimidines give very satisfactory results without extra additives (16). Solution-phase conditions with 4-chloropyrimidines give very satisfactory results without extra additives (16). LiCl is added only when triflate pyrimidines are used instead of halopyrimidines.

    Step 3. These preliminary positive results allowed the start of the third phase, a more comprehensive reaction optimization and scope determination program. In the first optimization round the influence of different equivalents of all reagents was studied using one stannane (allyltributylstannane) and one 4-chloropyrimidine moiety (Table 2). All reactions were performed in DMF at a temperature of 125°C using a double coupling procedure (12 h each). The reaction was monitored by HPLC analysis (ratio of starting material A to product B) after cleavage from the resin. This optimization round was performed on an automated synthesizer (Syro). The optimal conditions were found to be stannane (10 equiv), DIPEA (2.5 equiv), LiCl (4.5 equiv), and PdCl$_2$(PP$_3$)$_2$ (0.04 equiv) (entry 4).

    Using these conditions in a broader scope determination (nine stannanes) revealed that reaction conditions could hardly be improved further (see Table 3). Only three stannanes gave good results (entries 1–3) and two gave moderate conversion (entries 4 and 5). The remaining ones gave no or minor conversion. The lesson learned from these results was never to perform an optimization process using only a one-to-one combination of building blocks.

    A search of the literature on additional factors influencing the Stille reaction indicated that the additives PPh$_3$ and CuI as well as the amount of available catalyst often play an important role (16). Therefore, a second round of optimization was designed to study these parameters. Eight different conditions were si-

Table 2   First Optimization Round (Eight Experiments)

Entry	(Bu)$_3$SnR (equiv)	LiCl (equiv)	DIPEA (equiv)	PdCl$_2$(PPh$_3$)$_2$ (equiv)	Ratio A/B[a]
1	10	0	2.5	0.04	38/62
2	10	1.5	2.5	0.04	24/76
3	10	3	2.5	0.04	12/88
4	10	4.5	2.5	0.04	3/97
5	10	4.5	0	0.04	9/91
6	10	4.5	1.25	0.04	11/89
7	10	4.5	2.5	0.02	27/73
8	5	4.5	2.5	0.04	5/95

[a] Ratio of starting material A to product B determined by HPLC analysis after cleavage from the resin.

Table 3 Scope Determination Using First
Optimized Conditions

Entry	(Bu)$_3$SnR (R)	Ratio A/B[a]
1	Allyl	14/86
2	2-Furanyl	0/100
3	2-Thienyl	11/89
4	Phenylethynyl	57/43
5	Phenyl	42/58
6	Vinyl	100/0
7	EtO-vinyl	100/0
8	2-Pyridinyl	83/17
9	3-Pyridinyl	77/23

[a] Ratio of starting material A to product B determined
by HPLC analysis after cleavage from the resin.

multaneously studied in an automated fashion (Syro) including four different
stannanes (allyl-, phenylethynyl-, phenyl- and 2-pyridinyltributyltin). These stannanes were selected on the basis of their different reactivities (high, moderate,
and low; see Table 3). The number of equivalents of stannane (10 equiv), LiCl
(4.5 equiv), and DIPEA (2.5 equiv) found earlier were kept constant in this experiment. The results of these 32 experiments are described in Table 4.

The allytributyl- and phenyltributylstannane gave complete conversion under all conditions. The phenylethynyl- and 2-pyridinetributylstannanes gave satisfactory results only under the conditions listed in entries 16 and 32. Therefore,
these latter conditions are the most optimal ones found so far.

Thus, 55 experiments, performed in an automated fashion, were required to
identify robust reaction conditions for the Stille reaction on a 4-chloropyrimidine
moiety.

C. Successful Diketopiperazine Ring Formation

The last example is based on a frequently used scaffold in solid-phase chemistry,
the piperazinediones [diketopiperazines (DKPs)]. DKPs have been shown to exhibit a variety of biological activities (17). The ring system is rather rigid and
can be readily assembled from building block sets (i.e., amino acids and aldehydes), which are available in abundant variety. It is obvious that many publica-
tions in recent years have described a huge variety of resins, linkers, and/or cleavage methods (18–25) to synthesize DKP libraries. Based on this large number
of external experiences with this chemistry we decided to abandon steps 1 and

Table 4 Second Optimization Round (32 Experiments)

PdCl$_2$ (PPh$_3$)$_2$ (equiv)	PPh$_3$ (equiv)	CuI (equiv)	Entry	R = allyl, ratio A/B[a]	Entry	R = phenyl ethynyl, ratio A/B[a]	Entry	R = phenyl, ratio A/B[a]	Entry	R = 2-pyridinyl, Ratio A/B[a]
0.04	—	—	1	0/100	9	53/47	17	0/100	25	73/27
0.08	—	—	2	0/100	10	41/49	18	0/100	26	51/49
0.04	—	0.04	3	0/100	11	37/63	19	0/100	27	62/38
0.04	—	0.08	4	0/100	12	38/62	20	0/100	28	65/35
0.08	—	0.08	5	0/100	13	—	21	0/100	29	48/52
0.04	0.1	—	6	0/100	14	22/78	22	2/98	30	28/72
0.04	0.1	0.04	7	0/100	15	19/81	23	0/100	31	24/76
0.08	0.2	0.08	8	0/100	16	5/95	24	2/98	32	7/93

[a] Ratio of starting material A to product B determined by HPLC analysis after cleavage from the resin.

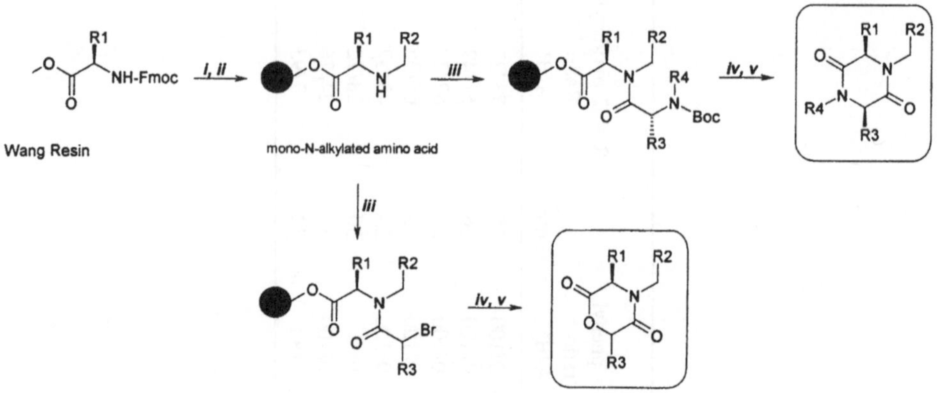

Figure 7 Synthetic protocol for DKP formation on solid support. (i) 40% Piperidine/ DMF, room temperature (RT). (ii) 1.2 equiv Aldehyde, NaBH(OAc)$_3$, DCM, RT, sonicate (2x). (iii) 1 equiv PyBrOP, DCM, 2 equiv DiPEA, 1 equiv Boc-AA-OH or 1 equiv bromoacetic acid. (iv) Neat TFA, RT. (v) Toluene, reflux, 5 h.

2 and immediately started with step 3. We selected the route according to the procedure of Gordon and Steele (18) (see Fig. 7) because we had had some preliminary in-house experience with a number of these steps.

Beforehand, we would have expected that this route using similar conditions could provide a diverse set of DKPs without any problems. It was also possible to make spin-off libraries of corresponding diketomorpholines. Perhaps one of the problems that could arise was the inability to control mono- and di-N-alkylation in step ii (Fig. 7). Therefore we changed the reaction conditions for this step for in-house optimized conditions. First we forced the imine formation using 10 equiv aldehyde in DCM/TMOF = 1:1 v/v at room temperature. After washing away the excess aldehyde the imine thus formed was reduced using 5 equiv NaBH(OAc)$_3$ in 1% HOAc/DMF. This two-step conversion could nicely be monitored by FTIR spectroscopy.

We then performed a scope determination for a representative set of DKPs, which had been selected on the basis of diversity parameters set for Organon good quality compounds.

After analysis it was shown that in more than 90% of the cases no DKP formation was observed, but only mono-N-alkylated product (see Fig. 8) was isolated. The need for reaction optimization on the acylation of the secondary amine obtained after the reductive amination was clear.

In Table 5 an overview is given of all the variations used in coupling reagents, solvents, and temperatures. The conclusion of that study was that the optimal acylation conditions are 3 equiv PyBrOP, 3 equiv Boc-AA-OH, and 6

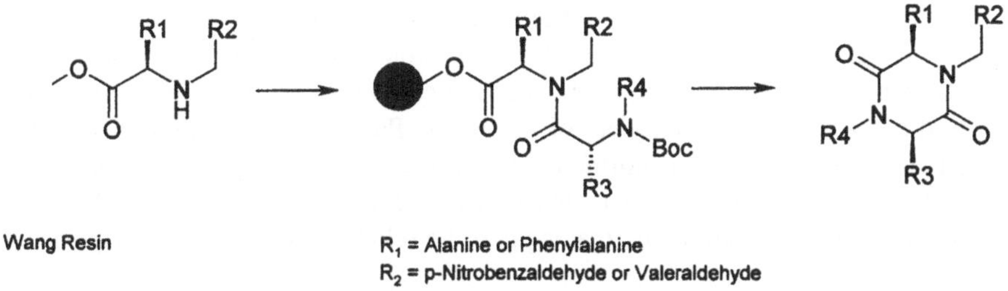
R1
R2
O
N
H
O
R1
R2
O
N
R4
O
O
N
Boc
R3
R1
R2
O
N
R4
N
O
R3
Wang Resin
R₁ = Alanine or Phenylalanine
R₂ = p-Nitrobenzaldehyde or Valeraldehyde

Figure 8 Optimization of the acylation with Boc-amino acid-OH.

equiv DiPEA in DCM/DMF = 1:2 v/v at room temperature for 4 h and repeat this step overnight.

Again, we performed a second round of scope determination and found that DKP formation was dependent on the side chains of the amino acids used. Moreover, in most cases the reaction was incomplete (monoalkylated product was still present) and therefore unusable for split-pool library synthesis.

A thorough study of the literature (back to step 1) revealed that failures of many established methods for acylation of difficult amines are rarely addressed.

One publication (26), based on solution-phase chemistry, mentioned this problem explicitly and gave the solution: the use of old-fashioned carbodiimides.

Step 2. A rough optimization showed that the use of 3 equiv Boc-Ala-OH and 1.5 equiv DiC in DCM at room temperature without base or HOBt gave a ratio of DKP to monoalkylated product formation of 2:1. An identical ratio was found when only 1.5 equiv Boc-Ala-OH and 2 equiv DiC were used. However, performing the reaction with 1.5 equiv of the preformed symmetric anhydride of Boc-Ala-OH under identical conditions gave the inverse ratio, 1:2. This suggests that the mechanism of this acylation is to proceed via an *O*-isoacylurea derivative and not via a symmetric anhydride formed in situ.

Step 3. A more complete reaction optimization round was started to identify the optimal acylation conditions for a combination of amino acids and alde-

Table 5 Factors Studied

Coupling reagent	Solvent	Temperature (°C)
PyBrOP	THF	RT
TFFH	DMF	50
HATU	DCM/DMF	
CiP	DCM	

hydes. The conditions of choice turned out to be 6 equiv amino acid and 8 equiv DiC in DCM at room temperature for 24 h (double coupling). The final scope determination using the "ultimate" reaction conditions showed DKP formation in all cases with an overall yield of 85–104%. Purity of the DKPs was >90%, and all compounds were confirmed with mass spectrometry. The lesson learned from this case study is that it is dangerous to skip steps 1 and 2 even if many literature references are found that support the reliability of a reaction sequence.

III. FACTORIAL DESIGN

In reaction optimization studies large numbers of variables (i.e., concentration, temperature, solvent, chemical reactants, and additives) are involved, and the importance of each variable needs to be understood. In solid-phase synthesis in general, additional factors such as the resin and the linker system might also play an important role. In the case study of the Stille reaction (see Sec. IIB), the first rough optimization round (Table 1) was done to evaluate the influence of the amount of equivalents of the reagents [stannane (two variations), Pd-cat (two variations)] and additives [LiCl (four variations), DIPEA (three variations)]. In this particular example a total of 48 combinations are possible. With the automated chemistry in hand, the chemist can decide to study all these combinations. However, in this case only eight combinations have been studied. With logical thinking most chemists would be able to reduce the number of experiments in this example and still get the required amount of information. If this chemist also would choose to vary all the remaining factors simultaneously (resin, linker, solvent, stannane, pyrimidine, and temperature), he or she would suddenly be confronted with thousands of combinations. It is impossible to study this in a comprehensive manner, and at the same time it will become more complicated to design an intelligent set of optimization reactions with the goal of performing a minimum number of experiments to obtain maximal information. As described in Section I, reaction optimization is still the most time-consuming factor in the entire process of library generation. Therefore, it is strange that not many efficacy-enhancing activities that focus on the optimization process are explored in combinatorial chemistry (27). Generally, it is observed that optimization is done on an ad hoc basis. In comparison, process chemistry has been facing a similar problem for a long time already, and there it is recognized that a more sophisticated approach is required (28). Statistical experimental design occupies itself with the question of how to conduct and plan experiments in order to extract the maximum amount of information in the smallest number of experimental runs (29,30). The basic idea is to vary all relevant factors simultaneously over a set of planned experiments and connect the results by means of a mathematical model. This model is then used for interpretation, prediction, and optimization.

One thing combinatorial chemistry certainly triggered is the awareness that chemical synthesis can be done more efficiently. Library production today is being performed in a (semi-) automated parallel fashion. However, the parallel approach is also suitable for performing reaction optimization, because many factors can be studied simultaneously. Doing this in combination with factorial design would certainly make the optimization process more efficient.

IV. FUNCTIONAL GROUP TRANSFORMATION

As pointed out earlier, the actual production of a library usually takes only a fraction of the time that is spent on optimization of the reaction conditions. This makes it worthwhile to investigate whether it is possible to increase the returns on the time spent on the latter aspect.

Methods to improve the effectivity of invested time were described earlier. The concept ''libraries from libraries'' (31,32) was introduced to capitalize on the initial effort expended in preparing a mixture-based library (see Fig. 9A). However, most of the examples given are based on a multiple conversion of several identical functionalities (e.g., N-permethylation or reduction of amide bonds). A similar principle can be applied by transforming a single functional group as demonstrated by the Chiron group (33) and has been termed ''postsynthesis modification'' (Fig. 9B). This basic strategy involves the introduction of a functional group that is used as the cornerstone for elaboration in a variety of heterocyclic ring systems. Both transformation methods result in compounds with alternative properties.

We advocate a further extension (34) of the Chiron approach. After having established the optimized reaction conditions for a given combinatorial library, this first-generation library is brought into production. A second-generation library is prepared by making use of most of the same building blocks and chemistry already optimized for the first one. However, in one position a functional group is introduced that is suitable for chemical transformation and should give access to a new region in chemical diversity space. The time spent on the preparation of this second library should be a small fraction of that needed for the first library. Only the introduction of the functional group and its transformation have to be optimized.

It can be argued that the second (transformed) library will be too similar to the first one and that duplication of properties will be the result. To check for this effect, a first-generation library was compared (*in silico*, as the coverage of diversity space is the criterion of interest) both to a transformed library and to a new combinatorial library of a totally different class. The benzodiazepine library published by Ellman's group (35) (see Fig. 10A) can serve as an example. The ''virtual transformation'' for this experiment was the following: In the R_1 position

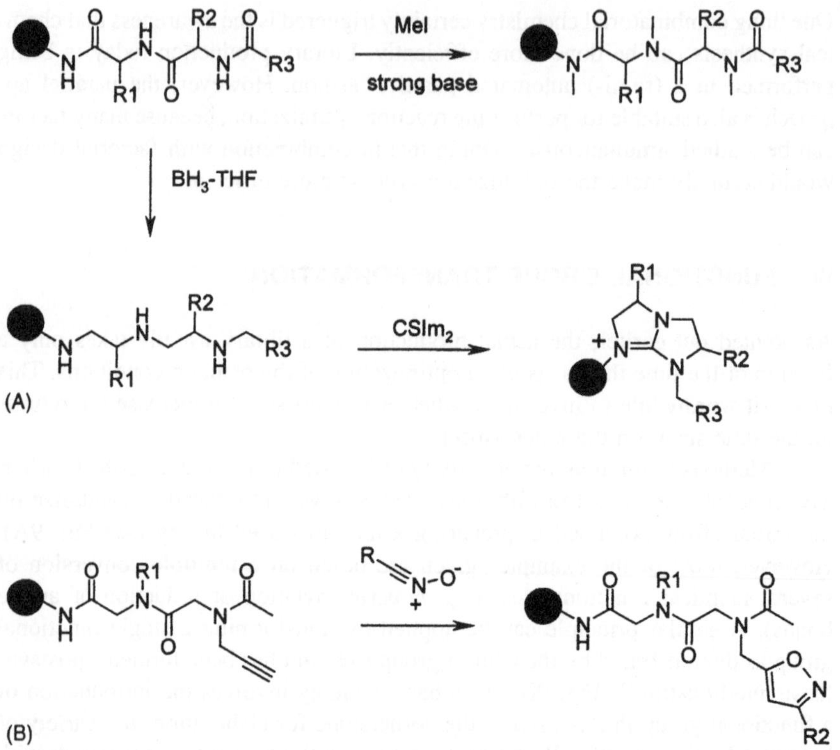

Figure 9 Methods for expanding existing libraries. (A) Libraries from libraries. (B) Postsynthesis modification.

of the benzodiazepine an aldehyde moiety is introduced that can be converted to nitrones. By means of a 1,3 dipolar cycloaddition with alkenes or alkynes, isoxazolidine and isoxazoline systems, respectively, can be obtained (see Fig. 10B). The functional group transformation aldehyde → nitrone → isoxazolidine on solid support was demonstrated earlier (36).

To demonstrate the validity of the concept of functional group transformation, we compare the original benzodiazepine library as described by Ellman with the proposed functional group–transformed benzodiazepine library. At the same time we compare it with a library that incorporated totally different chemistry. Arbitrarily, we chose the β-lactam library (see Fig. 11) published by Ruhland et al. (37) as an example of such a library.

Many methods have been described to compare the diversity content of sets of molecules (38). Here we give the comparison based on whole-molecule

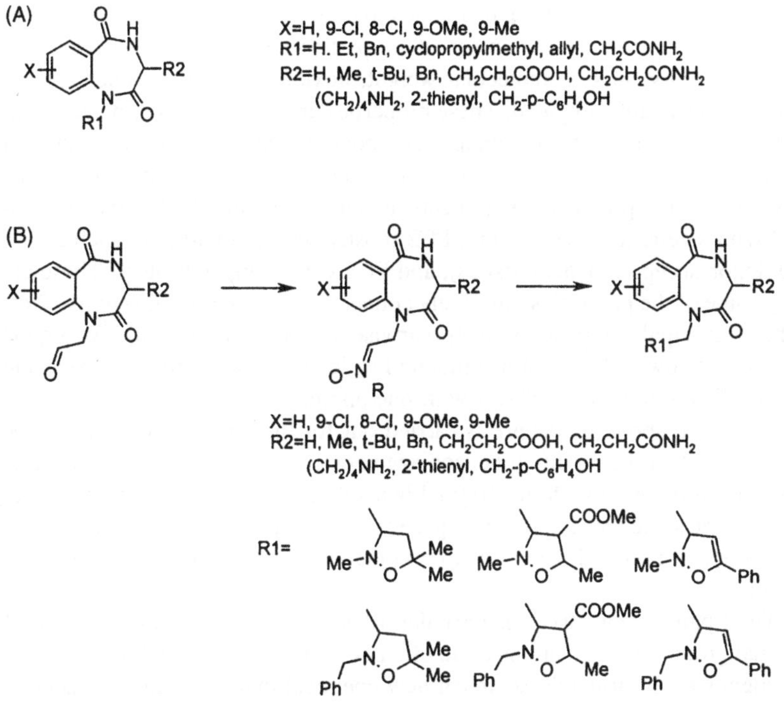

(A)

X=H, 9-Cl, 8-Cl, 9-OMe, 9-Me
R1=H. Et, Bn, cyclopropylmethyl, allyl, CH$_2$CONH$_2$
R2=H, Me, t-Bu, Bn, CH$_2$CH$_2$COOH, CH$_2$CH$_2$CONH$_2$
(CH$_2$)$_4$NH$_2$, 2-thienyl, CH$_2$-p-C$_6$H$_4$OH

(B)

X=H, 9-Cl, 8-Cl, 9-OMe, 9-Me
R2=H, Me, t-Bu, Bn, CH$_2$CH$_2$COOH, CH$_2$CH$_2$CONH$_2$
(CH$_2$)$_4$NH$_2$, 2-thienyl, CH$_2$-p-C$_6$H$_4$OH

R1=

Figure 10 Functional group transformation approach. (A) Original benzodiazepine library. (B) Benzodiazepine library after FGT.

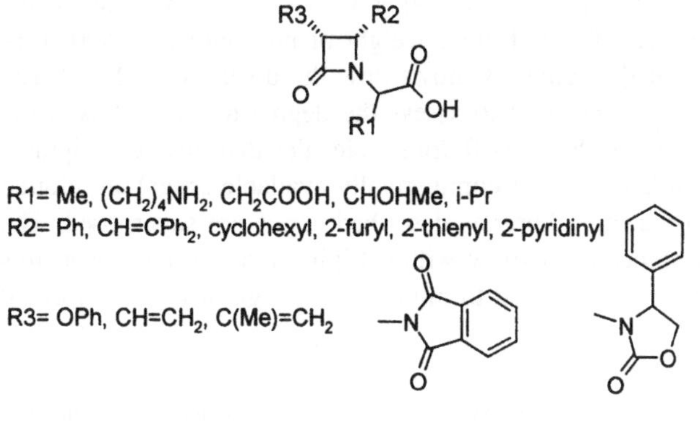

R1= Me, (CH$_2$)$_4$NH$_2$, CH$_2$COOH, CHOHMe, i-Pr
R2= Ph, CH=CPh$_2$, cyclohexyl, 2-furyl, 2-thienyl, 2-pyridinyl

R3= OPh, CH=CH$_2$, C(Me)=CH$_2$

Figure 11 β-Lactam library.

properties. Our method has been developed in house and briefly works as follows (39): For every molecule, a number of properties considered to be related to biological activity, such as molecular weight, flexibility, log P, and percent hydrophilic surface, are computed. These properties are subjected to principal component analysis; the first 10 principal components explain more than 85% of observed variance, and the first two explain nearly 60%. By inspection of the composition of the principal components, it can be seen that PC1 is mostly connected with size-related descriptors, PC2 mostly with lipophilicity (such as log P and donor/acceptor characteristics), and PC3 with shape indicators. These are exactly some of the properties intuitively connected with biological activity. Plotting these principal components either pairwise or three at a time will give a good impression of how well a certain compound collection covers property space and will also allow comparison of sets with one another.

When the above procedure is carried out for the three libraries, it can be seen (Fig. 12) that the modified benzodiazepines cover a set of properties that is almost nonoverlapping with the original benzodiazepines. In contrast, the β-lactams cover the same area in property space as the original benzodiazepine library. The same result is found when other combinations of principal components are inspected.

The "paper chemistry" exercise described above indicates the potential of functional group transformations as an efficacy-enhancing approach in combinatorial chemistry. In principle, as much new chemical diversity can be generated with a transformed library as with an entirely new chemical series. The potential of this approach is increased further by the fact that a large number of transformations have become available for application in a combinatorial chemistry setting over the last eight years (1–6). Of course, there is no guarantee that the additional diversity generated by this approach will be as large for every functional group transformation as the example presented in this chapter.*

One of the dangers of this concept is that huge molecules can easily be generated (two scaffolds attached to each other with several side chains) that exceed the proposed criteria for molecular weight of not more than 500 [Lipinski's "rule of five" (40)]. Figure 13 shows how the diversity method based on molecular properties can be used to assess the degree to which this is the case. Here, the properties for about 8000 drug molecules that meet the Lipinski rules (taken from our in-house file of commercially available drugs) are plotted in property space, and the three libraries described above are superposed. The transformed benzodiazepines are largely within Lipinski space; moreover, this visualization tool could easily be used to modify the design of the transformed

---

* The functional group transformation of a carboxylic acid to oxazoles was applied to different (proprietary) chemical series and led to the same conclusion.

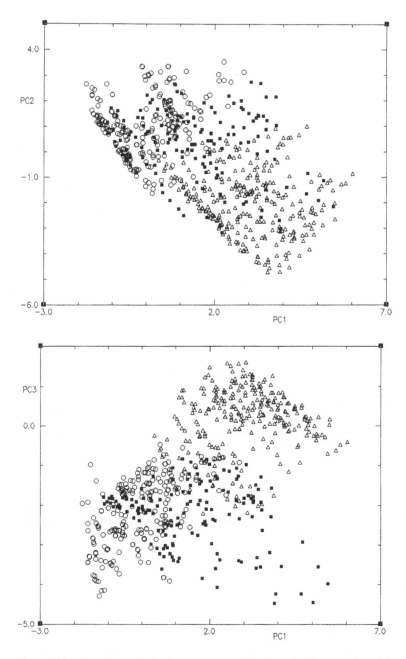

Figure 12 First three principal components of whole-molecule properties of the follow-ing libraries: original benzodiazepines (O), transformed benzodiazepines (Δ), and β-lactams (■). Top panel PC1 vs. PC2; bottom panel PC1 vs. PC3. The overlap between the original benzodiazepines and the β-lactams in property space is obvious, whereas the modified benzodiazepines are more or less completely separated.

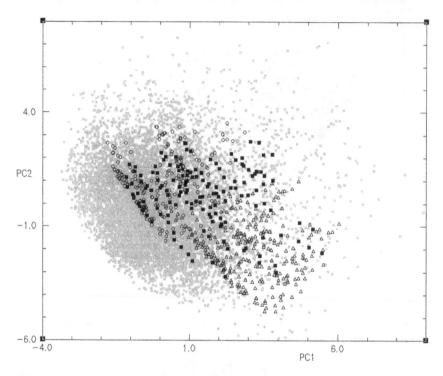

Figure 13 First two principal components of whole-molecule properties of the three libraries: original benzodiazepines (◯), transformed benzodiazepines (△), and β-lactams (■). Plotted on top of the properties for drug molecules satisfying the Lipinski criteria (■).

library so that it matches the Lipinski rules even better. Besides, it is debatable whether the "rule of five" should be used as an absolute criterion for hit-finding libraries.

V. CONCLUSIONS

The leitmotiv of this chapter is efficacy-enhancing tools in the process of identifying optimal reaction conditions on solid support. As about 70% of the time is spent on reaction optimization, these tools will mainly be focused on this part of the process. The first tool discussed is the four-step approach in which a filter (pinpointing phase) is introduced to eliminate the bad apples. Only those ideas that pass the pinpointing phase enter the time-consuming optimization step. An important message in this phase is "kill good, kill early." In the optimization phase the chemist should make use of (miniaturized) automated parallel chemistry to find the most optimal conditions. To cope with the overwhelming number of reaction conditions that can be studied when many factors can be varied, one should also consider the use of statistical design tools to assist in the optimization process. Finally, we discussed the power of reusing in-house optimized chemistry conditions. In a paper chemistry exercise we demonstrated the potential of functional group transformation to produce libraries that share a large part of their underlying chemistry but span different regions of property space.

Taken together, we have found that our approach has substantially increased our output of libraries of compounds with druglike properties.

ACKNOWLEDGMENT

We thank M. C. A. van Tilborg and I. Ilgün for their contributions to the Stille reaction and diketopiperazine formation, respectively.

REFERENCES

1. PHH Hermkens, HCJ Ottenheijm, DC Rees. Solid-phase organic reactions: A review of the recent literature. Tetrahedron 52:4527–4554, 1996.
2. PHH Hermkens, HCJ Ottenheijm, DC Rees. Solid-phase organic reactions. II: A review of the literature Nov 95–Nov 96. Tetrahedron 53:5643–5678, 1997.
3. S Booth, PHH Hermkens, HCJ Ottenheijm, DC Rees. Solid-phase organic reactions III: A review of the literature Nov 96–Dec 97. Tetrahedron 54:15385–15443, 1999.
4. F Balkenhöhl, C von dem Bussche-Hunnefeld, A Lansk, C Zechel. Combinatorial synthesis of small organic molecules. Angew Chem Int Ed Engl 35:2288–2337, 1996.
5. BA Bunin. The Combinatorial Index. San Diego: Academic Press, 1998.
6. SS Booth, CM Dreef-Tromp, PHH Hermkens, APA de Man, HCJ Ottenheijm. Survey of solid-phase organic reactions. In: G Jung, ed. Combinatorial Chemistry—Synthesis, Analysis, Screening. Weinheim: Wiley-VCH, 1999, pp 35–76.
7. AR Brown, PHH Hermkens, HCJ Ottenheijm, DC Rees. Solid phase synthesis. SynLett 817–827, 1998.
8. M Yokoyama, Y Menjo, M Watanabe, H Togo. Synthesis of oxazoles and thiazoles using thioimidates. Synthesis 1467–1470, 1994.
9. PHH Hermkens, MCA van Tilborg. Solid-phase organic chemistry: How to arrive at the best results. J Heterocyclic Chem 36:1595–1597, 1999.
10. LL Gundersen, AK Bakkestuen, AJ Aasen, H Overaa, F Rise. 6- Halopurines in palladium-catalyzed coupling with organotin and organozinc reagents. Tetrahedron 50:9743–9756, 1994.

11. J Solberg, K Undheim. Palladium-catalyzed coupling of organotin reagents and al-kenes with 4-iodopyrimidines. Acta Chem Scand, Ser B B41:712–716, 1987.

12. T Benneche. Acta Chem Scand 44:927, 1990.

13. Y Kondo, R Watanabe, T Sakamoto, H Yamanaka. Chemical studies on pyrimidine derivatives XLI. Palladium-catalyzed cross-coupling reaction of halopyrimidines with aryl and vinylstannanes. Chem Pharm Bull 37:2814–2816, 1989.

14. Y Kondo, R Watanabe, T Sakamoto, H Yamanaka. Condensed heteroaromatic ring systems XVI. Synthesis of pyrrolo[2,3-d]pyrimidine derivatives. Chem Pharm Bull 37:2933–2936, 1989.

15. J Solberg, K Undheim. Regiochemistry in palladium-catalyzed organotin reactions with halopyrimidines. Acta Chem Scand 43:62–68, 1989.

16. V Farani, V Krishnamurthy, WJ Scott. In: LA Paquette, ed. Organic Reactions, Vol. 50. Los Angeles: Wiley 1997, pp 3–633.

17. SD Bull, SG Davies, RM Parkin, F Sanchez-Sancho. The biosynthetic origin of diketopiperazines derived from D-proline. J Chem Soc, Perkin Trans 1:2313–2320, 1998.

18. DW Gordon, J Steele. Reductive alkylation on a solid phase: Synthesis of a pipera-zinedione combinatorial library. Bioorg Med Chem Lett 5:47–50, 1995.

19. BO Scott, AC Siegmund, CK Marlowe, Y Pei, KL Spear. Solid phase organic syn-thesis (SPOS): A novel route to diketopiperazines and diketomorpholines. Mol Di-versity 1:125–134, 1995.

20. VS Goodfellow, CP Laudeman, JI Gerrity, M Burkard, E Strobel, JS Zuzack, DA McLeod. Rationally designed non-peptides: Variously substituted piperazine librar-ies for the discovery of bradykinin antagonists and other G-protein-coupled receptor ligands. Mol Diversity 2:97–102, 1996.

21. AK Szardenings, TS Burkoth, HH Lu, DW Tien, DA Campbell. A simple procedure for the solid phase synthesis of diketopiperazine and diketomorpholine derivatives. Tetrahedron 53:6573–6593, 1997.

22. I Ugi, W Hörl, C Hanusch-Kompa, T Schmid, E Herdtweck. MCR6: Chiral 2,6-piperazinediones via Ugi reactions with α-amino-acids, carbonyl compounds, isocy-anides and alcohols. Heterocycles 47:965–975, 1998.

23. RA Smith, MA Bobko, W Lee. Solid-phase synthesis of a library of piperazinediones and diazepinediones via Kaiser oxime resin. Bioorg Med Chem Lett 8:2369–2374, 1998.

24. M Del Fresno, J Alsina, M Royo, G Barany, F Albericio. Solid-phase synthesis of diketopiperazines, useful scaffolds for combinatorial chemistry. Tetrahedron Lett 39:2639–2642, 1998.

25. AK Szardenings, V Antonenko, DA Campbell, N DeFrancisco, SI Ida, L Shi, N Sharkov, D Tien, Y Wang, M Navre. Identification of highly selective inhibitors of collagenase-1 from combinatorial libraries of diketopiperazines. J Med Chem 42: 1348–1357, 1999.

26. PHH Hermkens, TG van Dinther, CW Joukema, GN Wagenaars, HCJ Ottenheijm. Peptide backbone-to-backbone cyclisation as an avenue to β-turn mimics. Tetrahe-dron Lett 35:9271–9274, 1994.

27. DR Pilipauskas. Can the time from synthesis design to validated chemistry be short-ened? Med Res Rev 19:463–474, 1999.

28. R Carlson, A Nordahl. Exploring organic synthetic experimental procedures. Topics Curr Chem 166:1–64, 1993.
29. AC Atkinson, AN Donev. Optimum Experimental Designs. Oxford, UK: Clarendon Press, 1992.
30. DL Massart, L Buydens, S de Jong, P Lewi, J Smeijers, B Vandeginste. Chemometrics. Amsterdam: Elsevier, 1977.
31. JM Ostresh, GM Husar, SE Blondelle, B Dorner, PA Weber, RA Houghten. "Libraries from libraries": Chemical transformation of combinatorial libraries to extend the range and repertoire of chemical diversity. Proc Natl Acad Sci USA 91:11138–11142, 1994.
32. JM Ostresh, CC Schoner, VT Hamashin, A Nefzi, J-P Meyer, RA Houghten. Solid-phase synthesis of trisubstituted bicyclic guanidines via cyclization of reduced N-acylated dipeptides. J Org Chem 63:8622–8623, 1998.
33. Y Pei, WH Moos. Post-modifications of peptoid side chains: [3 + 2] Cycloaddition of nitrile oxides with alkenes and alkynes on the solid phase. Tetrahedron Lett 35:5825–5828, 1994.
34. PHH Hermkens, H Hamersma. Functional group transformation: An efficiency-enhancing approach in combinatorial chemistry. J Comb Chem 1:307–316, 1999.
35. CG Boojamra, KM Burow, LA Thompson, JA Ellman. Solid-phase synthesis of 1,4-benzodiazepine-2,5-diones. Library preparation and demonstration of synthesis generality. J Org Chem 62, 1240–1256, 1997.
36. WJ Haap, D Kaiser, TB Walk, G Jung. Solid phase synthesis of diverse isoxazolidines via 1,3-dipolar cycloaddition. Tetrahedron 54:3705–3724, 1998.
37. B Ruhland, A Bhandari, EM Gordon, MA Gallop. Solid-supported combinatorial chemistry of structurally diverse β-lactams. J Am Chem Soc 118:253–254, 1996.
38. PM Dean, RA Lewis (eds.). Molecular Diversity in Drug Design. Dordrecht: Kluwer Academic, 1999.
39. DM Bayada, H Hamersma. Compound diversity and representativity in chemical databases. J Chem Inf Comput Sci 39:1–10, 1999.
40. CA Lipinski, F Lombardo, BW Dominy, PJ Feeney. Experimental and computational approaches to estimate solubility and permeability in drug discovery and development settings. Adv Drug Delivery Rev 23:3–25, 1997.

5
The Choice of Solid Support

Wim Van Den Nest
DiverDrugs SL, L'Hospitalet de Llobregat, Barcelona, Spain

Fernando Albericio
University of Barcelona, Barcelona, Spain

I. INTRODUCTION

There is no doubt that combinatorial chemistry owes a great debt to R. Bruce Merrifield for the establishment of the solid-phase method for the synthesis of peptides (1). Nowadays, it is possible to affirm that Merrifield's prediction, "[I]t seems quite clear that a gold mine is awaiting the organic chemist who would look to solid supports for controlling and directing his synthetic reactions" (2), is a reality. The power of the solid-phase approach is of such importance that it has recently become possible to find scientific papers dealing with the preparation of libraries of organic compounds where the title clearly indicates that the library was made in the solution phase* rather than in the solid phase (3). The most significant difference between the solid and solution phases is the presence of the solid support in the former. Thus, the success of a synthetic process is strongly related to the choice of support and its performance.

Whereas the *handle* or *linker* should be thought of as a protecting group,†

---

* In polymer-supported chemistry, the term "solution-phase" has often been used to describe chemistry carried out on a polymer, such as polyethylene glycol (MW 2000–4000), which is soluble in the solvent in which the reaction is performed and that is precipitated at the end of the reaction (72).

† Some authors tend to confuse the terms "*handle*" or "*linker*" and "solid support." The term *handle* or *linker* should be used to indicate the way in which the solid support has been functionalized. In some cases, such in the chloromethyl/Merrifield resin, the linker is already incorporated into the resin (the chloromethyl groups). In other cases, the *handle* or *linker* is a bifunctional spacer molecule that becomes attached *permanently* to a functionalized resin at one end, often through a

the solid support should be considered as being analogous to a cosolvent (4). Thus, when a reaction is carried out on a polystyrene solid support, it can be considered that toluene is a cosolvent in the reaction. The solid support is therefore an integral part of the reaction and, in addition to being suitable for the reaction, needs to be optimized for any particular reaction. For this reason, the translation of organic reactions from the solution phase to the solid phase often requires some work to optimize the overall process. In general, reactions that tolerate excesses of reagent can be transferred more easily to the solid phase than those that require stoichiometric amounts of reagents. The latter situation rarely works well on solid supports (5) due to the fact that the solid-phase approach is based on the use of large excesses of reagents to drive the reactions to completion.

An optimal solid support should have the following characteristics:

1. Mechanically robust: batchwise and flow-continuous modes
2. Stable to variation in temperature
3. Mobile, well-solvated, and reagent-accessible sites
4. Acceptable loadings
5. Good swelling in a broad range of solvents, if applicable
6. Acceptable bead sizes, if applicable
7. Stable in acidic, basic, reducing, and oxidizing conditions
8. Compatible with radical, carbene, carbanion, and carbenium ion chemistry
9. Biocompatible and swelling in aqueous buffers
10. Little nonspecific binding to biomolecules

However, it is unrealistic to expect a solid support to have all these features, and therefore the most suitable support should be chosen for a specific purpose.

In contrast to the majority of chemical reagents, which tend to be of the same quality regardless of the supplier, the source of the solid support is extremely important. For instance, different polystyrene resins obtained from different manufacturers, or even from the same source, can have markedly different properties. The choice of the solid support source can have a significant influence on the performance of the synthetic process. If the preparation of a library is attempted with a specific batch of solid support, it is important that production of the library be carried out with the same batch.

Several classifications of solid supports have been proposed based on their physical properties (6,7). However, the solid supports used most widely today can be classified into just two groups: gel-type supports and modified surface type supports.

---

stable amide bond, and is linked *temporarily* to a growing molecule at the other end. The term "solid support" should be reserved exclusively to describe the polymer used. Thus, the solid support in the chloromethyl/Merrifield resin is polystyrene.

II. GEL-TYPE SUPPORTS

Gel-type supports are the most commonly used. The polymer network is flexible and can expand or exclude solvent to accommodate the growing molecule within the gel. Thus, the chemistry takes place within the well-solvated gel, which contains mobile and reagent-accessible chains (8). Three types of gel resins are commercially available and therefore these are the most used: polystyrene, polyethylene glycol, and hydrophilic PEG-based resins.*

A. Polystyrene Resins

Polystyrene (PS) resins are without doubt the most commonly used solid supports. Polystyrene systems consist of 1% cross-linked hydrophobic resins obtained by suspension polymerization from styrene and divinylbenzene.† They can be substituted with a wide variety of functionalities, although chloromethylation or aminomethylation are mostly used. Loadings can vary from 0.3 to 3 mmol/g.‡ For solid-phase peptide synthesis (SPPS), the loading should be kept at around 0.5 mmol/g to ensure a successful outcome for the synthesis (8). The resulting product is a hydrophobic swelling bead whose best swelling properties are seen in nonpolar solvents such as toluene or dichloromethane (DCM) (9). However, these systems can also be used in combination with other, more polar, solvents such as N,N-dimethylformamide (DMF), dioxane, and tetrahydrofuran. In SPPS, the use of acetonitrile is incompatible with a good quality final product (10). The bead size can be controlled very well, and beads that are virtually monosized have been obtained. Resins with average bead sizes of 10–200 µm are available. However, the use of seeding procedures has provided beads as large as 600 µm for grafting and library synthesis (11). It should be pointed out that the rate of diffusion in these larger beads is slower than that in the smaller ones, and the larger beads therefore show worse performance.

B. Polyethylene Glycol Resins

Based on the early work of Mutter and coworkers (12), the first commercially available polyethylene glycol-polystyrene (PEG-PS) resins were developed inde-

---

* Polyacrylamide resins were introduced in the 1980s for the solid-phase synthesis of peptides using the Fmoc-*t*Bu strategy and were developed based on the concept that the insoluble support and peptide backbone should have comparable polarities (73).

† PS with 2% cross-linking, which is mechanically more stable than those with less cross-linking, was used in the early years of solid-phase peptide synthesis. However, due to problems observed during the synthesis of difficult sequences, the cross-linking of this resin was then reduced to 1%.

‡ Full derivatization of polystyrene would give a substitution of approximately 10 mmol/g.

pendently in the mid-1980s by Zalipsky, Albericio, and Barany (10,13,14) (PEG-PS) and Bayer et al. (15) and Rapp (16) (TentaGel). These systems can be synthesized either by reaction of preformed oligooxyethylenes with aminomethylated polystyrene beads (PEG-PS) or by graft polymerization on polystyrene beads (TentaGel). More recently, other PEG-PS-based resins have been developed and commercialized. For example, Hudson and coworkers (17) developed Champion I and II (NovaGel), which are prepared from PEG blocks, and Dendrogel, which is prepared from branched PEG chains. In addition, Argogel has been prepared by PEG attachment through branched diol supports (18). Bradley and coworkers used the Dendrogel methodology and extended it to a dendrimer constructed with polyamidoamine (PAMAM) monomers on a TentaGel resin (19).

The first idea to involve grafting PEG onto PS was aimed at combining, on the same support, a hydrophobic PS core with hydrophilic PEG chains. Furthermore, in some of the aforementioned PEG-PS resins, the PEG unit might act as a *spacer* separating the starting point of the solid-phase synthesis from the PS core (20). This early hypothesis has not been corroborated in any other formulations of PEG-PS resins, and it is therefore possible to conclude that the *environment* effect of the PEG is far more critical than any spacer effect (20,21). PEG-PS resins are compatible with both polar and nonpolar solvents due to the unique conformational flexibility of PEG chains (6). The high degree of swelling provides the beads with a more firm and flow-stable character and makes PEG-PS resins physically stable in flow systems. The differences among different batches of PEG-PS resins are more accentuated than among PS resins. The content of PEG varies significantly between the different resins, and therefore their swelling properties can be markedly different. Furthermore, some of these systems can lose PEG during treatment with TFA. PEG-PS resins usually show lower loadings (0.15–0.3 mmol/g) than the PS ones. Regular PEG-PS resin beads have a particle size in the range of 35–150 μm, which implies a 50–200 pmol capacity. Rapp polymers have introduced a macrobead version of TentaGel that has a particle size between 300 and 700 μm with a narrow size distribution. This increase in particle size raises the capacity per bead by a factor of 100–1000 into the nanomolar range (10–100 nmol/bead) (22), which makes this resin particularly suitable for the preparation of chemical libraries following the "mix and split" approach (23). PEG-PS resins, depending of the PEG content, can be more difficult to dry than PS resins. Thus, in some cases washings with diethyl ether are required.

C. Hydrophilic PEG-Based Resins

Hydrophilic PEG-based resin is composed either exclusively of a PEG network or of a combination of PEG with a small amount of polystyrene or polyamide [developed by Meldal and coworkers (6)] or of acrylate with polymerizable vinyl groups [developed by Barany and Kempe (24)].

The Meldal resins are rationalized on the basis of two properties of PEG:

the high solvation potential of PEG in many solvents due to its amphipathic nature and the limited range of conformer populations of the PEG chains due to the preference of gauche–gauche interactions of vicinal carbon–oxygen bonds (6,21). The first member of this family was the PEGA resin (25), which was obtained by inverse suspension radical polymerization of various sizes of linear bis- and branched tris-2-aminopropyl-PEG samples with acryloyl chloride. Variation of the size and size distribution of the incorporated PEG as well as the addition of copolymerizing additives allows fine-tuning of the resin properties (26). The uniform beads swelled in all solvents ranging from toluene to aqueous buffers, a fact that makes PEGA resins especially suitable for assays of enzyme specificity and inhibition (6,21). To avoid the presence of secondary amide bonds, which can interfere with reactions involving carbon and carbenium ions, a new family of PEG-based resins (POEPOP/POEPS) was developed by partial derivatization of linear PEG (MW 1500) with vinylphenylmethyl chloride or vinylphenylpropyl chloride, followed by inverse suspension radical polymerization (27,28). The methyl version of this resin is not stable to Lewis acids owing to the benzylic linkage between the PEG and the PS. In addition, the preparation of vinylphenylpropyl chloride is not trivial, and even the small amount of PS (5–10%) present in the polymer led to a slight decrease in the favorable swelling observed for PEGA resins in polar solvents. Alternatively, polyoxyethylene cross-linked polyoxypropylene (POEPOP), which contains only ether bonds, was developed from a polymerization of PEG that was partially derivatized with chloromethyloxirane (29). Although the POEPOP resin is mechanically robust, with a relatively high loading (primary and secondary alcohols), and with good performance for organic transformations such as the Horner–Wadsworth–Emmons, nitroaldol, and Sakurai reactions, the presence of secondary ether bonds means that this solid support is not totally stable to strong Lewis acids (30). To overcome this problem, the SPOCC resin, in which all ether bonds and functional alcohol groups are primary, was developed (31). The preparation of this solid support, which retains all the swelling properties of the former PEGA resin, is based on the cross-linking by cationic polymerization of long-chain PEG that is terminally substituted with oxetane. These solid supports are all suitable for solid-phase enzyme assays, as demonstrated with proteolytic enzymes, glycosyl transferases, and protein disulfide isomerase (32,33). Members of the PEGA family of solid supports are usually fragile, and it is convenient to keep them slightly wet in a solvent such as methanol.

Another approach to PEG-containing supports is the CLEAR family, developed by Barany and Kempe (20,24). The different CLEAR resins are based on the copolymerization of branched PEG-containing cross-linkers such as trimethylolpropane ethoxylate triacrylate, which contains various ethylene oxide units, with amino-functionalized monomers such as allylamine or 2-aminoethyl methacrylate. These amino groups constitute the starting points for the solid-phase synthesis. The loadings of these solid supports are affected by the amount of functionalized monomer used for polymerization. Typical loadings are in the range 0.1–0.3 mmol/g. The CLEAR supports obtained both by bulk polymerization, which leads to irregularly shaped particles, and by suspension polymerization, which affords the preferred spherical beads, swell in a broad range of hydrophobic and hydrophilic solvents and have excellent physical and mechanical properties for both batchwise and continuous-flow systems.

CLEAR supports and the majority of the PEGA family supports are free of aromatic rings. This situation makes these solid supports very suitable for a broad range of reactions where the presence of such rings can cause problems because they can react with reagents or/and can jeopardize the solid-phase NMR control of the reactions.

III.   MODIFIED SURFACE TYPE SUPPORTS

Although a large number of materials exist that have been used for surface functionalization for solid-phase synthesis (5–7), the most commonly used, which are commercially available, are membranes and pellicular solid supports, where a mobile polymer is grafted to a rigid and inert plastic.

A.   Cellulose Membranes

Cellulose membrane systems are the basis of the SPOT concept, a highly parallel and technically simple arrangement that was first developed by Frank (34,35). SPOT is very flexible and economical relative to other multiple solid-phase procedures, especially regarding miniaturization and array geometries. Beaded cellulose was one of the first carriers tested by Merrifield for solid-phase synthesis but was found to be unsuitable (1). Later, cellulose paper was used for the solid-phase synthesis of both oligonucleotides (36) and peptides (37). Cellulose, usually Whatman 50 or 540, can be easily modified by O-acylation with protected amino acids to form amino acid esters (38). Although these functionalized membranes have been used for peptide synthesis, the lability of the ester bond limits their use in a broad range of chemical reactions. To circumvent this problem, the hydroxy functions of cellulose have been alkylated with epoxides containing a protected amino function (39). However, the preparation of the N-protected epoxypropylamines is laborious, and therefore this method has proven to be impractical for the functionalization of cellulose membranes. Recently, optimized methods have been developed that involve the incubation of cellulose membranes with epibromohydrin in the presence of perchloric acid, followed by reaction with 4,7,10-trioxa-1,13-tridecanediamine or diaminopropane (40,41).

Although cellulose membranes have proven to be very useful for SPOT syntheses and screening of peptide libraries (42), they suffer from some limitations when they are used for the preparation of libraries based on small organic molecules (35). For example, chemical degradation of cellulose can occur under the reaction conditions used for more conventional organic reactions. For instance, the glycosidic bonds of the cellulose can be labile under nucleophilic and acidic conditions. Furthermore, the large number of hydroxy functionalities can cause instability and interfere with subsequent reactions.

B. Polyolefinic Membranes

Polyalkanes (polyethylene and polypropylene) and their fluorinated derivatives [polyvinylidene fluoride and polytetrafluoroethylene (PTFE)] have been successfully used for solid-phase synthesis in reactors (43–45) as well as in the SPOT technique (35). The polypropylene membranes are highly stable, show low expansion in the presence of most common solvents, and display acceptable loadings. Functionalization can be performed by photoinduced graft copolymerization with functional acrylates by the following sequence: (a) Coating with photoinitiator (benzophenone) and selective UV excitation, causing hydrogen abstraction and radical formation; (b) application of an acrylate monomer solution, resulting in addition of radical sites to the double bond of the monomer; and (c) free radical polymerization. If acrylic acid and its methyl ester are used as monomers, the final functionalization in the form of amine groups can be achieved by amidation with 4,7,10-trioxa-1,13-tridecanediamine after activation of the carboxylic acids with $SOCl_2$, $(COCl)_2$, PCl_3, or PCl_5.

C. Pellicular Solid Supports

In systems with pellicular solid supports, a mobile polymer is grafted onto a rigid and inert plastic. This idea was first developed by Geysen, who initiated the "multipin concept" (46,47). This design has several handling advantages in that it can be fabricated into any particular shape or form and can be adapted to any desired array. Furthermore, the size of the inert plastic, together with the efficiency of the grafting, will determine the quantity of product that can be synthesized. Thus, unlike gel-type supports, it is the surface area and not the volume that determines the loading capacity. A further advantage of this kind of support is the consistency of the kinetics between grafted supports of different sizes and shapes. Such a correlation is not possible with gel-type supports because the rate of diffusion must change with size and therefore the reaction rates change.

The original pin support was poly(acrylic acid) grafted onto inert polyethylene, where the carboxylic acid moieties were capped with mono Boc-protected ethylenediamine (46). Polystyrene and the extremely hydrophilic polyhydroxyethyl methacrylate and poly(methacrylic acid/dimethylacrylamide) were also grafted on (48). The last grafted support in the "multipin concept" is the so-called Lanterns (47), where the actual solid support is polystyrene, which can be functionalized similarly to PS gel-type supports.

Comparison studies carried out in our laboratory between pellicular supports and PS resins have demonstrated that kinetic reactions are usually slower with the pellicular solid supports.

IV. LOADING DETERMINATION

Before starting a synthetic process, the initial loading and capacity of the solid support should be determined. It is on the basis of this loading that the quantities of reagents and the amount of final product that can be expected after cleavage will be calculated. Furthermore, the loading itself is also a parameter that is strongly related to the success of a synthesis. Thus, if the loading is too high, the "pseudodilution" phenomenon (49,50) associated with the solid-phase approach can be partially lost. This can be important, for instance, in those processes where a cyclization step is involved (51).

As the number of synthetic steps carried out on the solid support increases, so does the importance of the initial loading. For example, in SPPS the loadings that are considered optimal are around 0.5 mmol/g. Higher loadings can favor hydrophobic interactions between the peptide chains as well as the formation of hydrogen bonds, and these situations increase the risk that some of the reactive sites of the growing chain will become inaccessible (6,52). In the solid-phase organic synthesis (SPOS) of small molecules, the molecular weights of the target compounds are discrete in comparison with those of biomolecules, and larger amounts of compounds are required for proper characterization and biological tests. In these cases it is important to reach a balance between the initial loading, the amount of solid support, and the amount of product desired. There is no doubt that the use of solid supports with loadings greater than 1 mmol/g can be detrimental to the quality of the final product. Loadings indicated by the commercial suppliers should be confirmed before starting a synthesis, because suppliers usually give a broad range and the real value is often slightly out of the quoted range.

Several methods for the determination of loading are known, and most of those described in the literature are based on peptide chemistry. However, it is important to point out that the use of more than one method very often leads to a variety of results (53). The most accurate methods are those that allow a real quantification of the first building block anchored to the solid support. This is possible only in the case of SPPS, where the initial loading can be determined by incorporation of an amino acid and subsequent amino acid analysis of the acid-hydrolyzed amino acid/solid support. Alternatively, if the α-amino function is protected with the fluorenylmethyloxycarbonyl (Fmoc) group, the loading can be calculated by quantitative spectrophotometric monitoring following deblocking by treatment with piperidine. Methods based on the titration of amino groups usually give higher values, because all basic sites are titrated and the method is therefore not selective. For solid supports that contain heteroatoms such as Cl or Br in the reactive site, both elemental analysis and a modified Volhard titration are also very convenient. When the method used is based on the incorporation of a building block and its subsequent analysis, the solid support should be extensively and carefully washed to remove any building block that is noncovalently trapped on the solid support, because the presence of such compounds can interfere with the determination.

When a substituent is present for the determination of the loading, it is necessary to correct for the weight change during the reaction. Thus, the loading of a resin is expressed in millimoles per gram of resin according to the formula (54).

$$f = \frac{a}{1 - a \times e}$$

where f is the loading in millimoles per gram, a is the analytical concentration of substituent on the resin in millimoles per gram, and e is the equivalent weight of the substituent added.

A. Amino Acid Analysis

The amino acid to be incorporated in the resin should be stable to acidic conditions. Thus, Ser, Thr, Cys, Trp, Tyr, and Ile (the racemization that occurs during the acid hydrolysis implies its conversion to D-*allo*-Ile) should be avoided. Pro is not very convenient because its determination in a ninhydrin-based instrument at 540 nm is less accurate than for other systems. Thus, bifunctional amino acids such as Gly, Ala, Leu, or Phe, which are not β-branched and therefore show good reactivity, are the most suitable for this method. If the amino function is protected with an Fmoc group, its removal with piperidine–DMF (3:7) (1 × 1 min + 1 × 10 min) is advisable because Fmoc-amino acyl solid support can, depending on the conditions, be partially resistant to the acid hydrolysis.

Amino acyl resins (4–8 mg) are hydrolyzed in sealed tubes in propionic acid–12 N HCl (1:1) for 12 h at 130°C or 90 min at 150°C. For greater accuracy, the acid mixture may contain a known amount of an internal amino acid standard.

B. Ultraviolet Determination of the Fulvene-Piperidine Adduct (for Amines and Carboxyl Groups)

Amino groups are quantified by the UV determination of the fulvene-piperidine adduct formed by treatment of an Fmoc-amino-containing solid support with piperidine (55,56).

The Fmoc-amino acyl resin (4–8 mg) in a 10 mL polypropylene syringe fitted with a polyethylene disk is stirred or shaken in piperidine–DMF (3:7) (1 mL) for 30 min. MeOH (5 mL) is then added, and the solution is filtered and diluted with MeOH up to a total volume of 25 mL. The resultant fulvene-piperidine adduct has a characteristic UV absorption at 301 nm ($\varepsilon = 7800\ M^{-1}\ cm^{-1}$). For reference, a piperidine–DMF–MeOH mixture (0.3:0.7:24) is prepared, and the result is compared to a free Fmoc-amino acid solution of known concentration treated under identical conditions.

$$\text{Loading (mmol/g or }\mu\text{mol/mg)} = A_{301} \times \frac{0.025\,\text{L} \times 10^6\,\mu\text{mol/mol}}{7800\,\text{M}^{-1}\text{cm}^{-1} \times 1\,\text{cm} \times \text{mg resin}}$$

Similarly, carboxyl groups can be quantitatively monitored through the formation of the corresponding fluorenylmethyl ester and the UV determination of the fulvene-piperidine adduct formed after treatment with piperidine (56).

The carboxyl-containing solid support (4–8 mg) in a 10 mL polypropylene syringe fitted with a polyethylene disk is washed (1 mL portions) with DMF (3 × 0.5 min). The support is then treated with 9-fluorenylmethanol (5 equiv with respect to the expected functionalization), DiPCDi (5 equiv), and DMAP (0.1 equiv) in 0.5–1 mL of DMF (2 × 60 min) and washed with DMF (5 × 0.5 min). Finally, treatment with piperidine and UV determination are carried out as indicated above.

C. Quantitative Ninhydrin Test

Amino groups are quantified on the basis of their reaction with ninhydrin to produce Ruhemann's purple (55,57,58). Testing for Pro and Cys using this procedure should be avoided because they give a different response to ninhydrin.

Three reagent solutions are required: Solution A is phenol–EtOH (7:3); solution B is 0.2 mM KCN in pyridine; and solution C is 0.28 M ninhydrin in EtOH. The amino resin (4–8 mg) is placed in a test tube and incubated with 75 μL of solution A, 100 μL of solution B, and 75 μL of solution C for 7 min at 100°C. Five milliliters of EtOH–H$_2$O (6:4) is added, and the sample is vigorously mixed in a vortex. The solution is filtered and diluted with EtOH–H$_2$O (6:4) to give a total volume of 50 mL. The absorbance is measured at 570 nm ($\varepsilon = 15,000$ M^{-1} cm^{-1}). For reference, an EtOH–H$_2$O (6:4) mixture is prepared, and the result is compared to a free amino acid solution of known concentration treated under identical conditions.

$$\text{Loading (mmol/g or }\mu\text{mol/mg)} = A_{570} \times \frac{0.050\,\text{L} \times 10^6\,\mu\text{mol/mol}}{15,000\,\text{M}^{-1}\text{cm}^{-1} \times 1\,\text{cm} \times \text{mg resin}}$$

D. Picric Acid Titration (Gisin Method)

The Gisin method (59) is based on the formation of an ammonium picrate salt, its subsequent displacement by DIEA, and the UV determination of the diisopropylethylammonium picrate salt.

The amino acyl resin (4–8 mg) is placed in a 10 mL polypropylene syringe fitted with a polyethylene disk. The resin is washed (1 mL portions) with DCM (3 × 0.5 min), DIEA–DCM (1:19) (3 × 0.5 min), DCM (3 × 0.5 min), 0.1 M picric acid (3 × 1 min) in DCM, and DCM (6 × 0.5 min). The resin is then washed with DIEA–DCM (1:19) (3 × 0.5 min) and DCM (7 × 0.5 min), and the washings are retained. The combined solution is diluted with EtOH up to a total volume of 50 mL. The absorbance is measured at 358 nm (ε = 14,500 M^{-1} cm^{-1}). For reference, EtOH–DCM (8:2) is prepared.

$$\text{Loading}\,(\text{mmol/g or}\,\mu\text{mol/mg}) = A_{358} \times \frac{0.050\,\text{L} \times 10^6\,\mu\text{mol/mol}}{14{,}500\,\text{M}^{-1}\text{cm}^{-1} \times 1\,\text{cm} \times \text{mg resin}}$$

E. Backtitration with $HClO_4$ in HOAc

The amino resin (10–20 mg) is treated with 0.01 M $HClO_4$ in glacial HOAc (5 mL), and the mixture is shaken for 100 h. Excess $HClO_4$ is backtitrated with 7 × 10^{-3} M potassium biphthalate in glacial HOAc using crystal violet as an indicator (53,60).

F. Halide Analysis by the Modified Volhard Method

The resin (250 mg) is quaternarized by heating in pyridine (2.5 mL) for 1 h at 100°C, and the suspension is then diluted with HOAc–H_2O (1:1) (25 mL). The halide is displaced by the addition of HNO_3 (5 mL) and precipitated with a measured excess of standard $AgNO_3$ solution. The AgCl that is formed is coated with toluene, and the excess $AgNO_3$ is backtitrated with standard NH_4SCN solution, using ferric alum [$FeNH_4(SO_4)_2 \cdot 12\,H_2O$] as an indicator. A red color, due to the formation of $Fe(SCN)_3$, indicates that an excess of SCN^- is present and that the endpoint has been reached (61,62).

V. PRACTICAL CONSIDERATIONS

Gel-phase type solid supports cannot be magnetically stirred because fine particles will be formed. This will plug the filter disks and will cause a decrease in the overall yield. Thus, mechanical, orbital, or vortex stirring is advisable. In the case of PEGA solid supports, the stirring should be very slow because vigorous stirring can lead to the destruction of the beads.

Gel-phase type solid supports show the best performance when the bead size is not very disperse. This is because the rate of diffusion changes with the size of the bead and therefore reaction rates change. For this reason, it is convenient to examine the beads by optical microscopy to assess their homogeneity and integrity. Beads that are broken and/or are present in a broad size range should be discarded.

Spectroscopic analysis, mainly gel-phase ^{13}C NMR* (63) and FTIR, is a very convenient technique for the characterization of solid supports. The gel-phase ^{13}C NMR method has the great advantage of not requiring any special instrumentation besides a conventional FTNMR spectrometer and a conventional ^{13}C probe. In addition to ^{13}C NMR spectroscopy, ^{19}F NMR (64) and ^{31}P NMR (65,66) spectroscopy can also be used in cases where fluorine or phosphor are present in the handle or in the first amino acid. Furthermore, T_1 can be correlated to the mobility of the polymer chains and therefore to the performance of the solid support.

The best conditions for acquiring standard gel-phase ^{13}C NMR spectra from a variety of resins are well established, and working under optimal conditions, i.e., maximum swelling, gives rise to reasonably narrow lines for the pendant groups. (63–65).

The solid support is dried to constant weight under vacuum, and a sample (~300 mg) is weighed into a 10 mm NMR tube. The solvent is then introduced. The use of 5 mm tubes or microcells allows the experiment to be performed with a much smaller quantity of support (~80 mg and ~10 mg, respectively). Swelling of the polymer is crucial for the success of the experiment, and the choice of solvent must be carefully made in each case in order to optimize this property. Complete swelling of the polymer is usually achieved by allowing the resin to equilibrate with the solvent for 30 min after agitation with a vortex shaker. However, ultrasonic agitation is sometimes required to obtain homogeneous samples. The best results are obtained when the amount of solvent in the tube or in the microcell is the exact amount required to swell the resin. The ^{13}C NMR spectra can then be acquired using a relatively fast pulsing regime in order to take advantage of the fast relaxation properties of the resin. Again, the mathematical weighting of the FIDs before Fourier transformation can be easily optimized to discriminate the pendant group signals from the much faster-relaxing polymer backbone ^{13}C atoms. Sensitivity can also be substantially increased by using line-broadening routines (at least 1 Hz) during the weighting of the FID.

Infrared spectroscopy of solid supports has some value for their character-

---

* High resolution magic angle spinning (HRMAS) ^{1}H NMR is without doubt a superior method but has the limitation that specialized HRMAS probes are required (74–76).

ization and even for monitoring the loading, especially in those cases where the linkers have chemical functionalities, such as a carbonyl group, with strong IR absorption properties. The standard method consists of measuring difference spectra of the loaded resin relative to the unloaded one using either conventional IR or FTIR spectrometers (67). Samples of 2–10 mg are used to prepare standard KBr pellets. More recently, FTIR microspectroscopic methods have been described that allow single-bead measurements, a technique that opens the way for the analysis of solid-phase reactions during the course of the reaction (68,69). DRIFT (70) and photoacoustic (71) methods are also applied in monitoring reactions.

The best reaction performance is usually obtained at high concentrations. On the other hand, washing should be performed with large amounts of solvent (10 mL of solvent per gram of resin) to remove any excess building blocks or reagents that are not covalently attached to the solid support.

The quantity of product prepared in the solid phase is usually less than in a classical solution method. Therefore, NMR is not often the first choice to analyze the course of the reaction. Alternatively, HPLC/MS, in reverse phase, is a powerful technique for the first analysis of the reaction.

In SPPS the yields are calculated by amino acid analysis of the acid-hydrolyzed peptide-resin, and in solid-phase oligonucleotide synthesis, by UV determination. In SPOS, however, there is a tendency to calculate yields by weighing the final product. Although this method is totally valid in classical organic synthesis, where larger amounts of products are prepared, the use of such a method in SPOS warrants further discussion. Small amounts of product tend to be prepared, and the final cleavage to detach the compound from the solid support very often involves treatment with trifluoroacetic acid, which can clean the solid support and liberate other impurities accumulated during the synthesis. This process can subsequently lead to errors in the calculation of the yield.

REFERENCES

1. RB Merrifield. Solid phase peptide synthesis I. The synthesis of a tetrapeptide. J Am Chem Soc 85:2149–2154, 1963.
2. RB Merrifield. Solid phase synthesis. Adv Enzymol 32:221–296, 1969.
3. DL Boger, J Kyoo Lee. Development of a solution-phase synthesis of minor groove binding bis-intercalators based on Triostin A suitable for combinatorial synthesis. J Org Chem 65:5996–6000, 2000.
4. AW Czarnik. Solid-phase synthesis supports are like solvents. Biotechnol Bioeng 61:77–79, 1998.
5. D Hudson. Matrix assisted synthetic transformations: A mosaic of diverse contributions. I. The patterns emerge. J Comb Chem 1:333–360, 1999.

6.   M Meldal, Properties of solid supports. In: GB Fields, ed. Methods in Enzymology, Vol. 289. Solid-Phase Peptide Synthesis. Orlando, FL: Academic Press, 1997, pp 83–104.

7.   P Forns, GB Fields. The solid support. In: SA Kates, F Albericio, eds. Solid-Phase Synthesis: A Practical Guide. New York: Marcel Dekker, 2000, pp 1–77.

8.   VK Sarin, SBH Kent, RB Merrifield. Properties of swollen polymer networks: Solvation and swelling of peptide-containing resins in solid-phase peptide synthesis. J Am Chem Soc 102:5463–5470, 1980.

9.   KC Pugh, EJ York, JM Stewart. Effects of resin swelling and substitution on solid phase synthesis. Int J Peptide Protein Res 40:208–213, 1992.

10. S Zalipsky, JL Chang, F Albericio, G Barany. Preparation and applications of polyethylene glycol-polystyrene graft resin supports for solid-phase peptide synthesis. React Polym 22:243, 1994.

11. F Svec, JMJ Frechet. New designs of macroporous polymers and supports. Science 273:205–211, 1996.

12. H Becker, HW Lucas, J Maul, VN Pillai, H Anzinger, M Mutter. Polyethyleneglycols grafted onto crosslinked polystyrenes: A new class of hydrophilic polymeric supports for peptide synthesis. Makromol Chem Rapid Commun 3:217–223, 1982.

13. S Zalipsky, F Albericio, G Barany. Preparation and use of an aminoethyl polyethylene glycol-crosslinked polystyrene graft resin support for solid-phase peptide synthesis. In: CM Deber, VJ Hruby, KD Kopple, eds. Peptides 1985. Proc 9th Am Peptide Symp. Rockford, IL. Pierce, 1986, pp 257–260.

14. SA Kates, BF McGuinness, C Blackburn, GW Griffin, NA Sole, G Barany, F Albericio. "High-load" polyethylene glycol-polystyrene (PEG-PS) graft supports for solid-phase synthesis. Biopolymers (Peptide Sci) 47:365–380 1998.

15. E Bayer, B Hemmasi, K Albert, W Rapp, M Dengler. Immobilized polyoxyethylene, a new support for peptide synthesis. In: VJ Hruby, DH Rich, eds. Peptides 1983. Proc 8th Am Peptide Symp. Rockford, IL: Pierce, 1983, pp 87–90.

16. W Rapp. PEG grafted polystyrene tentacle polymers: Physico-chemical properties and application in chemical synthesis. In: G Jung, ed. Combinatorial Peptide and Nonpeptide Libraries: A Handbook. Weinheim, Germany: VCH, 1996, pp 425–464.

17. JH Adams, RM Cook, D Hudson, V Jammalamadaka, MH Lyttle, MFA Songster. A reinvestigation of the preparation, properties, and applications of aminomethyl and 4-methylbenzhydrylamine polystyrene resins. J Org Chem 63:3706–3716, 1998.

18. OW Gooding, S Baudart, TL Deegan, K Heisler, J Labadie, WS Newcomb, JA Porco, P van Eikeren. On the development of new poly(styrene-oxyethylene) graft copolymer resin supports for solid-phase organic synthesis. J Comb Chem 1:113–122, 1999.

19. NJ Wells, A Basso, M Bradley. Solid-phase dendrimer synthesis. Biopolymers (Peptide Sci) 47:381–396, 1998.

20. G Barany, F Albericio, SA Kates, M Kempe. Poly(ethylene glycol)-containing supports for solid-phase synthesis of peptides and combinatorial organic libraries. In: JM Harris, S Zalipsky, eds. Poly(ethylene glycol): Chemistry and Biological Applications. ACS Symp Ser 680. Washington, DC: Am Chem Soc, 1997, pp 239–264.

21. D Hudson. Matrix assisted synthetic transformations: A mosaic of diverse contributions. II. The pattern is completed. J Comb Chem 1:403–457, 1999.

22. W Rapp, KH Wiesmüller, B Fleckenstein, V Gnau, G Jung. New tools for libraries: Macrobeads of 750 μm diameter and selective orthogonal functionalization of inner sites and outer surface of beads. In: HLS Maia, ed. Proceedings of the 23rd European Peptide Symposium. Leiden: ESCOM, 1995, pp 87–89.

23. KS Lam, M Lebl. Combinatorial library based on the one-bead–one-compound concept. In: G Jung, ed. Combinatorial Peptide and Nonpeptide Libraries. Weinheim, Germany: VCH, 1996, pp 163–201.

24. M Kempe, G Barany. CLEAR: A novel family of highly cross-linked polymeric supports for solid-phase peptide synthesis. J Am Chem Soc 118:7083–7093, 1996.

25. M Meldal. PEGA: A flow stable polyethylene glycol dimethyl acrylamide copolymer for solid phase synthesis. Tetrahedron Lett 33:3077–3080, 1992.

26. FI Auzanneau, M Meldal, K Block. Synthesis, characterization and biocompatibility of PEGA resins. J Peptide Sci 1:31–34, 1995.

27. M Renil, M Meldal. POEPOP and POEPS: Inert poly(ethyleneglycol) cross-linked polymeric resin supports for solid-phase synthesis. Tetrahedron Lett 37:6185–6188, 1996.

28. J Buchardt, M Meldal. A chemically inert hydrophilic resin for solid-phase organic synthesis. Tetrahedron Lett 38:8695–8698, 1998.

29. J Rademan, M Meldal, K Bock. Solid-phase synthesis of peptide isosters by nucleophilic rections with N-terminal peptide aldehydes on a polar support tailored for solid-phase organic chemistry. Chem Eur J 5:1218–1225, 1999.

30. J Rademan, M Meldal, K Bock. Novel polar resins for solid phase organic chemistry and enzymatic on-bead assays. Towards new peptide isoster libraries. In: S Bajusz, F Hudecz, eds. Proceedings of the 25th European Peptide Symposium. Budapest: Akadémiai Kiadó, 1999, pp 38–39.

31. J Rademan, M Grötli, M Meldal, K Bock. SPOCC: Resin for solid-phase organic chemistry and enzyme reactions. J Am Chem Soc 121:5459–5466, 1999.

32. M Meldal, I Svendsen. Direct visualization of enzyme inhibitors using a portion mixing inhibitor library containing a quenched fluorogenic peptide substrate. 1: Inhibitors for subtilisin Carlsberg. J Chem Soc Perkin Trans 1:1591–1596, 1995.

33. M Meldal, I Svendsen, L Juliano, MA Juliano, E Del Nery, J Scharfstein. Inhibition of Cruzipain visualized in a fluorescence quenched solid-phase inhibitor library. D-Amino acid inhibitors for cruzipain, cathepsin B and cathepsin L. J Peptide Sci 4: 83–91, 1997.

34. R Frank. SPOT synthesis: An easy technique for the positionally addressable, parallel chemical synthesis on a membrane support. Tetrahedron 48:9217–9232, 1992.

35. H Wenschuh, R Volkmer-Engert, M Schmidt, M Schulz, J Schneider-Mergener, U Reineke. Coherent membranes supports for parallel micro-synthesis and screening of bioactive peptides. Biopolymers (Peptide Sci) 55:188–206, 2000.

36. R Frank, W Heikens, G Heisterberg-Moutsis, H Blöcker. A new general approach for the simultaneous chemical synthesis of a large number of oligonucleotides: Segmental solid supports. Nucl Acid Res 11:4365–4377, 1983.

37. R Frank, R Döring. Simultaneous multiple peptide synthesis under continuous flow conditions on cellulose paper discs as segmental solid supports. Tetrahedron 44: 6031–6040, 1988.

38. B Blankemeyer-Menge, M Nimtz, R Frank. An efficient method for anchoring

Fmoc-amino acids to hydroxyl-functionalised solid supports. Tetrahedron Lett 31: 1701–1704, 1990.

39. R Volkmer-Engert, B Hoffman, J Schneider-Mergener. Stable attachment of the HMB linker to continuous cellulose membranes for parallel solid phase spot synthesis. Tetrahedron Lett 38:1029–1032, 1997.

40. T Ast, N Heine, L Germeroth, J Schneider-Mergener, H Wenschuh. Efficient assembly of peptomers on continuous surfaces. Tetrahedron Lett 40:4317–4318, 1999.

41. K Licha, S Bhargava, C Rheinländer, A Becker, J Schneider-Mergener, R Volkmer-Engert. Tetrahedron Lett 41:1711–1715, 2000.

42. A Kramer, U Reineke, L Dong, B Hoffmann, U Hoffmüller, D Winkler, R Volkmer-Engert, J Schneider-Mergener. Spot synthesis: Observations and optimizations. J Peptide Res 54:319–327, 1999.

43. RH Berg, K Almdal, W Batsberg Pederson, A Holm, JP Tam, RB Merrifield. Long-chain polystyrene-grafted polyethylene film matrix: A new support for solid-phase peptide synthesis. J Am Chem Soc 111:8024–8026, 1989.

44. SB Daniels, MS Bernatowicz, JM Coull, H Köster. Membranes as solid supports for peptide synthesis. Tetrahedron Lett 30:4345–4348, 1989.

45. Z Wang, RA Laursen. Multiple peptide-synthesis on polypropylene membranes for rapid screening of bioactive peptide. Peptide Res 5:275–280, 1992.

46. HM Geysen, SJ Rodda, TJ Mason, G Tribbick, PG Schoofs. Strategies for epitope analysis using peptide synthesis. J Immunol Methods 102:259–274, 1987.

47. F Rasoul, F Ercole, Y Pham, CT Bui, Z Wu, SN James, RW Trainor, G Wickham, NJ Maeji. Grafted supports in solid phase synthesis. Biopolymers (Peptide Sci) 55: 207–216, 2000.

48. NJ Maeji, RM Valerio, AM Bray, RA Campbell, HM Geysen. Graft supports used with the multipin method of peptide synthesis. J Reactive Polym 22:203–212, 1994.

49. S Mazur, PJ Jayalekshmy. Chemistry of polymer-bound o-benzyne frequency of encounter between substituents on cross-linked polystyrene. J Am Chem Soc 101: 677–693, 1979.

50. G Barany, RB Merrifield. Solid-phase peptide synthesis. In: E Gross, J Meienhofer, eds. The Peptides, Vol 2. New York: Academic Press, 1979, pp 1–284.

51. C Blackburn, SK Kates. In: GB Fields, ed. Methods in Enzymology, Vol. 289. Solid-Phase Peptide Synthesis. Orlando, FL: Academic Press, 1997, pp 175–198.

52. C Chiva, M Vilaseca, E Giralt, F Albericio. An HPLC-ESMS study on the solid-phase assembly of C-terminal proline peptides. J Peptide Sci 5:131–140, 1999.

53. E Giralt, F Albericio. Free amino content of benzhydrylamine resins used as solid supports for peptide synthesis. An Quím 77C:134–137, 1980.

54. V Najjar, RB Merrifield. Solid phase peptide synthesis. VI. The use of the o-nitrophenylsulfenyl group in the synthesis of the octadecapeptide bradykinylbradykinin. Biochemistry 5:3765–3770, 1966.

55. GB Fields, Z Tian, G Barany. Principles and practice of solid-phase peptide synthesis. In: GA Grant, ed. Synthetic Peptides: A User's Guide. New York: WH Freeman, 1992, pp 77–183.

56. M Bassedas, F Albericio, F López-Calahorra. Quantitative monitoring of carboxyl groups in polymers. Anal Chim Acta 219:161–163, 1989.

57. VK Sarin, SBH Kent, JP Tam, RB Merrifield. Quantitative monitoring of solid-

phase peptide synthesis by the ninhydrin reaction. Anal Biochem 117:147–157, 1981.

58. E Kaiser, RL Colescott, CD Bossinger, PI Cook. Color test for detection of free terminal amino groups in the solid-phase synthesis of peptides. Anal Biochem 34: 595–598, 1970.

59. BF Gisin. The monitoring of reactions in solid-phase peptide synthesis with picric acid. Anal Chim Acta 58:248–249, 1972.

60. S Ehrlich-Rogozinski. Nonaqueous determination of the tert-butyloxycarbonyl group in amino acid and peptide derivatives. Isr J Chem 12:31–34, 1974.

61. JM Stewart, JD Young. Solid Phase Peptide Synthesis. San Francisco: WH Freeman, 1969.

62. B Gutte, RB Merrifield. The synthesis of ribonuclease A. J Biol Chem 246:1922–1941, 1971.

63. E Giralt, J Rizo, E Pedroso. Application of gel-phase ^{13}C NMR to monitor solid-phase peptide synthesis. Tetrahedron 40:4141–4152, 1984.

64. F Albericio, M Pons, E Pedroso, E Giralt. Comparative study of supports for solid-phase coupling of protected peptide segments. J Org Chem 54:360–266, 1989.

65. F Bardella, R Eritja, E Pedroso, E Giralt. Gell-phase ^{31}P NMR. A new analytical tool to evaluate solid-phase oligonucleotide synthesis. Bioorganic Med Chem Lett 3:2793–2796, 1993.

66. CR Johnson, B Zhang. Solid-phase synthesis of alkenes using the Horner–Wadsworth–Emmons reaction and monitoring by gel phase ^{31}P NMR. Tetrahedron Lett 36:9253–9256, 1995.

67. JR Hauske, P Dorf. A solid phase Cbz chloride equivalent: A new matrix specific linker. Tetrahedron Lett 36:1589–1592, 1995.

68. B Yan, G Kumaravel, H Anjaria, A Wu, RC Petter, CF Jewell, JR Wareing. Infrared spectrum of a single resin bead for real-time monitoring of solid-phase reactions. J Org Chem 60:5736–5738, 1995.

69. B Yan, G Kumaravel. Probing solid-phase reactions by monitoring the IR band compounds on a single "flattened" resin bead. Tetrahedron 52:843–848, 1996.

70. TY Chan, R Chen, MJ Sofia, BC Smith, D Glennon. High throughput on-bead monitoring of solid phase reactions by diffusive reflectance in infrared Fourier transform spectroscopy (DRIFTS). Tetrahedron Lett 38:2821–2824, 1997.

71. F Gosselini, M Direnzo, TH Ellis, WD Lubell. Photoacoustic FTIR spectroscopy, a non-destructive method for sensitive analysis of solid-phase organic chemistry. J Org Chem 61:7980–7981, 1996.

72. PH Ton, KD Janda. Soluble polymer-supported organic synthesis. Acc Chem Res 33:546–554, 2000.

73. E Atherton, RC Sheppard. Solid Phase Peptide Synthesis: A Practical Approach. Oxford, UK: IRL Press, 1989.

74. WL Fitch, G Detre, CP Holmes, NJ Shoolery, PA Keifer. High resolution ^{1}H NMR in solid-phase organic synthesis. J Org Chem 59:7955–7956, 1994.

75. PA Keifer, L Baltusis, DM Rice, AA Tymiak, JN Shoolery. J Magn Reson A 119: 65–75, 1996.

76. C Dhalluin, C Boutillon, A Tartar, J Lippens. J Am Chem Soc 119:10494–10500, 1997.

6

Applications of Fluorine-19 NMR Spectroscopy to Combinatorial Chemistry

Jill Hochlowski
Abbott Laboratories, North Chicago, Illinois

I. INTRODUCTION

Fluorine-19 nuclear magnetic resonance (^{19}F NMR) spectroscopy offers an excellent tool for the investigation of combinatorial chemistry processes, from the optimization of reaction conditions via quantitative techniques and characterization of reaction products to the determination of the reaction kinetics including the cleavage of library compounds from solid supports. Nuclear magnetic resonance spectroscopy in general is an extremely information-rich technique for structural investigation, and ^{19}F NMR spectroscopy in particular offers advantages over both carbon and proton NMR specific to combinatorial chemistry applications. Most resins upon which libraries are synthesized do not contain fluorine, so there is no background signal due to the solid support.

Fluorine is also a sensitive nucleus to detect, owing to the high gyromagnetic ratio and the fact that the spin ½ nucleus is present at 100% natural abundance, resulting in a sensitivity of 83% relative to that of proton NMR. Fluorine has a wide chemical shift dispersion, where small changes in structure result in large movements of signal. For example, methyl fluoride and ethyl fluoride resonate at $\delta = -272$ ppm and $\delta = -213$ ppm, respectively. (Chemical shifts for fluorine are generally recorded against fluorotrichloromethane as standard assigned at $\delta = 0$ ppm.) The sensitivity of chemical shift to structural changes is observed even when these changes are fairly remote from the fluorine atom being observed. This feature makes fluorine a good nucleus to use for the investigation of chemical or environmental changes occurring during synthesis or cleavage. Finally, the fluorine–carbon bond is extremely stable under most conditions employed during combinatorial chemistry manipulations and hence can serve as an excellent "tag" for the investigation of reactions, kinetics, or library encoding strategies.

Much of what has been learned to date in the techniques for the application of proton and carbon NMR spectroscopy to combinatorial chemistry can be directly applied to ^{19}F NMR. Solid-phase NMR spectra are acquired on samples in the gel-phase or solvent-swelled state. For highly flexible resins such as TentaGel, standard NMR spectra give acceptable linewidths for attached library compounds. For compounds bound to the less flexible polystyrene supports, magic angle spinning (MAS) techniques (1–4) have been shown to yield spectra with good linewidth by eliminating the line-broadening influences due to magnetic susceptibility.

Attaining narrow linewidths is an important criterion in achieving good ^{19}F NMR data. An excellent study done by Keifer on the influence of resin characteristics and spectral acquisition solvent on the linewidths for proton NMR signals for attached library compounds acquired under MAS conditions is found in Ref. 5. In this study, spectra were acquired for nine different commercially available resins in seven different solvents, and the linewidths were recorded for a *tert*-butyl singlet located on a resin-bound amino acid. The structure of the resin was found to exert the predominant influence on linewidth, with solvent having a lesser, though significant, effect. Dimethylformamide in general provided the best data, followed by dichloromethane, then dimethylsulfoxide. These data, although acquired for proton signals, can be directly applied to provide useful guidelines for optimization of ^{19}F NMR conditions as well. A review on the application of NMR spectroscopy to combinatorial chemistry details studies with various nuclei and methodologies is found in Ref. 6.

II. REACTION CHARACTERIZATION BY FLUORINE-19 NMR SPECTROSCOPY

The first application of ^{19}F NMR spectroscopy to the study of solid-phase synthesis was made by Manatt et al. in 1980 (7). Although not a practical application for the purpose of reaction monitoring, this work laid the foundation and demonstrated the potential of gel-phase fluorine-19 NMR spectroscopy in combinatorial chemistry. These workers synthesized a series of resin-bound fluorine-containing peptides, including fluoroaryloxycarbonyl-protected amino acid derivatives with 2-, 3-, and 4-fluorobenzylchlorocarbonates and peptides with terminal 2-fluorobenzyloxycarbonyl and 4-fluorobenzyloxycarbonyl forms of various amino acid residues. Using fluorobenzene as an internal standard and hexafluorobenzene as a lock signal, proton-decoupled spectra were acquired of chloroform-swelled beads in a 5 mm NMR tube on a probe tuned for fluorine and proton, at 94.1 MHz and 100 MHz, respectively. Exemplary of the ¹⁹F NMR data obtained under these conditions, linewidths for signals of compound bound to beads ranged from 26 to 39 Hz for the terminal 2-fluoro amino acid derivatives, whereas chemical shift dispersion observed between structures such as 2-F-Gly-PS, 2-F-Phe-PS, 2-F-Ala-PS, and 2-F-Pro-PS varied from one another by a maximum of only 4 Hz. The 4-fluoro amino acid derivatives showed signals of linewidths from 30 to 35 Hz, whereas chemical shift dispersion observed between structures such as 4-F-Phe-PS and 4-F-Ala-PS varied by only 1 Hz between them but by ~400 Hz between 2-F and 4-F derivatives. Therefore 2-fluoro and 4-fluoro benzene derivatives could be distinguished under these conditions, whereas the presence of different amino acids could not be established.

An interesting observation was made relative to the fluorobenzene standard, which appeared as two distinct signals in the NMR spectrum. This was attributed to material in differential environments within the swollen gel versus that in the outside solution, the determination being based upon varying the concentration of beads in a given solution volume. The higher field signal was determined in this manner to be due to solvent diffused into the polymer matrix.

In another study, the reaction progress for a series of fluorinated aromatic compounds bound to TentaGel resin was followed by Svensson et al. (8) using ¹⁹F NMR spectroscopy. Gel-phase spectra were acquired on 50–200 mg samples of beads swelled in either deuterochloroform or deuterodimethylsulfoxide at 375.5 MHz. Good signal-to-noise ratios were obtained in 10 min of spectral acquisition time under these conditions with acceptable linewidths. Changes in the chemical shift of the fluorine are observed (see Fig. 1) as the *m*-fluoro ester **1** is converted into the related acid **2** and amide **3**, respectively.

A subsequent series of compounds were synthesized with an *n*-butyl group replacing the Tentagel resin, thus allowing a comparison of spectra acquired for structures in solution-phase vs. gel-phase modes. Chemical shift correlations between these two structural types differed by at most 0.1 ppm. The authors note that better correlation was observed in DMSO than in chloroform in cases where hydrogen bonding could occur on functional groups near the fluorine atom being observed.

This same group (9) demonstrated the applicability of ¹⁹F NMR spectroscopy to the optimization of the cleavage of final product from solid support. This was accomplished by employing fluorine-containing linkers analogous to those commonly used in solid-phase synthesis, and a series of acid-, base-, and nucleophile-labile resins were investigated (see Fig. 2).

In one example, reaction optimization data were reported for a peptoid-type structure synthesized on TentaGel resin and attached through a fluorine-containing TFA-labile linker (see Fig. 3). Fluorine-19 NMR spectra were ac-

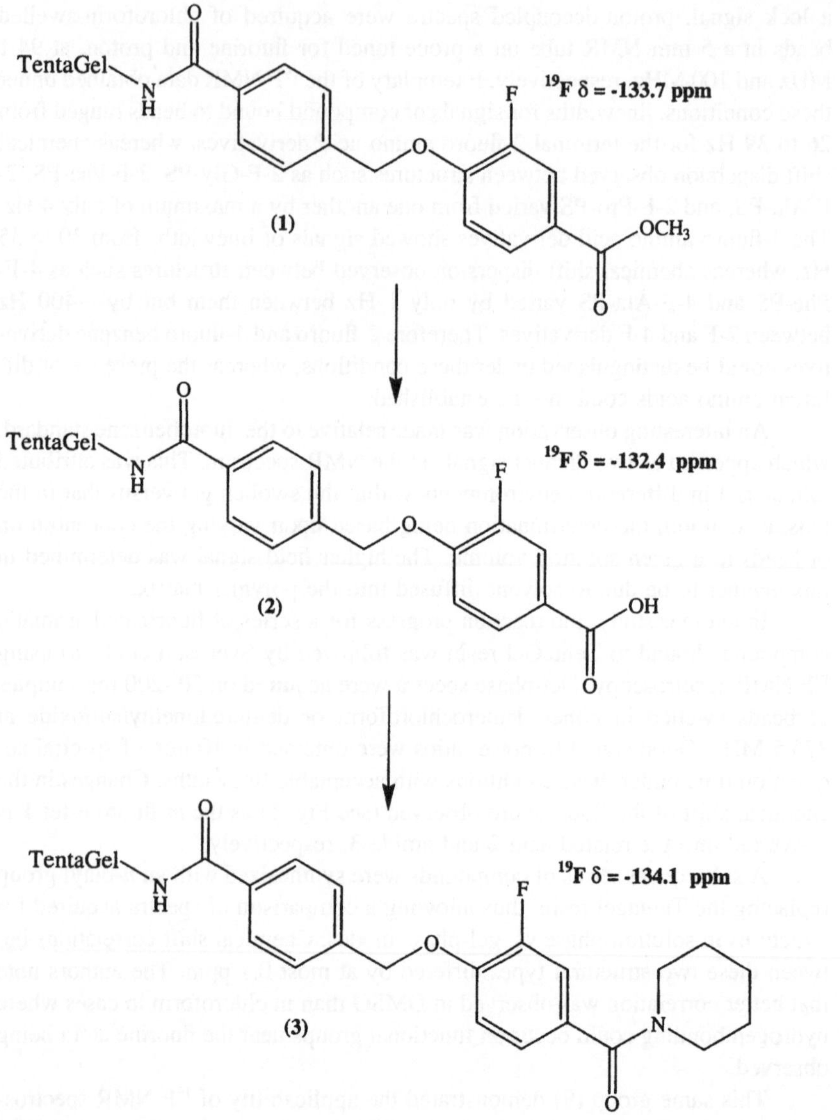

Figure 1 Fluorine-19 NMR shifts for a reaction sequence on solid support. (From Ref. 8.)

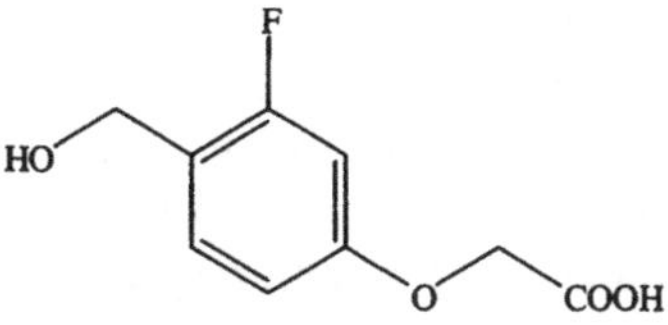

Figure 2 Fluorinated linkers for solid-phase library synthesis. (From Ref. 9.)

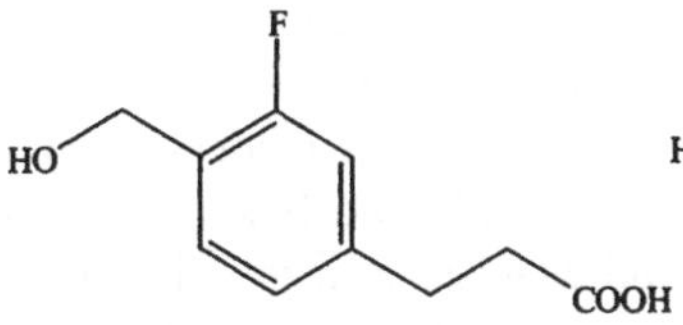

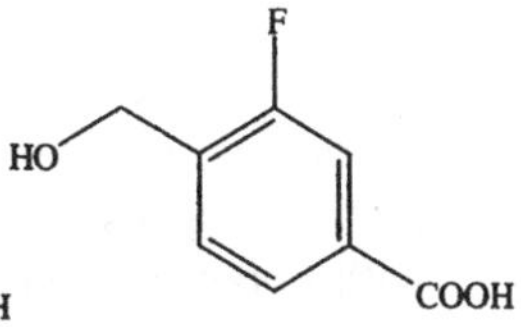

quired of deuterochloroform-swelled beads in a standard NMR tube at 375.5 MHz. Measurable differences in the chemical shift of the fluorine signal were observed during the bromoacylation of **7** to form **8** with signals at $\delta = -117.3$ and -115.1 ppm, respectively. The ability to distinguish and simultaneously record signals corresponding to starting material and product was useful in that it indicated incomplete reaction under the initial conditions studied, when the signal for the starting material was observed at 20% relative intensity. The ¹⁹F NMR spectrum acquired after a second bromoacylation condensation sequence indicated, by the absence of signal attributed to starting material, that complete conversion had occurred at that time.

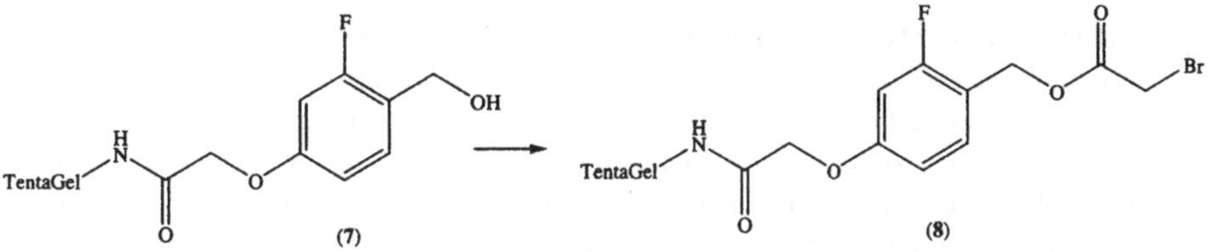

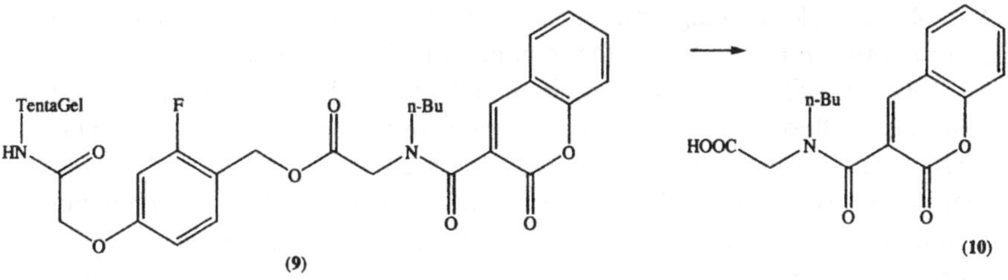

Figure 3 Library intermediates and product attached to TentaGel via a fluorinated linker. (From Ref. 9.)

Cleavage conditions for the release of compound **10** from the resin-bound state **9** were also optimized by ^{19}F NMR. Under aqueous TFA (1:2) cleavage conditions, incomplete release was evidenced by the presence of two distinct fluorine signals for the bead-bound linker with library compound cleaved and with library compound attached at $\delta = -117.3$ ppm and -115.3 ppm, respectively. In contrast, treatment of this same sample with 1 M LiOH for 2.5 h resulted in complete release of compound from the beads as evidenced by the single fluorine linker signal observed at $\delta = -117.3$ ppm. Loss of material from the resin was also evidenced at an earlier stage in the reaction sequence by the appearance in the ^{19}F NMR spectrum of a compound detached linker signal with characteristic shift at $\delta = -117.3$ ppm.

III. QUANTITATION OF COMBINATORIAL LIBRARY COMPOUNDS BY ^{19}F NMR

Fluorine-19 NMR spectroscopy can be used as a quantitative as well as qualitative technique in solid-phase synthesis by the incorporation of a fluorine-containing standard within the sample being analyzed. This standard can be attached either directly to the solid-phase support or as an external standard of known concentration within the sample being analyzed.

Drew et al. (10) achieved quantitative determinations by the incorporation of a fluorine standard directly into the polymeric material being used for library synthesis. A fluorinated chloromethylpolystyrene resin is generated by the copolymerization of styrene, 1,4-divinylbenzene, 4-vinylbenzyl chloride, and 4-fluorostyrene. Initial studies varying the amount of 4-fluorostyrene indicated an optimal content of this internal standard to be 0.3 mmol/g resin, which gave an acceptable signal-to-noise ratio in a reasonable amount of spectral acquisition time, as balanced against the cost of materials used. For this loading level, ~50 mg of resin swollen in deuterochloroform in a standard NMR tube gave good results with spectra acquired at 188.2 MHz within 6 min or at 470.1 MHz within 1 min. Magic angle spinning experiments improved spectral resolution and hence signal-to-noise, further reducing the minimal sample requirements.

This standardized resin could subsequently be used to study the attachment to resin of any fluorine-containing library element by the relative integration of ^{19}F NMR signals due to the support and library compound being studied. In a demonstrative experiment, a Michael reaction was optimized by the quantitation of a fluorinated reactant being attached. 3-Fluorothiophenol was reacted with resin-bound cinnamic acid ester with varying amounts of several different base catalysts under parallel conditions. An integration of the resin signal at $\delta = -117.3$ ppm vs. the Michael adduct signal at $\delta = -111.9$ ppm provided data on conditions of maximal yield. This methodology represents a simple approach to the optimization of virtually any reaction where a fluorinated reactant analog could be employed.

An external standard was used for quantitation experiments by Stones et al. (11) for the determination of loading on Merrifield resin. 2-Fluorophenol was attached directly to the resin, and a known concentration of fluorobenzene was used as a cosolvent, generating a mixed-phase sample whose ¹⁹F NMR spectrum was then acquired. It was determined that a solvent system of deuterochloroform–benzene–fluorobenzene (50:49:8:0.2) was optimal for the studies, because it gave a neutral buoyancy of bead samples swelled in this mixture. Weighed samples (30 mg) of beads were swelled in this standardized solvent and provided a mixed-phase solution–gel fluorine NMR spectrum in which the fluorobenzene signal at $\delta = -113$ ppm was integrated against the Merrifield attached 2-fluoroether signal at $\delta = -134$ ppm to provide quantitative data on resin loading. Prior to the quantitation experiments, inversion recovery measurements were acquired and indicated a spin–lattice (T1) relaxation rate of over 4 s for the Merrifield resin-attached 2-fluorophenyl sample; hence the authors used an interpulse relaxation delay of 25 s to ensure good quantitative data. No data are given on linewidths obtained under these conditions in which magic angle spinning was not employed. However, owing to the wide variance observed in chemical shifts, a very broad signal is obviously acceptable. Although it is not obvious from the article how the authors ensure that consistent amounts of solution and gel are presented to the magnetic field, good correlation is demonstrated between observed and predicted values in a blind experiment that acquired spectra on blended mixtures of known ratios of functionalized and unfunctionalized resin.

Our group has also applied mixed-phase ¹⁹F NMR spectroscopy to the quantitation of loading for various linkers, on various resins, and for various library structures by the attachment of a fluorine-containing moiety, either directly to the resin itself or as a monomer attached to a diversity position on specific cores (12). The ¹⁹F NMR signal due to library compound is then integrated against a solution standard in mixed-phase NMR experiments. We also found, as has been reported by Wehler and Westman (13) for proton NMR applications, that the magic angle spinning technique improves the linewidth for samples on most solid-phase resins. For the acquisition of ¹⁹F NMR spectra of TentaGel-bound samples, magic angle spinning is not required owing to the relative flexibility of the attachment tether.

In order to ensure that small-volume samples were located entirely within the active volume of the observed coil, we had fabricated a special small-volume NMR tube (Wilmad Glass Co., Buena, NJ) with a liquid capacity of 10 µL that could be fitted into a Varian MAS-NMR nanoprobe (Varian Instruments, Palo Alto, CA). Polypropylene plugs were made to replace the Teflon plugs normally used with microtubes in order to eliminate ¹⁹F NMR signal interference from this fluorinated source. In a typical quantitation experiment, 10–30 randomly selected

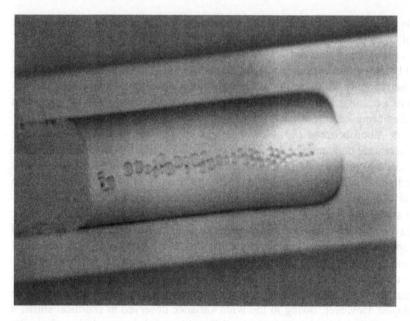

Figure 4 A 10 µL custom designed NMR tube loaded with beads and solution standard.

beads are counted, placed into the microtube as shown in Figure 4, and swelled in a measured volume of deuterodimethylformamide containing a fixed concentration of a fluorinated standard. Spectra were acquired at 470.1 MHz at a spin rate of 2000 Hz with an interpulse delay of 4.7 s with spectral acquisition times of 5–30 min. Figure 5 shows the results of such an experiment for 20 beads of resin sample **11** swelled in 5 µL of **12** at a concentration of 10 nmol/µL.

A relative integration of the two signals observed (the broader at δ = −61.6 ppm due to the resin sample and the sharper at δ = −61.8 ppm due to the solution

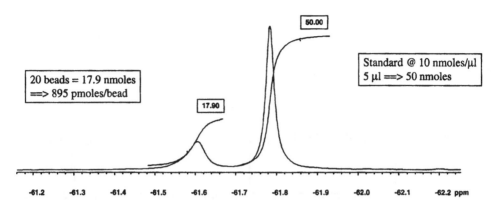

Figure 5 Determination of bead loading by MAS ¹⁹F NMR.

standard), indicates the average relative load of library compound, which was determined in this case to be 895 pmol per bead of **11**. It is possible under these conditions to determine the absolute amount of compound on a single bead sample, but this required spectral acquisition times of up to 4 h to achieve a sufficient signal-to-noise ratio.

In conjunction with experiments on the quantitative analysis of library compounds during synthesis, we realized that ¹⁹F NMR spectroscopy might also assist in providing data on their behavior during subsequent bioassay. A common method for the assay of solid-phase mix-and-split libraries involves their partial cleavage and diffusion into an agar matrix containing assay components (14). A question arose as to what was the actual local concentration of materials cleaved from beads and available for assay at the lower limits of aqueous solubility. Although it would not be practical to obtain quantitative information for material diffusing into agar from a partially cleaved, single-bead sample, the amount of material could be determined for representative members of a library with varying log *P* values in a bulk format experiment. Experiments were therefore undertaken (12) wherein samples of 20–50 beads were subjected to standard gas-phase TFA partial cleavage conditions, then placed on a film poured with a known volume of agar, made with assay buffer dissolved in deuterium dioxide containing a fixed concentration of a fluorinated standard.* After allowing 1 h for diffusion of TFA-cleaved library compound into the agar, beads were removed and the agar was melted, then transferred to a 5 mm standard NMR tube. Quantitative transfer of

* I also thank Dr. Robert Gentles, Dr. Peter Dandliker, and Mr. Martin Voorbach for advice and assistance in this work.

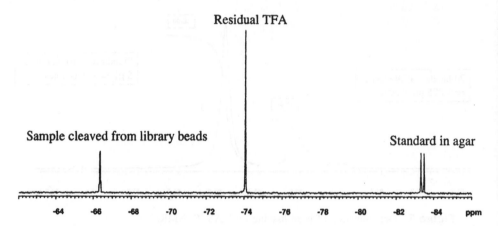

Figure 6   Fluorine-19 NMR spectrum acquired in agar.

agar to the NMR tube was not required owing to the fact that a relative integration at this point would yield quantitative information on the amount of material diffusing into the agar matrix. Standard solution-phase NMR spectral acquisition techniques of the resolidified agar were sufficient to give good linewidths. Quantitative data for the diffusion of any fluorinated compound into agar could be obtained by integration against the fluorine-containing standard in the buffer solution. A ^{19}F NMR signal for TFA used in the gas-phase cleavage always gave a residual peak in the spectrum. Figure 6 shows such an agar-gel ^{19}F NMR spectrum for the bead-cleaved compound **13** with 4-difluoromethoxybenzoic acid as the internal buffer standard within the agar.

(13)

IV.   KINETICS STUDIES IN COMBINATORIAL LIBRARY COMPOUNDS BY ^{19}F NMR

It is a simple step from obtaining quantitative data by ^{19}F NMR spectroscopy to deriving kinetics information, because one simply needs to take quantitative readings over time. This can be done to study either the kinetics of reaction or the kinetics of library cleavage. Shapiro et al. (15) used ^{19}F NMR spectroscopy to study the kinetics of the SNAr reaction on RinkAmide resin, as outlined in Figure 7. Kinetics information was obtained by halting the reaction at different time points by removing a portion of the resin and washing with excess DMF. Gel-phase ^{19}F NMR spectra were acquired in deuterobenzene on washed 0.5 mL aliquots of resin with 1–1.5 h of spectral acquisition time on a Bruker AC-300 MHz instrument in 5 mm tubes. The authors took advantage of the broad peak arising from the fluorine-containing probe components as an external standard. As an alternative to sampling aliquots of the reaction, the progress was monitored under magic angle spinning conditions at a spin rate of 4000 Hz in a Bruker-400 wide-bore instrument with a 7 mm rotor. The latter technique resulted in dramatically decreased spectral acquisition time requirements, with 4 min being adequate to obtain data with satisfactory signal-to-noise.

Our group has used ^{19}F NMR spectroscopy for the investigation of cleavage kinetics from solid supports (10). By the relative integration of on-bead vs. off-bead signals, we could determine the cleavage kinetics for various linkers, the kinetics for various cleavage conditions for a specific linker, or the cleavage rates for diverse compounds under identical conditions. For these studies, all experiments were carried out under magic angle spinning conditions at a spin rate of 2000 Hz on a Varian Unity 500 MHz NMR tuned to fluorine at 470.1 MHz. A typical spectrum would be acquired in a few minutes with 50–100 beads, swelled in 10 µL of cleavage solution. The spectral acquisition times were generally short enough that a deuterium lock signal was unnecessary. Figure 8 shows the ^{19}F NMR spectra obtained at two different time points for the cleavage of **14** to **15** in 15% TFA in dichloromethane (Fig. 9). The signals due to on-bead and off-bead fluorinated compound were easily identified (Fig. 9), both from their chemical shifts at $\delta = -71.6$ ppm and $\delta = -71.2$ ppm, respectively, and from their relative linewidths of 57 Hz and 19 Hz, respectively. A plot of the relative integration over time of these two signals gives a precise picture of the cleavage kinetics for the reaction. These experiments can be carried out sequentially with

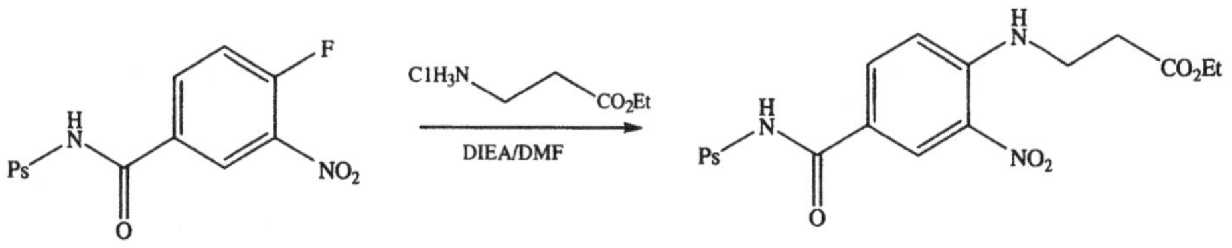

Figure 7 Reaction scheme for SNAr displacement. (From Ref. 15.)

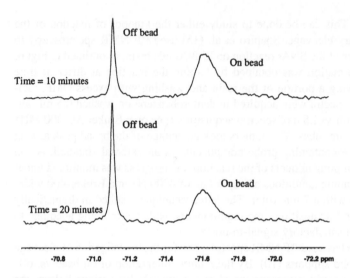

Figure 8 Fluorine-19 NMR spectra acquired during the cleavage in 15% TFA–DCM of **14** to **15**.

different percentages of TFA to investigate the cleavage profiles of a specific linker. Figure 10 depicts the relative cleavage kinetics in 2%, 10%, and 20% TFA in dichloromethane. These studies have obvious application in investigating new linker technologies.

Another useful application for the study of relative cleavage kinetics is to investigate the response of different library structures to a specific set of conditions. It is important that all compounds have at least similar cleavage behaviors under the conditions used when the assay technique employed involves the partial cleavage of library compounds. This ensures that specific cleavage conditions

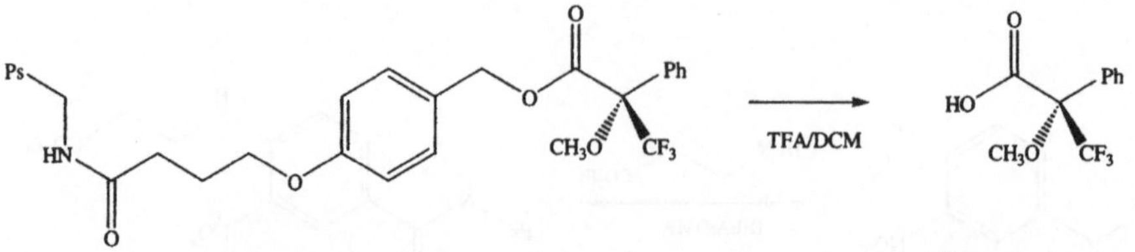

(14)
Chemical shift: δ = -71.6 ppm
Linewidth = 57 Hz

(15)
Chemical shift: δ = -71.2 ppm
Linewidth = 19 Hz

Figure 9 On-bead and off-bead chemical shifts and linewidths for Moshers acid.

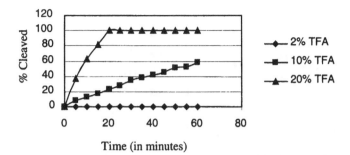

Figure 10   Cleavage kinetics for the conversion of **14** to **15** in varying amounts of TFA in DCM.

(of time and percent TFA) provide similar concentrations of material to assay. Using fluorinated model compounds with different electronic and steric characteristics, such as Moshers acid (**15**), Norfloxacin (**16**), 2,6-difluorobenzoic acid (**17**), and 4,4,4-trifluorobutyric acid (**18**), relative cleavage properties can be investigated.

(**16**)

(**17**)

(**18**)

Figure 11 depicts the relative cleavage conditions for these compounds from a Wang-type linker (of the type depicted in structure **14**) in 10% TFA–DCM. For this particular linker, although some differences in cleavage rates for

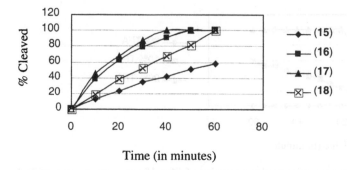

Figure 11 Relative cleavage kinetics for **15, 16, 17,** and **18** in 10% TFA–DCM.

these diverse compounds were observed, a time of 30 min could be chosen when materials available for assay varied by just over twofold in concentration.

V. ENCODING OF COMBINATORIAL CHEMISTRY LIBRARIES BY ^{19}F NMR

The same characteristics of ^{19}F NMR spectroscopy that present good methodologies for reaction monitoring, quantitation, and kinetics also enable a single-bead library encoding technique on solid phase. Specifically, since ^{19}F NMR chemical shifts are extremely sensitive to their environment, a wide range of distinguishable "tags" can be found in the many commercially available fluorine-containing compounds with various chemical structures. The sensitivity of detection of fluorine in NMR means that single-bead samples can be read, or decoded by their ^{19}F NMR spectra in a reasonable amount of time. Further, the C—F bond is stable to the range of conditions typically used to cleave library compounds from beads for the purpose of assay or analysis. Thus, a fluorine code attached via a noncleavable linkage will remain readable on beads from which the library entity for which they are encoding has been removed. We exploited these characteristics and developed a library encoding method (16) wherein combinations of fluorinated compounds are orthogonally attached to solid support beads and used to represent monomers in mix-and-split combinatorial libraries.

Our group usually synthesizes large mix-and-split libraries containing three diversity sites in a format of 25–100 diversity elements per site. Because libraries are screened using pools of compounds generated by avoiding mixing after attachment of third diversity elements, any given bead selected by assay has an uncertainty of only first and second positions. Assay cleavage conditions remove only a certain percentage of library compound from each bead, leaving residual library compound on the bead for subsequent analysis. A hybrid approach is used to deconvolute the structures on beads selected by assay, wherein a mass spectrum is first obtained from the remaining material cleaved from the bead. Owing to the mass redundancy of structures in screened pools, a molecular weight alone cannot always provide the structural assignment. However, a scheme for the encoding of first position monomer by ^{19}F NMR provides the additional information required to ascertain complete structural information in all cases.

First position encoding was achieved by the attachment to bulk aminomethylpolystyrene resin of lysine linker as shown in structure **19**.

(19)

Commercially available fluorine-containing carboxylic acids, such as those shown in Table 1, are attached to the ε-amino group of the linker to generate bulk pools of beads containing a characteristic fluorine signature. A variety of single fluorine-containing tags are commercially available, but to enhance the signal-to-noise ratio achieved per unit of spectral acquisition time, we use only those compounds that contain two or three identical fluorine atoms.

Large batches of bulk encoded resin are generated for subsequent use in the synthesis of multiple libraries. Typically, 100 differentially coded pools are generated using two-compound combinations of tags such as those listed in Table 1. Multibead ^{19}F NMR spectra are acquired for each pool and retained as a decoding data set. Selected typical spectra from this initial acquisition are shown in Figure 12. For library synthesis, to each of these bulk-encoded pools a TFA-labile linker, library core, and first position monomers will be sequentially added at the orthogonal amine, after which a standard mix-and-split library synthesis is carried out. Libraries are assayed in an agar format after partial library cleavage with gas-phase TFA. Each active bead is recovered and subjected to liquid-phase cleavage of the remainder of library compound to generate material for mass spectral analysis. Finally, the "spent" bead, which retains the fluorine code attached via the noncleavable amide linkage, is subjected to ^{19}F NMR spectral acquisition. This acquisition is achieved by swelling the bead in deuterodimethylformamide in a specially designed micro NMR tube (12), 10 μL in volume, along with a single control bead, as shown in Figure 13. The control bead has a single fluorine carboxylic acid compound attached to aminomethylpolystyrene, and its structure is different from those used in library encoding. Spectral acquisi-

Table 1 Fluorine-Carboxylic Acid Tags for Library Encoding

$\delta = 114.7$ ppm $\qquad$ $\delta = 113.2$ ppm $\qquad$ $\delta = 110.3$ ppm $\qquad$ $\delta = 108.9$ ppm

$\delta = 81.6$ ppm $\qquad$ $\delta = 81.4$ ppm $\qquad$ $\delta = 80.8$ ppm $\qquad$ $\delta = 70.9$ ppm

$\delta = 69.0$ ppm $\qquad$ $\delta = 63.0$ ppm $\qquad$ $\delta = 61.7$ ppm $\qquad$ $\delta = 61.3$ ppm

$\delta = 66.8$ ppm $\qquad$ $\delta = 57.5$ ppm $\qquad$ $\delta = 56.8$ ppm $\qquad$ $\delta = 42.4$ ppm

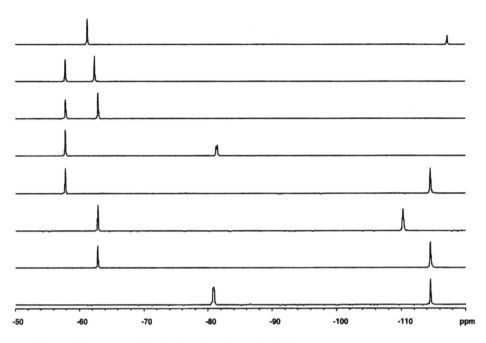

Figure 12 Selected ¹⁹F NMR codes. (From Ref. 16.)

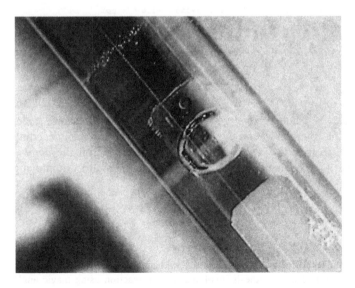

Figure 13 A 10 µL tube for gel-phase MAS-NMR containing a single bead to be decoded (dark in color) and a control bead (light in color). (From Ref. 16.)

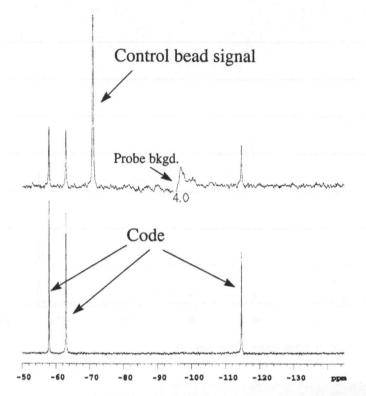

Figure 14 Comparison of ^{19}F NMR spectra of single bead to be decoded (upper spectrum) and bulk encoded resin pool to which it has been assigned (lower spectrum). (From Ref. 16.)

tion generally takes 1–4 h to achieve good signal-to-noise for a readable code. Figure 14 depicts the comparison of a single-bead spectrum acquired for decoding and the multibead spectrum to which the code has been matched from the archive. Numerous libraries have been encoded and decoded in our laboratory employing this technique.

REFERENCES

1. IE Pop, CF Dhalluin, BP Deprez, PC Melnyk, GM Lippens, AL Tartar. Monitoring of a three-step solid phase synthesis involving a Heck reaction using magic angle spinning NMR spectroscopy. Tetrahedron 52(37):12209–12222, 1996.
2. T Wehler, J Westman. Magic angle spinning NMR: A valuable tool for monitoring

the progress of reactions in solid phase synthesis. Tetrahedron Lett 37(27):4771–4774, 1996.

3. MJ Shapiro, J Chin, RE Marti, MA Jarisinski, Enhanced resolution in MAS NMR for combinatorial chemistry. Tetrahedron Lett 38(8):1333–1336, 1997.

4. SK Sarkar, RS Garigipati, JL Adams, PA Keifer. An NMR method to identify nondestructively chemical compounds bound to a single solid-phase bead for combinatorial chemistry applications. J Am Chem Soc 118:2305–2306, 1996.

5. P Keifer. Influence of resin structure, tether length, and solvent upon the high-resolution 1-H NMR spectra of solid-phase synthesis resins. J Org Chem 61:1558–1559, 1996.

6. MJ Shapiro, JS Gounarides. NMR methods utilized in combinatorial chemistry research. Prog Nucl Magm Resonance Spectrom 35:153–200, 1999.

7. SL Manatt, CF Amsden, CA Bettison, WT Frazer, JT Gugman, BE Link, JF Lubetich, EA McNelly, SC Smith, DJ Templeton, RP Pinnell. A fluorine-19 NMR approach for studying Merrifield solid-phase peptide synthesis. Tetrahedron Lett 21:1397–1400, 1980.

8. A Svensson, T Fex, J Kihlberg. Use of 19-F NMR spectroscopy to evaluate reactions in solid phase organic synthesis. Tetrahedron Lett 37(42):7649–7652, 1996.

9. A Svensson, K-E Bergquist, T Fex, J Kihlberg. Fluorinated linkers for monitoring solid-phase synthesis using gel-phase 19-F NMR spectroscopy. Tetrahedron Lett 39:7193–7196, 1998.

10. M Drew, E Orton, P Krolikowski, JM Salvino, NV Kumar. A method for quantitation of solid-phase synthesis using ^{19}F NMR spectroscopy. J Comb Chem 2:8–9, 2000.

11. D Stones, DJ Miller, MW Beaton, TJ Rutherford, D Gani. A method for the quantitation of resin loading using ^{19}F gel phase NMR spectroscopy and a new method for benzyl ether linker cleavage in solid phase chemistry. Tetrahedron Lett 39:4875–4878, 1998.

12. J Hochlowski, D Whittern C Woodall, T Sowin. Applications of ^{19}F-NMR spectroscopy to combinatorial chemistry. Proceedings of NMHCC Combinatorial Chemistry, San Diego, 1998.

13. T Wehler, J Westman. Magic angle spinning NMR: A valuable tool for monitoring the progress of reactions in solid phase synthesis. Tetrahedron Lett 37(27):4771–4774, 1996.

14. B Beutel. Discovery and identification of lead compounds from combinatorial mixtures. Annu Rep Med Chem 32:261–268, 1997.

15. MJ Shapiro, G Kumaravel, RC Petter, R Beveridge. 19-F NMR monitoring of SNAr reactions on solid support. Tetrahedron Lett 37(27):4671–4674, 1996.

16. JE Hochlowski, DN Whittern, TJ Sowin. Encoding of combinatorial chemistry libraries by fluorine-19 NMR. J Comb Chem 1(4):291–293, 1999.

7

Optimization of Solid-Phase Organic Syntheses by Monitoring Reactions on Single Resin Beads Using FTIR Microspectroscopy

Bing Yan
ChemRx Advanced Technologies, Inc., South San Francisco, California

Robert F. Dunn, Qing Tang, Gary M. Coppola, and Isidoros Vlattas
Novartis Pharmaceuticals Corporation, Summit, New Jersey

I. INTRODUCTION

This chapter describes the application of a single-bead FTIR microspectroscopic method in the optimization stage of solid-phase organic synthesis (SPOS) (1–10) and combinatorial chemistry. This method provides much-needed on-support reaction information at each stage of a synthetic sequence. Examples on the method and reaction monitoring directly on resin beads are presented to demonstrate the value of the method.

A. Reaction Optimization in Combinatorial Chemistry

In order to synthesize a library with high purity, optimization of reaction conditions is critical. Because of the complexity of the chemistry and the diversity of building blocks, the optimization of a synthetic protocol is very time-consuming. Furthermore, the lack of adequate analytical methods for on-support analysis is often brought up as a problem in this process. If libraries are prepared without adequate refinement of conditions, they are of poor quality. Because libraries of the "one-bead, one-compound" type are so large that their members cannot all be fully characterized, well-optimized reaction conditions are essential.

B. Qualitative and Quantitative FTIR Analysis on Solid Phase

In SPOS, both qualitative and quantitative analyses are required to determine whether the reactions work and whether they proceed to completion. FTIR has been used for both purposes. In principle, any chemical transformation will produce some spectral changes that can be detected to give "yes or no" information on the reaction. If IR signals of a functional group in the starting material convert to an IR band in another functional group in the product, the quantitative conversion percentage can be determined. A survey of 721 solid-phase organic reactions reported (11) before 1998 indicates that 80% of these reactions can produce detectable changes in the IR spectrum and 50% can be quantitatively analyzed.

C. Typical Characteristics of Solid Supports

Although there are several classes of solid-phase supports that are most used in SPOS, the microsporous polystyrene (PS) resin cross-linked with 1–2% of divinylbenzene (DVB) is the most popular. This kind of resin is uniformly loaded with chemical reacting groups in a density range of 0.5–1.0 mmol/g. Higher loading density resins have been produced for only a few resins. PS resin absorbs solvents in a process called swelling. Swollen beads constitute a gel phase in which chemical reactions are carried out. The gel phase shows selectivity in absorbing reagent or reactant and therefore dictates the reaction rate and yield. The swelling property of solid supports in various solvents is important considering that the solvent requirements differ for different reaction steps. This dynamic solvation also affects the reaction kinetics, the yield, and the quality of the final library. Dry beads with most solvent molecules removed remain in a gel state and are easily flattened for transmission single-bead FTIR measurement. Reactions in various solvents and reagent conditions can be monitored using the single-bead FTIR method with only a brief wash with DCM.

II. AN OVERVIEW OF APPROACHES AND METHODS

This section discusses FTIR methods with an emphasis on single-bead FTIR (12–15) and data analysis methods.

A. FTIR and FT Raman Methods in General

A variety of FTIR and FT Raman methods have been used in the analysis of solid-phase synthesis products. Beam condensers (16), attenuated total reflection (ATR) (16), and macro FT Raman spectroscopy (16–18) are methods requiring approximately 0.2 mg of resin beads. The diffuse reflectance infrared Fourier transform spectroscopy (DRIFT) (19), photoacoustic spectroscopy (20), and KBr pellet (21–29) methods have been used to study resin-bound compounds requiring 5–10 mg of resin samples. Although spectral quality is not superior for these methods, they are useful for qualitative reaction monitoring. In the KBr pellet method, light scattering and interference from stray light are two undesirable consequences. Another drawback of this method is that it is difficult to maintain KBr powder moisture-free. On the other hand, an advantage of this approach is that it requires a standard IR instrument that can often be found in nonspecialized laboratories.

B. Single-Bead FTIR Analysis

Most spectra shown in this chapter were collected on an FTIR spectrophotometer coupled with an IR microscope. No sample preparation is necessary. A few resin beads were compressed with a diamond compression cell or hand pressed between two NaCl windows. One side of the diamond or NaCl window was viewed under the microscope to locate a single bead. The transmission mode was used for the measurement. By switching the measuring mode to reflectance and using an ATR objective, spectra from the bead surface or a surface-functionalized support such as a multipin crown or microtube can also be obtained.

The single-bead FTIR method was first introduced in 1995 (12). The superior quality of the spectra obtained by the flattened single-bead method compared to that obtained by other FTIR methods from many more beads is attributable to the fact that (a) measurements using transmission mode, in general, can acquire a better spectrum than other modes of acquisition and (b) the single-bead method allows the minimum amount of stray light to pass through.

Several aspects of the single-bead FTIR method are responsible for its wide applicability. First, a single bead represents the whole population of beads in a properly mixed reaction mixture (12). Second, IR spectra taken from different beads (different sizes, different reaction times) can be compared quantitatively using polystyrene bands as an internal reference. Therefore, it is not necessary to examine the same bead for a reaction kinetics study in the course of a synthesis and data acquired from various beads are comparable. Third, a bead analyzed in the open air shows little moisture signal because of the hydrophobic nature of the resin.

The characteristics of combinatorial solid-phase organic synthesis have also contributed to the wide applicability of the FTIR method. Organic reactions selected to synthesize a combinatorial library are usually well optimized in solution so that the corresponding SPOS reactions would usually lead to the expected resin-bound products. Consequently, the analytical task in solid-phase organic synthesis is only to confirm the presence of the desired products. FTIR is particularly suited for this task, for several reasons. First, IR easily monitors functional group transformation. The functional group that is being monitored by IR does not need to be directly involved in the reaction. Building blocks used in synthesis can be selected to contain an IR-detectable group that remains unaffected in the reaction product. Direct monitoring of compounds on a solid phase is generally quicker and more convenient than methods requiring cleavage and analysis of intermediates, and this monitoring is particularly advantageous when synthetic intermediates are unstable to cleavage conditions. FTIR is a sensitive technique and requires only one bead for nondestructive analyses.

C. Analysis of Surface-Functionalized Polymers

The attenuated total reflection (micro- and macro-ATR) methods record spectra from the surface of a solid sample. These methods are effectively insensitive to sample thickness and are therefore ideally suited for analyses of surface-functionalized polymers such as multipin crowns and microtubes. The macro-ATR method has been used for qualitative studies of surface-functionalized polymers (30). We recently applied a micro-ATR method to obtain IR spectra from a polymer surface and used this method for quantitative kinetics analysis of reactions occurring on surface-functionalized polymers (see Sec. III.E).

D. Chemometrics in Spectral Analysis—A Case Study

Data analysis is an important aspect of bead FTIR studies, because the generated information is usually buried in the spectra and needs to be extracted. The complexity in an IR spectrum sometimes leads to severe spectral overlaps, particularly in the fingerprint region, making the extraction of quantitative information difficult. Instrument drift and inconsistencies in measurement, operation, and analysis procedures further accentuate the problem. When an IR band is isolated, a simple peak area integration can account for the relative changes in the concentration of the bound compound. A peak deconvolution program (such as PeakFit from Jandel Scientific) can also analyze slightly overlapped peaks (31,32). However, all of these methods are limited to high quality spectra with isolated or lightly overlapped bands. There is a need for a suitable method to analyze the severely overlapped spectra.

Chemometric and multivariate calibration methods can be used to extract quantitative information when there is an overlap of spectral information over all measured spectral regions. These methods can achieve increased precision from the redundant information in the spectra, account for baseline variations, and detect outliers (random spectral signals not related to observed reaction). The major loading factors obtained are close representations of the pure component concentration in a chemical transformation. They allow for a quantitative interpretation of the changes in spectral features.

To validate the chemometrics analysis method, the transformation in Scheme 1 was used for analysis because in product **2**, integration of an isolated band at 1730 cm^{-1} can be compared with chemometrics results obtained by analyzing overlapped spectral regions below 1670 cm^{-1}. Spectra of the starting resin **1** and the product **2** and spectra taken at various times during the course of the reaction were recorded. Spectra of **1** and **2** are shown as spectra a and b, respectively, in Figure 1.

The partial least squares (PLS) calculation method was performed using the fingerprint region of the IR spectra from 800 cm^{-1} to 1670 cm^{-1}. The primary loading factor (90% in weight) is given with spectrum c of Figure 1. This loading factor is a combination of all the correlated spectral changes involved in this transformation process. A number of key signature bands, which were previously buried due to overlapping, become unambiguously resolved and identified in the loading factor (Fig. 1c) from PLS analysis. The peak at 1622 cm^{-1} was assigned to the C=N double bond stretching vibration of **2**, the one at 1575 cm^{-1} to the $-NO_2$ asymmetrical stretching mode, and the one at 1159 cm^{-1} to the C$-$O$-$C asymmetrical stretching vibration. Therefore, the loading factor provides aug-

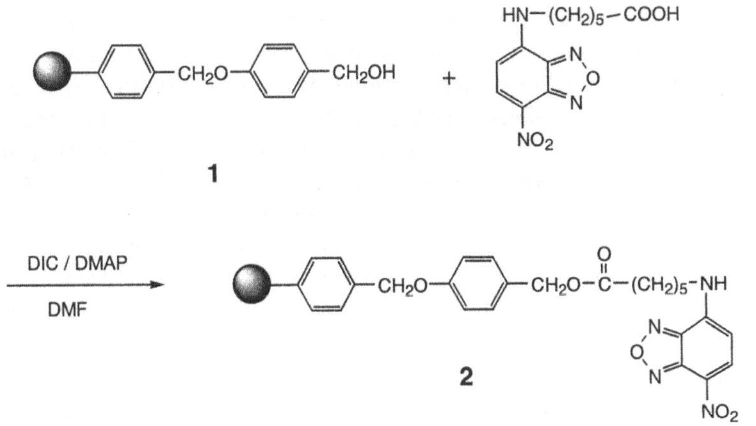

Scheme 1

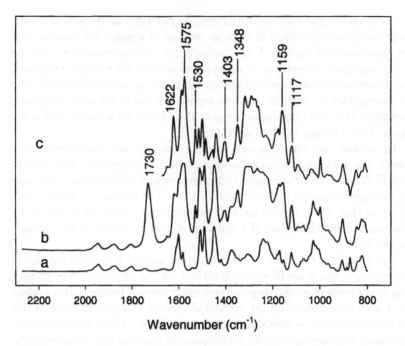

Figure 1 Infrared spectra of (a) **1** (starting material); (b) **2** (product); (c) the primary loading factor for the spectral changes during the synthesis of **2**.

mented and enriched information about the changes of all spectral features in an SPOS transformation.

The normalized score of the first loading factor is shown in Figure 2 as filled circles, which represent the time-dependent changes during the course of the synthesis of **2**. This score is obtained from the analysis of all the spectral intensities in the overlapped fingerprint region (800–1670 cm^{-1}). As a comparison, the time course that is also calculated from the peak area of the isolated $C=O$ band in the range of 1670–1780 cm^{-1} for **2** is shown in Figure 2 as open circles. This time course is used as a reference for the validation of the PLS calibration method. Time courses calculated from the two spectral regions are nearly identical, demonstrating that the multivariate calibration method is a powerful method for studying the kinetics of solid-phase organic reactions when spectral features are severely overlapped. Curve-fitting analysis to a pseudo-first-order reaction rate equation yielded a rate constant of 4.2×10^{-4} s^{-1} for the PLS calibration method (solid line in Fig. 2), which is very close to that derived from the single peak area calibration (3.7×10^{-4} s^{-1}; dashed line). Using these proce-

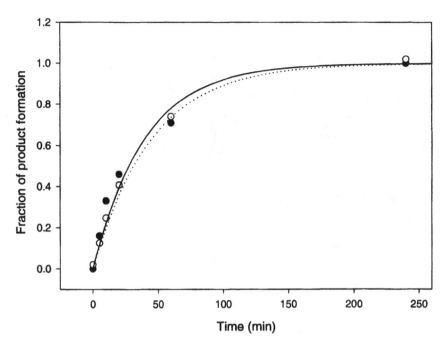

Figure 2 Time course for synthesis of **2** (●), (—) from PLS calibration method; (○), (···) from integration of peak area in the range 1670–1780 cm^{-1}.

dures, three other reactions including an ether synthesis were quantitatively analyzed (33).

III. RESULTS AND DISCUSSION

The single-bead FTIR method has been used in various stages of solid-phase reaction optimization. These include quality control (QC) of the starting material resin, resin loading determination, in-process reaction monitoring, solid-phase reaction kinetics, optimization of a diverse set of reaction parameters, and cleavage condition optimization. Examples of these applications are outlined here.

A. Quality Control of Starting Solid Supports

The quality of commercial resins that are used as starting materials for synthesis varies from manufacturer to manufacturer and even from batch to batch from the

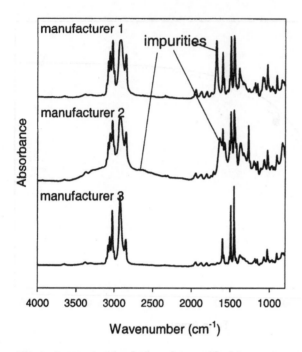

Figure 3 Single-bead IR spectra of aminomethylated polystyrene resins from three different commercial suppliers.

same source. Aminomethylated polystyrene resins from three different manufacturers were studied by single-bead FTIR (Fig. 3). Only the resin from manufacturer 3 has no impurities. In another study, surface aminomethylated polymers from two sources were compared (Fig. 4). Although both manufacturers claimed that the polymer they provided was aminomethylated polystyrene grafted on a polypropylene base, the one from manufacturer 1 was clearly contaminated with impurities.

Other quality problems from commercial resins also need to be checked. For example, the aldehyde functionality is not stable, and its loading decreases with storage time. Almost no chlorotrityl resin is hydrolysis-free as detected by single-bead FTIR.

B. Resin Loading Efficiency and Kinetics

The attachment of the linker or scaffold species to solid support is a fundamental step in SPOS. It is important that this step proceed to completion. Commonly used resins often utilize chloromethyl, aminomethyl, or hydroxymethyl groups

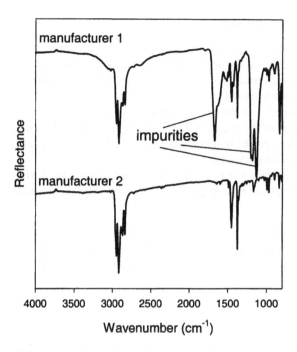

Figure 4 Micro-ATR FTIR spectra of aminomethylated polystyrene surface-grafted polymers from two different commercial suppliers.

as the starting reaction functional group in this initial step. The completion of the loading step on chloromethyl resin can be determined by detecting the remaining chlorine on the resin by using combustion elemental analysis (34) to determine the yield of the loading reaction. Using chlorine elemental analysis, the coupling yield of the first step in Schemes 3 and 4 (see next section) was determined satisfactorily. When aminomethyl resins are used as the starting material, qualitative and quantitative Kaiser tests are used to monitor completion of the reaction. The completion of the formation of an ester or ether with the hydroxymethyl groups can be confirmed reliably by using FTIR to monitor vibrations of the hydroxyl and carbonyl bonds. Other loading reactions that can be monitored by FTIR include ester or amide formation from a carboxylic acid resin which is monitored by confirming the disappearance of the carbonyl band in the IR spectrum. Similarly, reactions starting from aldehyde resins can be monitored by the disappearance of IR bands associated with the carbonyl and $C-H$ bonds of the aldehyde group. These on-resin analytical methods can be effectively used to monitor a wide range of resin loading reactions.

C. In-Process Reaction Monitoring

If the disappearance of an IR band from the starting material accompanies the formation of product, the decrease in the IR band can be used to estimate the percentage of conversion in this reaction step. This bears a close resemblance to the monitoring of a chemical reaction by thin-layer chromatography. The conversion of the starting material spot to the product spot with a different retention time indicates the completion of the reaction. The sophistication of FTIR is that the actual reaction information can be obtained directly without the complication and contamination of the cleavage reaction.

A single-bead FTIR analysis can be rapidly performed at any time during synthesis. The qualitative or quantitative in-process reaction monitoring is shown by the following examples.

Example 1. The synthesis of a resin-bound isonitrile (Scheme 2) according to the reported procedure (35) was monitored by single-bead FTIR. The disappearance of the IR band at ~3500 cm^{-1} from the starting resin indicated

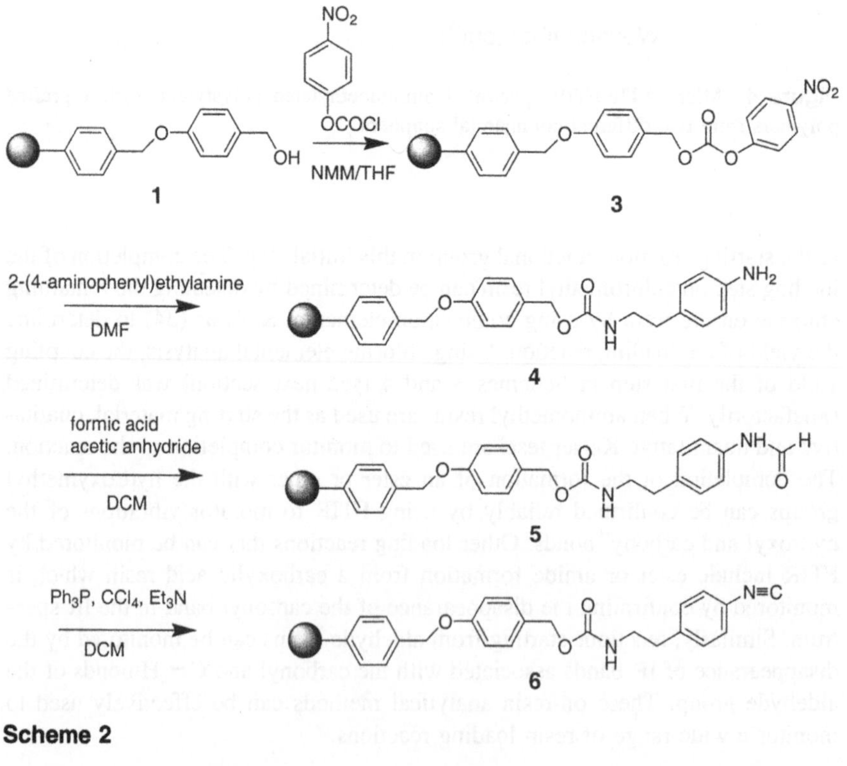

Scheme 2

the completion of step 1. The conversion of the band at 1765 cm^{-1} from **3** to the band at 1715 cm^{-1} for **4** indicated the completion of step 2. Finally, the formation of the formamide **5** was observed by the presence of the band at 1684 cm^{-1} in step 3. Subsequent conversion of **5** to the isonitrile **6** was again quantitatively monitored by observing the disappearance of the amide carbonyl band at 1680 cm^{-1} and the concomitant emergence of the band at 2150 cm^{-1}. The yield was ~85% for this step on the basis of FTIR analysis alone (Fig. 5).

Example 2. This example is related to the synthesis and application of a scavenger resin for amines (Scheme 3) (36). Both high loading (3.5 mmol/g) and low loading (1.7 mmol/g) Merrifield resins **7** were used to synthesize scavenger resins. High loading resin **7** displays an IR band with higher intensity at 1265 cm^{-1}, characteristic of the CH$_2$—Cl wagging vibration. Resin **7** was first converted to scavenger resin **8**, and this transformation was monitored by the formation of two IR bands at 1732 and 1786 cm^{-1} (Fig. 6). A variety of amines were then captured by the resin as shown in the case of 2-(4-bromophenyl)ethylamine. The completion of the quenching reaction was indicated by the disappear-

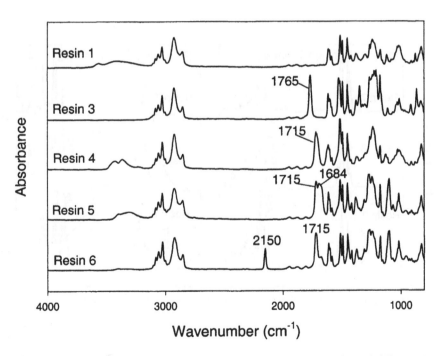

Figure 5 Single-bead IR spectra of the starting resin **1**, intermediates **3,4,5**, and the product **6** (see Scheme 2).

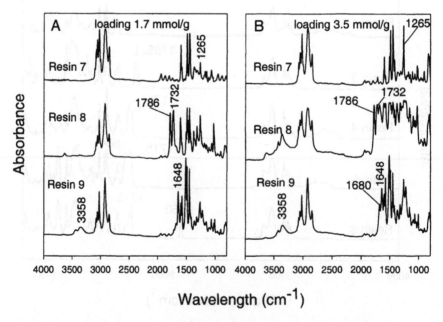

Scheme 3

Figure 6   Single-bead IR spectra of the starting resin **7**, scavenger resin **8**, and aminated product **9** (see Scheme 3) for both (A) low loading and (B) high loading series.

ance of the two carbonyl bands at 1732 and 1768 cm^{-1} and the formation of the amide carbonyl band at 1648 cm^{-1}.

Example 3. As shown in Scheme 4, the resin-bound aldehydes **10** were prepared by reacting Merrifield resin **7** with 4-hydroxy-2-methoxybenzaldehyde. The resin-bound aldehyde was converted in a two-step reductive amination process to the secondary amine **11**. The secondary amine **11** was then treated with a variety of electrophiles to form carbamates **12**, ureas **13**, amides **14**, and sulfonamides **15** (37). The conversion of **7** to **10** was confirmed by the new IR bands at 2769 and 1682 cm^{-1}, attributable to the aldehyde functionality (Fig. 7). The yield of this transformation was determined by combustion elemental analysis of chlorine. The chlorine content was 0.97 mmol/g for the starting resin **7** and <0.01 mmol/g for the product **10**, corresponding to a conversion yield of 99%. Alternatively, the conversion of **10** to **11** was monitored by single-bead FTIR,

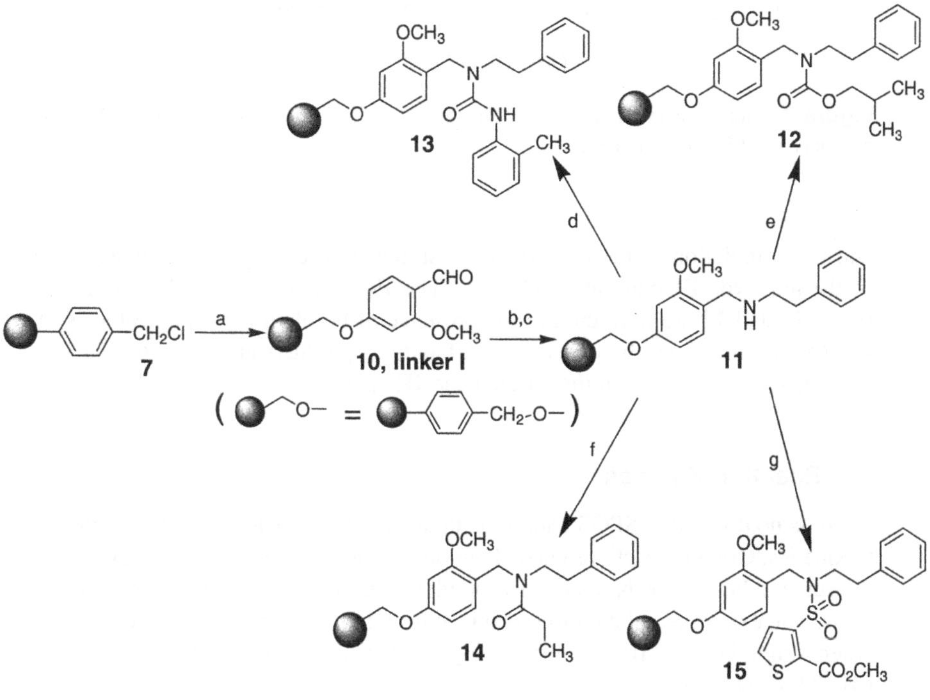

Scheme 4 Reaction conditions: (a) 4-Hydroxy-2-methoxybenzaldehyde, NaH, DMF; (b) CH(OMe)$_3$, Phenethylamine; (c) NaBH$_4$, THF, EtOH; then MeOH, reflux; (d) o-tolyl isocyanate, DMF, 60 °C; (e) isobutylchloroformate, DIEA, DCM; (f) propionyl chloride, DIEA, NMM, DCM; (g) carbomethoxythiophene-3-sulfonyl chloride, NMM, DCM.

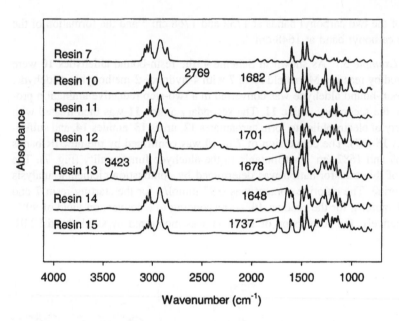

Figure 7   Single-bead IR spectra of the starting resin **7**, intermediates **10** and **11**, and products **12–15** (see Scheme 2).

and the completion of the reaction was estimated based on the area integration of the aldehyde IR band at 1682 cm^{-1}. The IR signals at 2273–2373 cm^{-1} in resins **11** and **12** are attributed to the presence of borohydride. The synthesis of **12–15** was confirmed by observing IR bands of products at 1701, 1678, 1648, and 1737 cm^{-1} (Fig. 7) in the single-bead IR spectra.

D.   Reaction Kinetics

There is no doubt that SPOS cannot be understood completely without reference to kinetics. Organic reactions carried out on polymer support have generally been expected to be slower than the corresponding homogeneous solution-phase reactions. The experimental data to test this expectation were obtained only recently when single-bead FTIR was effectively used in the study of reaction kinetics (31–33,37,41–45).

   An approach to the study of reaction kinetics on solid support is summarized below. A very small amount of resin beads is taken out of the reaction vessel at various times, rinsed briefly with solvent, flattened as described in Sec. II.B, and subjected to transmission measurement. Typical IR bands of the starting

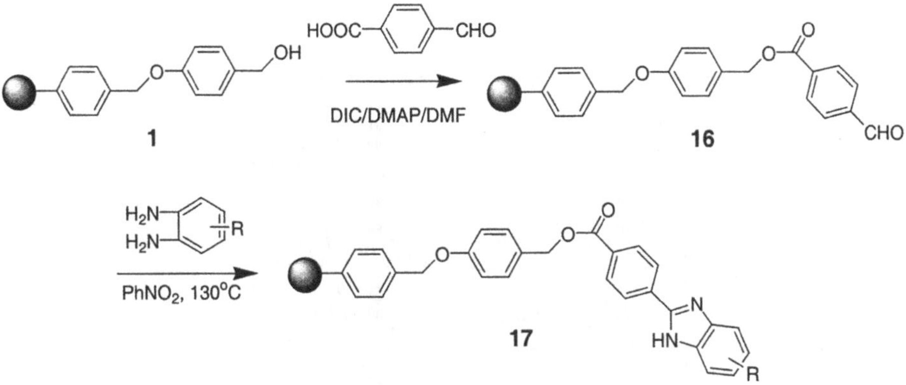

Scheme 5

material or the product are integrated, plotted against time, and fitted to a pseudo-first-order rate equation. The quality of curve fitting can be found in several published articles (31–33,37,41–44). The kinetics of the synthesis of **16** from **1** (Scheme 5) is determined (Fig. 8). The starting material band at 3470 cm^{-1} and the product bands at 1720 and 1703 cm^{-1} are integrated (Fig. 8A) and plotted against time with curve-fitting results (Fig. 8B).

An updated summary for the kinetics of a variety of solid-phase organic reactions is shown in Figure 9. This figure shows the best fit of the kinetic data for the reactions depicted in Schemes 1, 5–20, and 22–24 representing 20 solid-phase organic reactions with pseudo-first-order reaction rates ranging from 1.1 × 10^{-4} to 9.0 × 10^{-3} s^{-1}. Data shown here represent the first kinetic measurement before any serious optimization. These data demonstrate that most solid-phase organic reactions proceed at a higher rate than was generally expected previously. Many of these solid-phase organic reactions were complete in 2–3 h. Extended reaction times of 24 or 48 h, as have usually been used in SPOS, are unnecessary for most reactions and will lead to the formation of more side products.

E. Optimization of Reaction Conditions

From our own experience in SPOS, we have found that single-bead FTIR is an essential analytical tool. In addition to the above testimonies, its usefulness is further attested in the following examples.

1. Solid Support

Optimal reaction conditions and reaction rates vary when the same reaction is carried out on different solid supports. This is due to the different physicochemi-

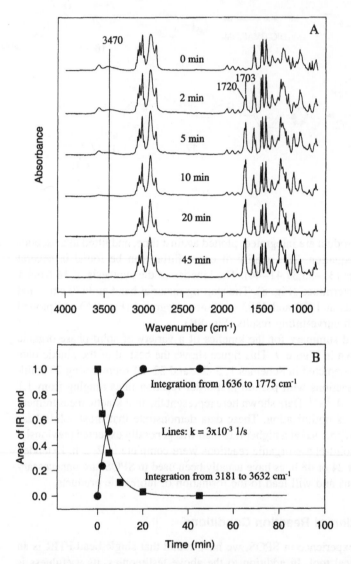

Figure 8 (A) Single-bead IR spectra of the reaction product at various times during the reaction in Scheme 5. (B) Time course of the reaction in Scheme 5 followed by integration of the IR band of the product at specified frequencies.

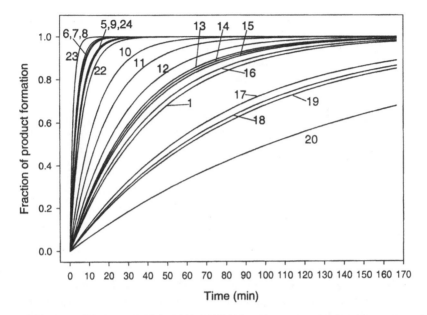

Figure 9 Time courses of reactions of Schemes 1, 5–20, and 22–24. The kinetics of product formation was analyzed as in Figure 8. The calculated time courses according to rate constants obtained from the best fit are displayed for each reaction. Pseudo-first-order rate constants (s^{-1}) determined for the reactions were (1) 3.8×10^{-4}; (5) 5.0×10^{-3}; (6) 4.6×10^{-3}; (7) 4.3×10^{-3}; (8) 3.1×10^{-3}; (9) 3.0×10^{-3}; (10) 2.5×10^{-3}; (11) 1.23×10^{-3}; (12) 6.0×10^{-4}; (13) 5.0×10^{-4}; (13b) 4.8×10^{-4}; (14) 4.6×10^{-4}; (15) 4.1×10^{-4}; (16) 2.2×10^{-4}; (17) 2.0×10^{-4}; (18) 1.9×10^{-4}; (19) 1.1×10^{-4}. * 3b represents the same reaction 3 carried out using 100-fold less starting resin and 50% of the resin loading as compared to the original reaction 3.

cal properties associated with various solid supports. Polystyrene, for instance, is highly hydrophobic. PS-PEG resin, on the other hand, is cross-linked 1% divinylbenzene-polystyrene grafted with 70% PEG linkers and therefore more hydrophilic. Reactive sites in PS-PEG resins are separated from the hydrophobic polystyrene framework by the long flexible PEG spacers. It is commonly assumed that reactions carried out on PS-PEG resins proceed faster than those carried out on polystyrene resins.

Although the attachment of a Knorr linker to solid supports proceeds faster on PS-PEG resins (TentaGel, Champion) than on microporous PS resins (38), the kinetics of four classes of reactions (in Schemes 7, 9, 14–17, and 19) were studied on both microporous PS resins and PS-PEG resins using single-bead FTIR (32). The catalytic oxidation of alcohol by tetra-n-propylammonium perru-

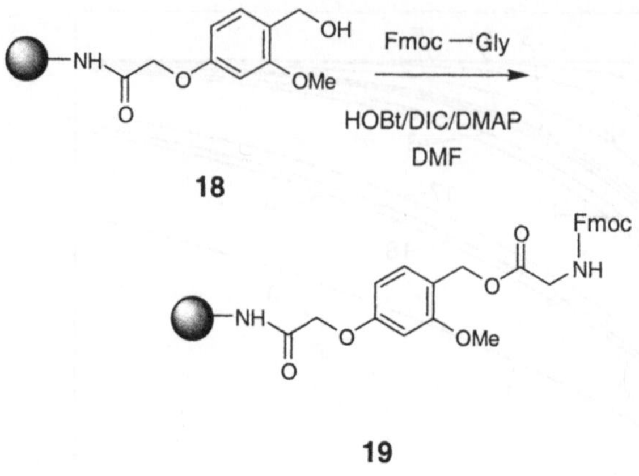

Scheme 6

thenate was clearly faster on PS-PEG resin. On the other hand, a series of esterification reactions proceeded with similar rates on both resins. A dansylhydrazone formation (Scheme 9) and a 5-oxazolidinone ring-opening reaction with amine (Scheme 19) were found to be faster on PS resins. These findings challenge the common perception that reactions on the PS-PEG resin are always faster than those on PS resins.

Surface-modified polypropylenes such as multipin crowns and microtubes constitute another kind of solid support. The reaction in Scheme 7 was carried out

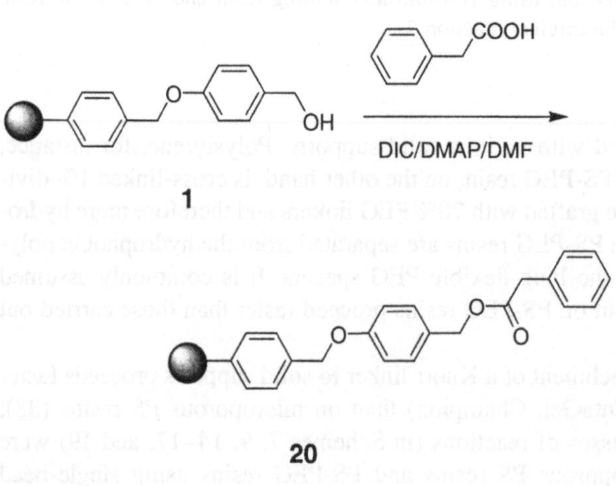

Scheme 7

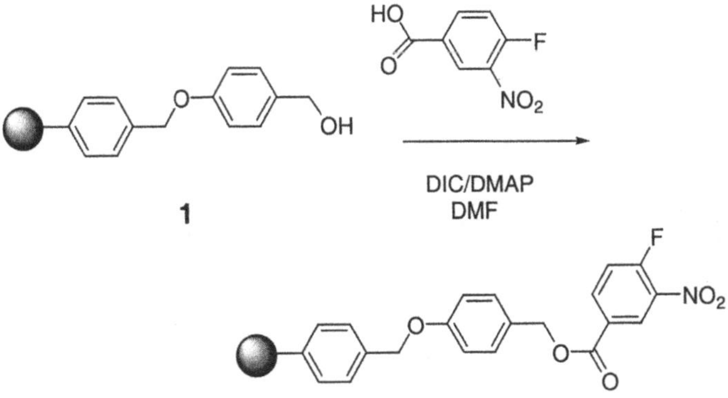

1

21

Scheme 8

on low and high cross-linked polystyrene resins (1% and >20% divinylbenzene, respectively), polystyrene-PEG resin, and surface-functionalized microtubes (39). Infrared spectra at various times on these various supports are shown in Figure 10. The best fits for kinetics data on these supports are compared in Figure 11. The reaction on the highly cross-linked polystyrene resin was slower than on other supports, and the reaction on the microtube was faster than on PS resins.

As expected, the effects of polymer matrix on the reaction rate were found to be at least as complex as the solvent effects encountered in solution-phase

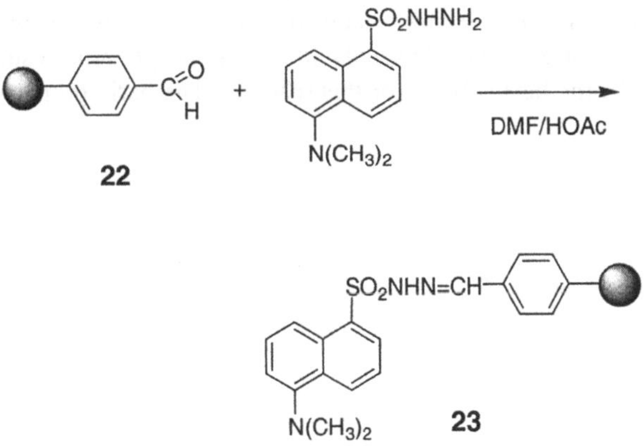

22

23

Scheme 9

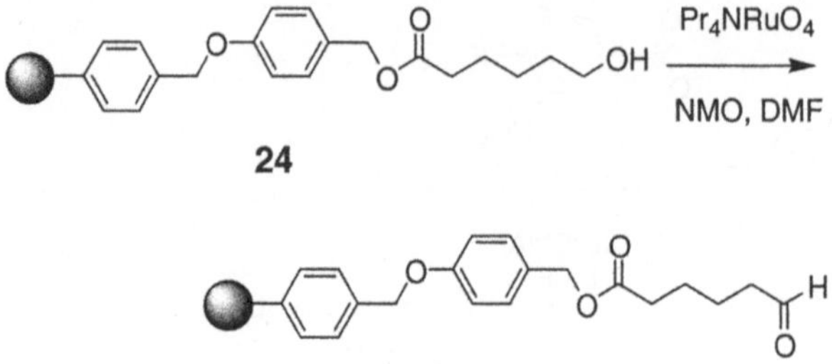

24

25

Scheme 10

reactions. Broad generalizations about the characteristic effects of any given resin for a series of different reactions should be cautiously considered. Reaction rates on supports depend on solvent swelling, selective absorption, hydrogen bonding, hydrophobicity, and polarity, and no single polymer support is best suited for all reactions or conditions.

2. Solvent Effects on SPOS

When the same solid support is used, solvent has multiple roles in solid-phase organic syntheses, including swelling the polymer support, stabilizing the reaction intermediate, and solubilizing the reagents and reactants. Solvent selection is often critical for solid-phase organic syntheses. Examples of this are shown in Figures 12 and 13. In these examples, the reaction in Scheme 20 was carried out on polystyrene resin in different solvents. Although DMF is a good swelling solvent, the reaction does not proceed to completion because of the poor solubility of the reagents in this solvent. On the other hand, pyridine, which is a better solvent for the reagents and a good swelling solvent for the polystyrene resin, was ultimately shown to be the best solvent for this reaction. The basic property

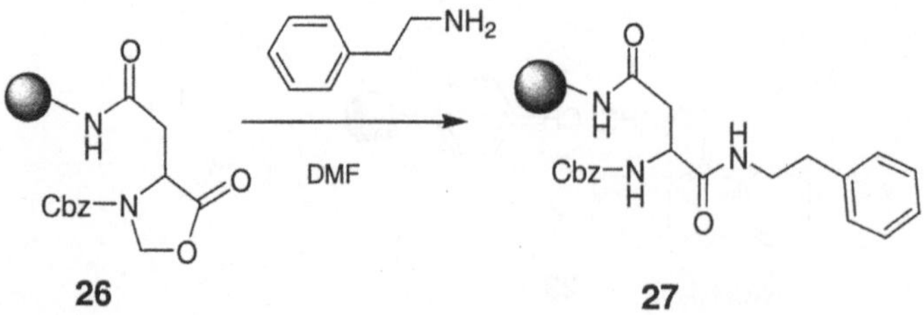

26   **27**

Scheme 11

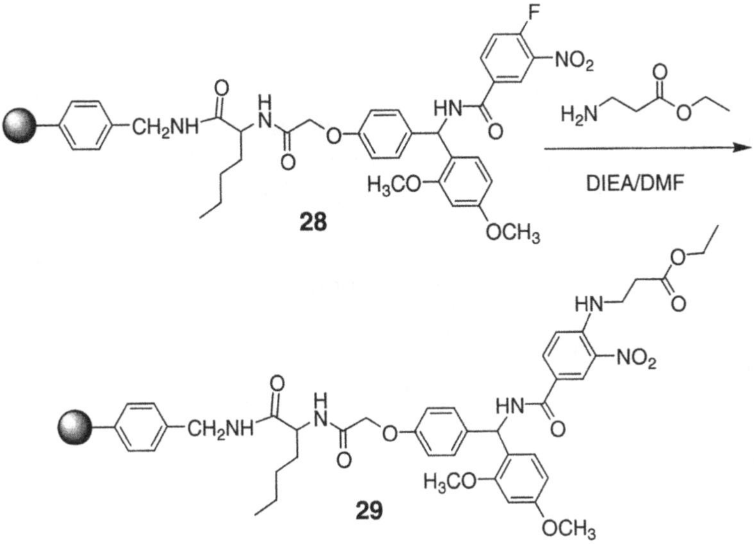

Scheme 12

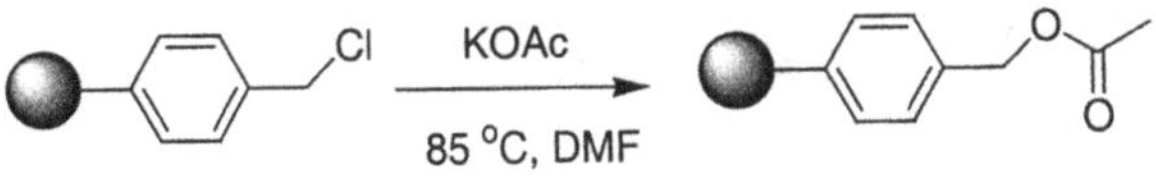

Scheme 13

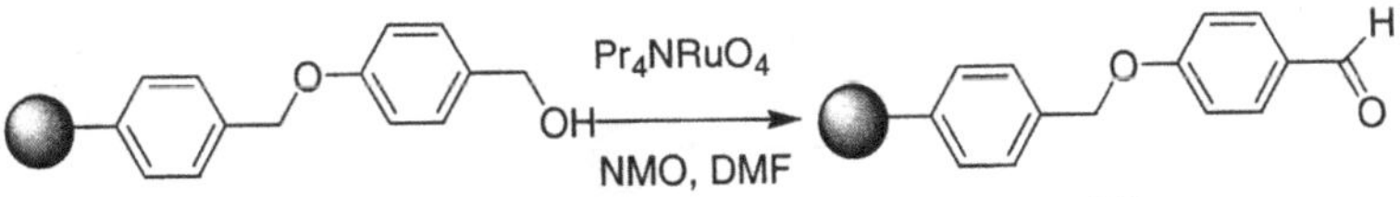

Scheme 14

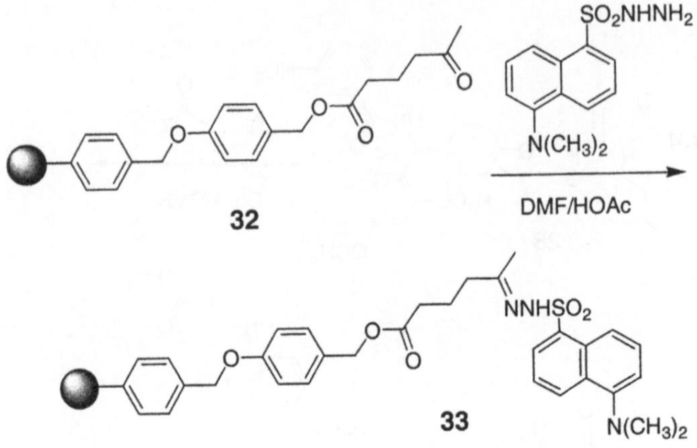

Scheme 15

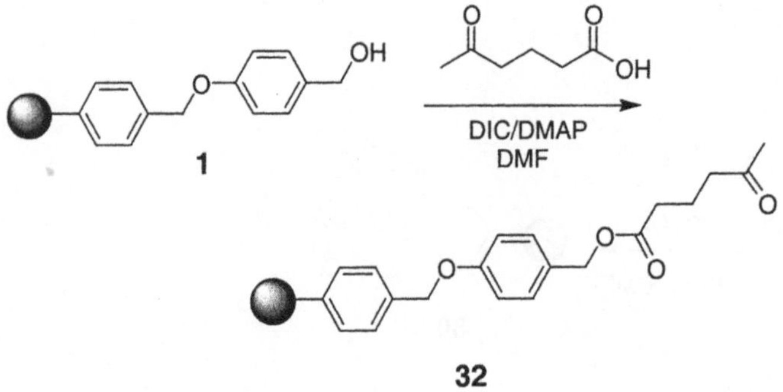

Scheme 16

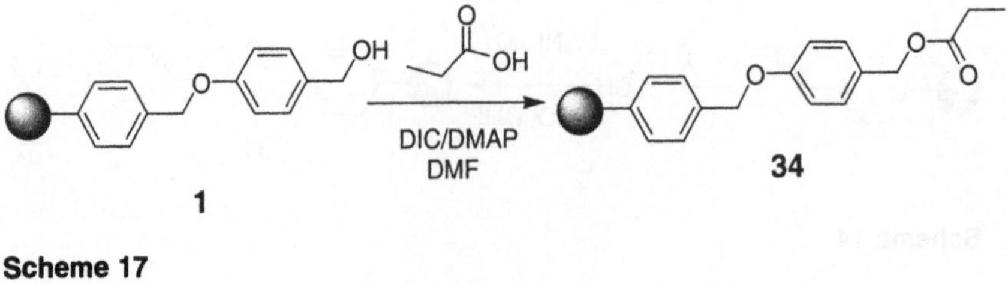

Scheme 17

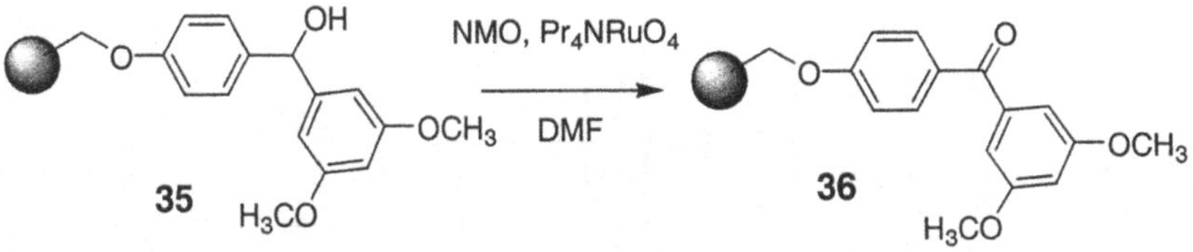

Scheme 18

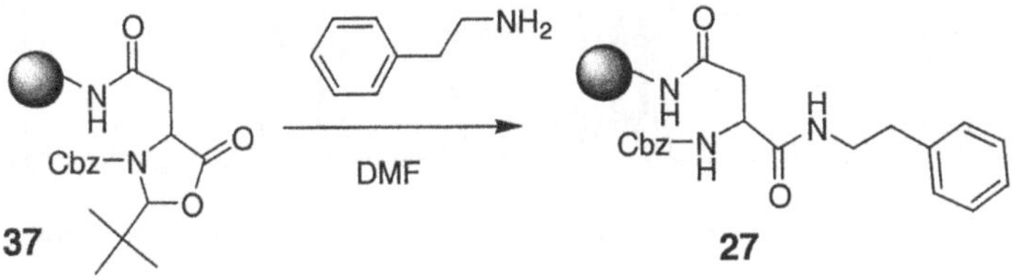

Scheme 19

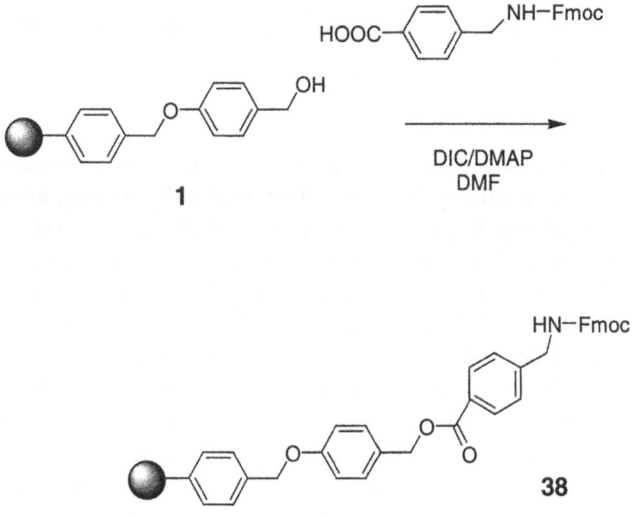

Scheme 20

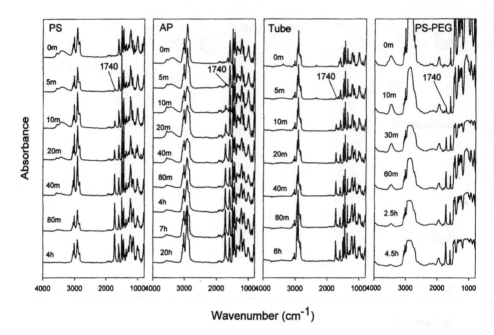

Figure 10 Single-bead IR spectra of the product for the reaction of Scheme 7 at various times on polystyrene (PS), Argopore (AP) microtube reactor (Tube), and polystyrene polyethyleneglycol resin (PS-PEG).

of pyridine is not the major factor, because the addition of triethylamine in DMF did not improve reaction efficiency.

3. Mixing and Agitation

Resin suspensions are typically agitated by shaking in reaction tubes mounted in orbit shakers (~45°), by angular rocking on a wrist shaker, by rotating tubes on a rotary device, by bubbling N_2 gas, or by conventional magnetic stirring.

The reaction in Scheme 9 was studied as a test case (twofold excess of dansylhydrazine in DMF, 30 min) to compare the mixing efficiency in six different mixing methods (40). Single-bead IR and the fluorescence quantitation method were used to quantitate the reaction. The rotation and nitrogen bubbling methods provided the highest mixing efficiencies under mild conditions. Magnetic stirring also gave high mixing efficiencies but caused fragmentation of the beads. The 180° rotation mixing experiment required longer reaction times. Mixing with an orbit shaker did not give satisfactory reaction yields compared with the other techniques. A tenfold excess of reagents and increased reaction times could sometimes compensate for the completion of slow reactions.

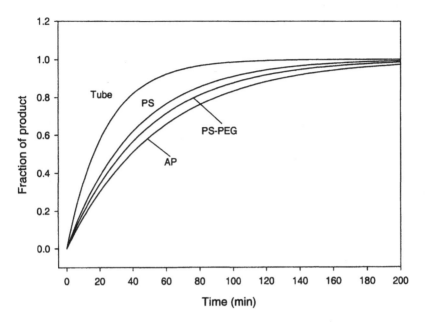

Figure 11 Time courses for reactions as in Figure 10 wherein IR peak areas at 1740 cm^{-1} for various time points were fitted to a pseudo-first-order rate equation. Rate constants were obtained by the best fit as in Figure 8.

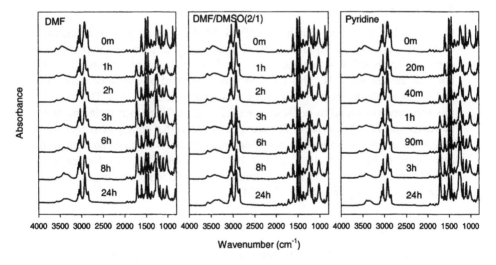

Figure 12 Single-bead IR spectra of the product at various times for the reaction of Scheme 20 in different solvents as shown.

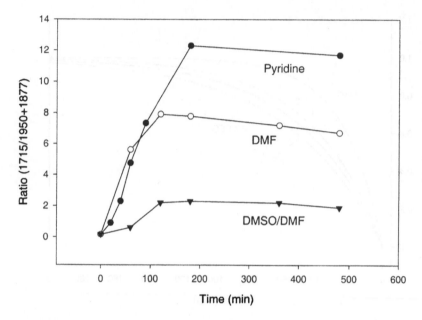

Figure 13 Time courses for the reaction in Figure 12 wherein IR peak areas at 1715 cm^{-1} relative to resin peaks are plotted against time.

4. Catalysts

Only soluble catalysts are suitable for SPOS. We have found that larger amounts of catalysts are usually required for solid-phase reactions than for solution-phase methods. The reaction in Scheme 14 is a catalytic oxidation of the benzylic alcohol **1** to the aldehyde **31**. This reaction was monitored by single-bead IR (41) and by an aldehyde spectrofluorometric quantitation method (33). The yield of this reaction was found to be 94% when 0.2 equiv of tetra-*n*-propylammonium perruthenate (TPAP) was used. The areas of IR bands that undergo changes were integrated and fitted to a pseudo-first-order rate equation. The rate constants were determined to be 4.61×10^{-4}, 1.64×10^{-4}, and 1.18×10^{-4} s^{-1} for reactions using 0.2, 0.1, and 0.05 equiv, respectively, of TPAP, demonstrating the importance of catalyst use specifically for SPOS.

5. Temperature

Reactions in Scheme 21 were monitored by single-bead FTIR. The conversion of the Rink acid to the corresponding chloride went to completion as judged by the disappearance of the hydroxyl band at ~3500 cm^{-1}. The reaction of **41** with

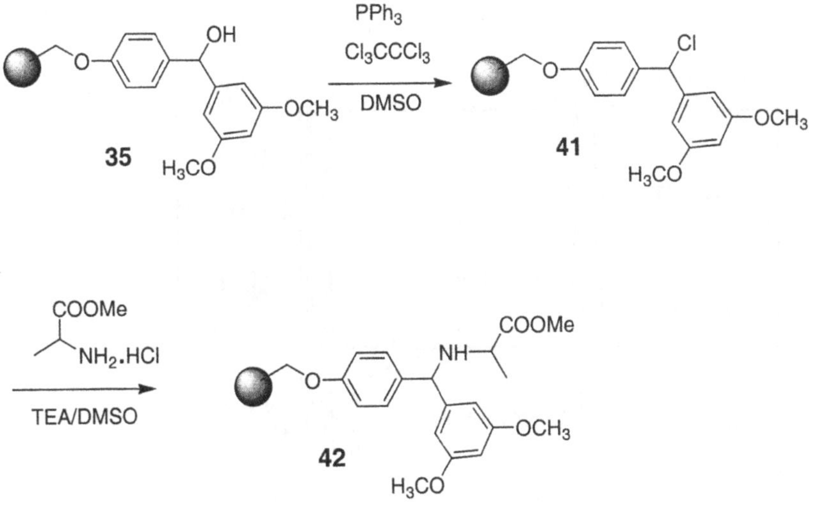

Scheme 21

an amino acid in the next step did not proceed at room temperature after 4 h of reaction time, whereas the same reaction went nearly to completion at 50°C (Fig. 14). Raising temperature is routine in organic synthesis to drive the reaction to completion. Single-bead FTIR is a simple method to select proper reaction temperature. PS resin can sustain temperatures as high as 130°C without any deformation (31).

F. Determination of Yield and Loading

Many quantitative applications of FTIR are known. They require external or internal standard and a calibration curve. A quantitative method for aldehydes and nonhindered ketones using dansylhydrazine (Scheme 9) has been described elsewhere (44). The method was validated by single-bead FTIR measurement. The better accuracy and higher speed of this method compare favorably to the Fmoc derivatization method. Hydroxyl and carboxyl groups on resin can be routinely quantitated by reaction with 9-anthroylnitrile and 1-pyrenyldiazomethane (45). In order to demonstrate the application of hydroxyl and carboxyl quantitation methods for the monitoring of SPOS reactions, we use the example depicted in Scheme 22. In this reaction, the starting hydroxyl resin had a loading of 1.17 mmol/g, as determined using the 9-anthroylnitrile method. The end product was a carboxyl resin with a loading of 1.09 mmol/g owing to the overall increase in the "molecular weight" of resin.

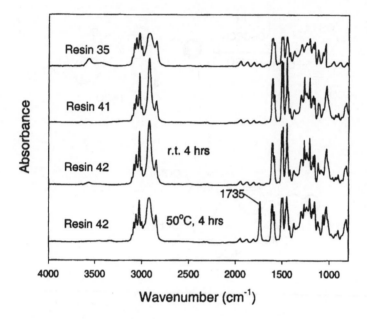

Figure 14 Single-bead IR spectra of the starting resin and products at various stage of synthesis (Scheme 21).

The reaction in Scheme 22 was monitored by single-bead FTIR (Fig. 15). As the reaction progresses, the hydroxyl stretching bands at 3570 and 3450 cm^{-1} disappear and two carbonyl stretching bands at 1735 and 1712 cm^{-1} as well as the broad feature from 3300 to 2500 cm^{-1} emerge. These single-bead FTIR measurements further confirmed the interconversion between the hydroxyl and carboxyl groups during the reaction. The peak areas of the carbonyl band in the product at various times were integrated and plotted against time (triangles in Fig. 16).

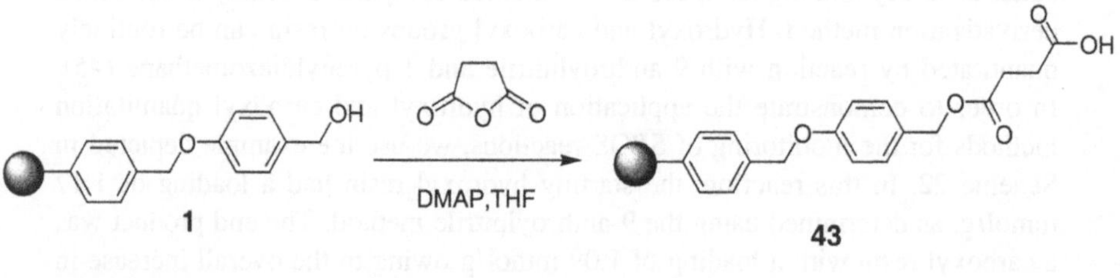

Scheme 22

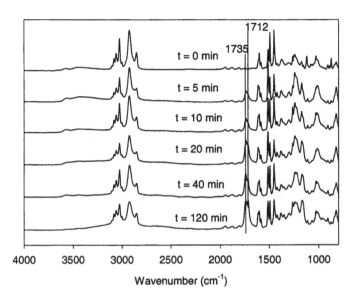

Figure 15 Single-bead IR spectra of the starting resin and the product at various times during the reaction of Scheme 22.

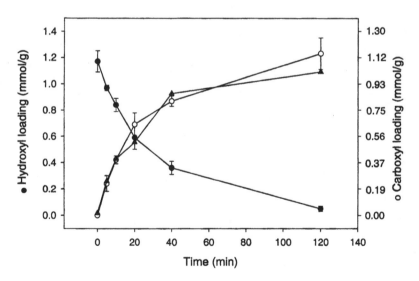

Figure 16 Time courses of the reaction in Scheme 23 obtained by three different measurements: (▲) IR peak area integration for double carbonyl bands; (●) the quantitative determination of the absolute amount of hydroxyl groups; and (○) the quantitative determination of the absolute amount of carboxyl groups.

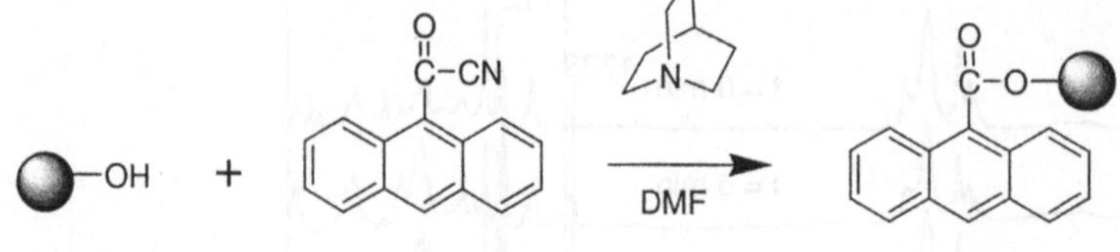

44   **45**

Scheme 23

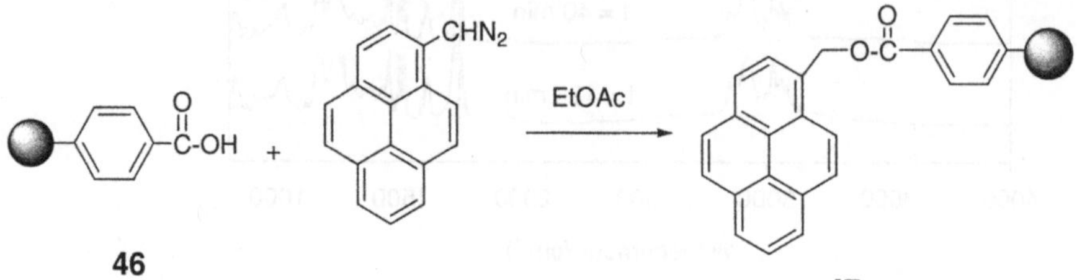

46

47

Scheme 24

Table 1

Linker	I	II	III	IV
R = 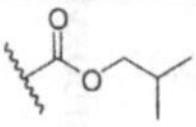	12	48	52	56
R = 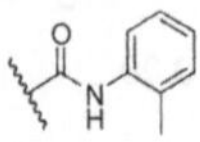	13	49	53	57
R = 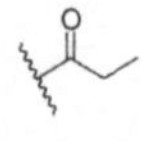	14	50	54	58
R = 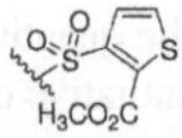	15	51	55	59

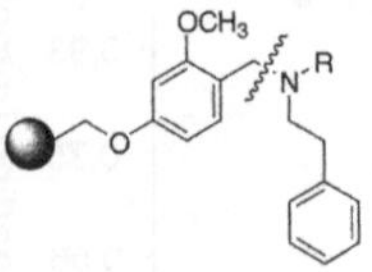

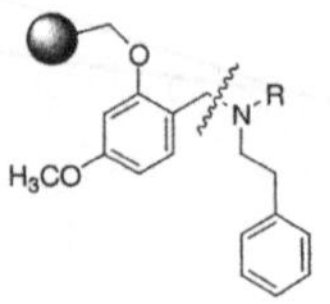

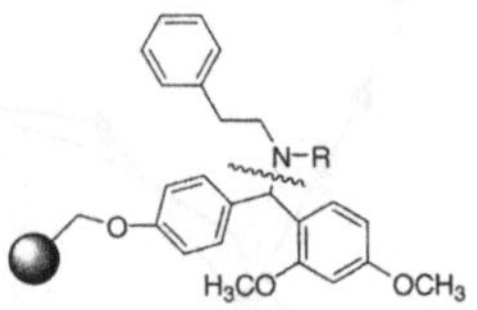

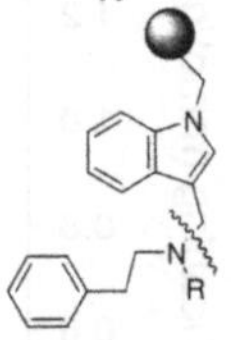

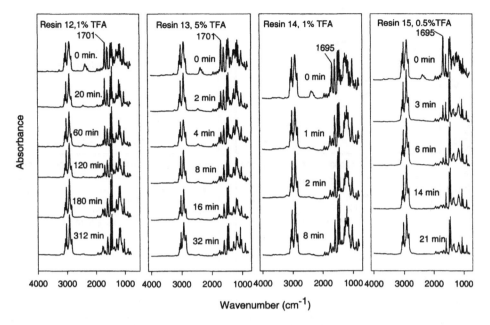

Figure 17   Single-bead IR spectra of resins **12–15** at various times during a TFA cleavage reaction.

The amounts of hydroxyl and carboxyl groups at various times during the reaction were then determined and plotted (closed and open circles in Fig. 16) using duplicate measurements. Note that loadings of starting resin and products are not the same scale because the loading of the product (mmol/g) is decreased after the addition of resin weight. These results also demonstrated that the accurate quantitation of yield can be achieved at various functional group loading levels. Hydroxyl groups could be detected at a level of 0.05 mmol/g resin loading. The sensitivity in these methods is extremely useful for monitoring yields in SPOS.

G. Cleavage Kinetics and Efficiency

The successful synthesis of an organic compound on the solid support accomplishes only part of the task in SPOS. The cleavage of the final product from the solid support, if not optimized, reduces the purity of final product. Therefore, it needs considerable attention.

Optimal cleavage conditions should cleave the product rapidly, in high yield, and under mild conditions. As an example, the TFA cleavage kinetics of

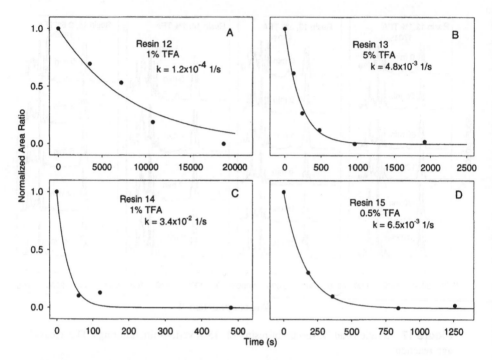

Figure 18 TFA cleavage time courses for reactions as in Figure 17 wherein IR peak areas at 1701, 1701, 1695, and 1695 cm^{-1} for resin **12–15** at various time points were integrated (●), plotted against time, and also fitted to a pseudo-first-order rate equation (lines). Rate constants show in the figure were obtained from the best fit as in Figure 8.

16 resin-bound carbamates, ureas, amides, and sulfonamides from four different acid-labile linkers (Table 1) are described below (43).

1. Monitoring of Cleavage Reactions of Resins **12–15** from Linker I

Resins **12–15** (~30 mg each) were reacted with 1% TFA in DCM. A droplet of the suspension was taken at various times for single-bead FTIR analysis after the beads were washed with THF and DCM. Resin **13** underwent only 50% cleavage in 3 h whereas resin **15** underwent 95% cleavage in 1 min. The cleavage conditions were further refined by using 5% TFA for resin **13** and 0.5% TFA for resin **15** (Fig. 17). The areas of the carbonyl bands were integrated after a peak deconvolution procedure using a PeakFit program from Jandel Scientific (San Rafael, CA). The data for the carbamates are plotted against time in Figures (Fig. 18A–18D). The time course was fitted to a pseudo-first-order reaction rate

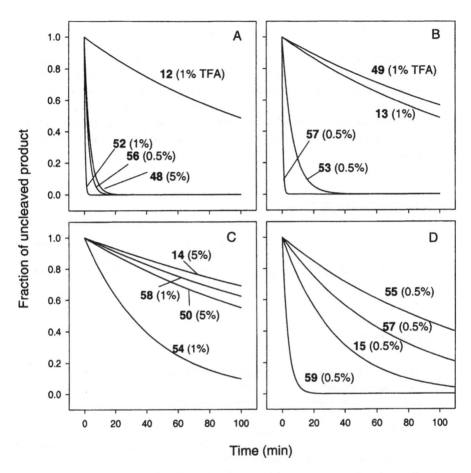

Figure 19 Time courses calculated from rate constants of TFA cleavage reactions for resins **12–15** and **48–59**. All rate constants were obtained as in Figure 18.

equation, and rate constants were determined to be 1.2×10^{-4} s^{-1} (1% TFA), 4.8×10^{-3} s^{-1} (5% TFA), 3.4×10^{-2} s^{-1} (1% TFA), and 6.5×10^{-3} s^{-1} (0.5% TFA) for resins **12–15**, respectively.

The results of a thorough kinetics study of cleavage reactions of 16 resin-bound compounds, including carbamates, ureas, secondary amides, and sulfonamides (Table 1), are shown in Figure 19. They show that cleavage conditions are generally milder than those commonly used in SPOS. Among various acid-labile linkers studied in this work, the indole linker (56–59) is found to be the most labile followed by the Rink-type linker (52–55).

IV. CONCLUSION

This chapter has demonstrated that single-bead FTIR plays a unique role in nearly every aspect of SPOS optimization with a large number of real-life examples. Single-bead FTIR is a simple, sensitive, and fast method for monitoring reactions on solid support without stopping them or cleaving product from the resin. It provides quantitative and kinetic information with respect to the process of transferring solution-phase organic reactions to solid phase and optimizing every synthesis step.

ACKNOWLEDGMENTS

We are very grateful to our coworkers for participating in the projects described and to the authors of Refs. 12, 13, 16, 31–34, and 37–45. We particularly thank Lina Liu, Jay B. Fell, and Drs. Qun Sun, Wenbao Li, Gnanasambandam Kumaravel, Bore Raju, and Roger E. Marti for their contributions.

REFERENCES

1. M Fridkin, A Patchornik, E Katchalski. J Am Chem Soc 88:3164, 1966.
2. CC Leznoff. Chem Soc Rev 3:65, 1974.
3. JI Crowley, H Rapoport. Acc Chem Res 9:135, 1976.
4. CC Leznoff. Acc Chem Res 11:327, 1978.
5. JMJ Frechet. Tetrahedron 37:663, 1981.
6. P Hodge. In: DC Sherrington, P Hodge, eds. Synthesis and Separations Using Functional Polymers. New York: Wiley, 1988, Chap 2.
7. JS Fruchtel, G Jung. Angew Chem Int Ed Engl 1996:35.
8. IW James. Mol Diversity 2:175, 1996.
9. PHH Hermkens, HCJ Ottenheijm, D Rees. Tetrahedron 52:4527, 1996.
10. R Brown. Contemp Org Syn 4:216, 1997.
11. S Booth, PHH Hermkens, HCJ Ottenheijm, DC Rees. Tetrahedron 54:15385, 1998.
12. B Yan, G Kumaravel, H Anjaria, A Wu, R Petter, CF Jewell Jr, JR Wareing. J Org Chem 60:5736, 1995.
13. B Yan, G Kumaravel. Tetrahedron 52:843, 1996.
14. DE Pivonka, K Russell, T Gero. Appl Spectrosc 50:1471, 1996.
15. DE Pivonka, TR Simpson. Anal Chem 69:3851, 1997.
16. B Yan, H-U Gremlich, S Moss, GM Coppola, Q Sun, L Liu. J Comb Chem 1:46, 1999.
17. J Hochlowski, T Sowin, J Pan. Cyprus '98. New Technologies and Frontiers in Drug Research, 1998.
18. SS Rahman, DJ Busby, DC Lee. J Org Chem 63:6196, 1998.

19. TY Chan, R Chen, MJ Sofia, BC Smith, D Glennon. Tetrahedron Lett 38:2821, 1997.
20. F Gosselin, M Di Renzo, TH Ellis, WD Lubell. J Org Chem 61:7980, 1996.
21. JM Frechet, C Schuerch. J Am Chem Soc 93:492, 1971.
22. JI Crowley, H Rapoport. J Org Chem 45:3215, 1980.
23. JR Blanton, JM Salley. J Org Chem 56:490, 1991.
24. MJ Kurth, LAA Randall, C Chen, C Melander, RB Miller, K McAlister, G Reitz, R Kang, T Nakatsu, C Green. J Org Chem 59:5862, 1994.
25. H-S Moon, NE Schore, MJ Kurth. Tetrahedron Lett 35:8915, 1994.
26. C Chen, LAA Randall, RB Miller, AD Jones, MJ Kurth. J Am Chem Soc 116:2661, 1994.
27. JK Young, JC Nelson, JS Moore. J Am Chem Soc 116:10841, 1994.
28. M Reggelin, V Brenig. Tetrahedron Lett 38:6851, 1996.
29. IEA Deben, THM van Wijk, EM van Doornum, PHH Hermkens, ER Kellenbach. Mikrochim Acta (suppl) 14:245, 1997.
30. H-U Gremlich, SL Berets. Appl Spectrosc 50:532, 1996.
31. Q Sun, B Yan. Bioorg Med Chem Lett 8:361, 1998.
32. W Li, B Yan. J Org Chem 63:4092, 1998.
33. B Yan, H Yan. Application of the single bead FTIR microspectroscopy and chemometrics in combinatorial chemistry: Multivariate calibration methods for monitoring solid phase organic synthesis. J Combi Chem 3:78, 2001.
34. B Yan, CF Jewell Jr, SW Myers. Tetrahedron 54:11755, 1998.
35. C Hulme, J Peng, G Morton, JM Salvino, T Herpin, R Labaudiniere. Tetrahedron Lett 39:7227, 1998.
36. GM Coppola. Tetrahedron Lett 39:8233, 1998.
37. B Yan, N Nguyen, L Liu, G Holland, B Raju. Kinetic comparison of TFA cleavage reactions of resin-bound carbamates, ureas, secondary amides and sulfonamides from benzyl and benzhydryl based linkers. J Combi Chem 2:66, 2000.
38. W Li, X Xiao, AW Czarnik. J Comb Chem 1:127, 1999.
39. B Yan, Q Tang. A comparison of solid-phase organic reaction kinetics on various solid supports. Submitted.
40. W Li, B Yan. Tetrahedron Lett 38:6485, 1997.
41. B Yan, Q Sun, JR Wareing, CF Jewell Jr. J Org Chem 61:8765, 1996.
42. B Yan, JB Fell, G Kumaravel. J Org Chem 61:7467, 1996.
43. B Yan, N Nguyen, L Liu, G Holland, B Raju. Kinetic comparison of trifluoroacetic acid cleavage reactions of resin-bound carbamates, ureas, secondary amides, and sulfonamides from benzyl-, benzhydryl-, and indole-based linkers. J Comb Chem 2(1): 66–74, 2000.
44. B Yan, L Wenbao. Rapid fluorescence determination of the absolute amount of aldehyde and ketone groups on resin supports. J Org Chem 62:9354–9357, 1997.
45. B Yan, L Liu, C Astor, Q Tang. Rapid determination of the absolute amount of resin-bound hydroxyl or carboxyl groups and qualitative "color" test for the optimization of solid-phase combinatorial and parallel organic synthesis. Anal Chem 71: 4564, 1999.

8

Spectroscopic Techniques for Optimizing Combinatorial Reactions on Solid Supports

Andrea M. Sefler and Douglas J. Minick
GlaxoSmithKline, Research Triangle Park, North Carolina

I. SOLUTION-PHASE VS. SOLID-PHASE REACTION OPTIMIZATION

Reaction optimization is an important part of solution-phase organic synthesis. Many conditions can affect the success of a reaction, including the solvent, pH, temperature, and concentration of reagents. In general, the goal of a reaction optimization is to modify these conditions to maximize the yield and purity of the desired product. Analytical chemistry is required to assess the yield and purity of any reaction. Chemists use many analytical techniques to follow solution-phase reactions; these include ultraviolet (UV) spectroscopy, thin-layer chromatography (TLC), gas chromatography (GC), high pressure liquid chromatography (HPLC), mass spectrometry (MS), infrared spectroscopy (IR), and nuclear magnetic resonance spectroscopy (NMR). The chemist typically uses these techniques in the optimization process to monitor the extent of completion of a reaction, to confirm the identity of a reaction product, and to determine the purity of the product.

Reaction optimization also plays a critical role in solid-phase organic synthesis (SPOS). Solid-phase reaction optimizations, however, are considerably more challenging than solution-phase optimizations, primarily because fewer analytical techniques are available for determining structures of complex organic molecules bound to solid supports. Classical analytical techniques such as electronic absorption spectroscopy (1) or simple colorimetric assays (2) can be used to detect common functional groups such as amines or carboxylic acids. However, SPOS encompasses a very broad range of reactions, many of which cannot be adequately monitored by these simple tests.

The solid-phase chemist can use all of the standard analytical techniques if the material is cleaved from the solid support. This approach is generally undesirable, however, because intermediates may simply not cleave from the support. Alternatively, they may degrade under the typically harsh cleavage conditions, introducing new impurities and side products. The interpretation of analytical data is often ambiguous when these problems are encountered, making it more difficult to optimize the reaction. Furthermore, cleaving material after each reaction step is inefficient in longer reaction sequences, especially if enough material is desired for full structural analysis. Because of these factors, we recommend using analytical techniques that are capable of determining structures of reaction products bound directly to solid supports.

Many analytical techniques have been used to study solid supports and attached compounds. Some of these include a mass spectrometric (MS) study of molecules bound to single beads (3), a fluorescence study of load distributions on various supports (4), and a study of physicochemical properties of combinatorial beads using electron paramagnetic resonance (EPR) spectroscopy (5). While these techniques have contributed to our knowledge of solid-phase organic synthesis, many of them are simply not practical in industrial research laboratories where chemists have become accustomed to easy-to-use analytical equipment, often operating in open-access environments. The techniques of IR and NMR spectroscopy, however, are fully capable of monitoring solid-phase organic reactions directly on the solid support (6–8). Moreover, these techniques are relatively easy to use and are familiar to the synthetic chemist.

II. BACKGROUNDS OF THE TECHNIQUES

A. NMR Analysis of Solid-Phase Reactions

Nuclear magnetic resonance is an indispensable analytical technique for the organic chemist. The synthetic chemist typically uses solution-phase NMR to follow reactions, to assist structure elucidation, or to provide purity analysis and/or quantification of samples, all of which are necessary parts of reaction optimization. With solid-phase chemistry advancing beyond the realm of peptide synthesis, solid-phase NMR (9–13) has also become an important tool for the synthetic chemist.

The spectra obtained for solid-phase NMR samples are similar to those obtained for liquid samples; however, the data do show some differences. The spectral resolution varies with the nature of the solid support and the length of the linker and is strongly dependent on the solvent used. Solvent studies have been performed for both resins (14) and Chiron SynPhase crowns (15) to aid in the proper choice of solvent for the best quality data. Although the peaks are generally too broad to see coupling information, chemical shifts of compounds bound to solid supports are nearly identical to the solution-phase values, allowing rough assignments to be made. The full complement of standard two-dimensional (2-D) spectra (i.e., COSY, NOESY, HMQC, HMBC) can also be obtained, if necessary, to aid the assignment of more complicated molecules (16–18). More typically, the spectra of the starting material and product are compared to determine whether or not the expected structural changes have occurred on the solid support and roughly, to what extent.

The hardware used for solid-phase NMR is also slightly different. For compounds attached to solid supports, some nuclei (^{13}C, ^{19}F, etc.) give reasonable spectra using conventional NMR probes. However, probes that use magic angle spinning (MAS) NMR give the most useful data for proton spectra (13). Although the spectra can be acquired with either a conventional solids cross polarization (CP)/MAS probe or with a high resolution magic angle spinning (HRMAS) probe such as the Varian Nanoprobe (9), Keifer et al. (19) showed that the HRMAS probes give higher resolution spectra. Bruker also manufactures a similar HRMAS probe for use with their spectrometers. These probes are relatively low cost accessories and provide good quality data on the conventional 300–500 MHz spectrometers used in synthetic chemistry labs.

B. Infrared Analysis of Solid-Phase Reactions

After reaching a state of apparent maturity in the 1960s, IR spectroscopy began a period of robust growth that has continued to the present day. Since the late 1980s, improvements in accessory design have been substantial, particularly in the field of Fourier transform infrared (FTIR) microspectroscopy, where advances have produced a series of simple-to-use, low cost, high sensitivity accessories for microanalysis. These advances have been accelerated by the burgeoning interest in combinatorial chemistry and the widespread recognition that vibrational spectroscopy is a powerful tool for studying solid-phase reactions. It is not surprising that vibrational spectroscopy is among the most useful analytical techniques in combinatorial chemistry given its long history in solid-state analysis (20). Most of the techniques within this branch of molecular spectroscopy (e.g., KBr pellets, diffuse reflectance, photoacoustic spectroscopy) have been used to study solid-phase reactions, and several articles describe the general utility of these techniques in combinatorial chemistry (21–24). Although both Raman scattering and near-infrared (NIR) spectroscopy continue to be investigated as viable tools for solid-phase reaction monitoring (25–28), IR has proven to be the most useful of these optical techniques. We attribute this primarily to the enormous volume of IR data found in the literature, a basic understanding of IR spectra among organic chemists, and the evolving availability of low cost, robust, user-friendly spectrometers and accessories. As a result, this technique has been applied in a wide range of studies involving many different aspects of combinatorial chemistry, including simple reaction monitoring (29–31), photochemical cleavage (32), the study of diffusion rates of solvents into resin (33), bead decoding (34), and quantitative analysis (35). Many of these in-depth studies have been performed with an IR microscope and a transmission flow cell like the one described by Pivonka et al. (36). Unfortunately, this equipment is complex and prevents IR from becoming the "TLC technique of combinatorial chemistry," as described by Yan and Gremlich (23). Instead, attenuated total reflectance Fourier transform IR (ATR FTIR) microspectroscopy performed with simple ATR microsampling accessories has emerged as the best technique for routine studies of solid-phase reactions (37).

Gremlich and Berets (38) reported one of the earliest applications utilizing a simple ATR microsampling accessory. In this work, a SplitPea accessory from Harrick Scientific was used to analyze chemical "megacrowns" directly. The SplitPea device is a single-bounce diamond ATR accessory that fits directly into the sampling chamber of most FTIR spectrometers and is capable of producing fairly high quality IR spectra using only the room temperature (DTGS) detector in the instrument. We purchased this accessory and found that although it was capable of providing fairly good quality data, it was difficult to use overall, primarily because of the crystal design and lack of video-capture capability. These limitations made it difficult to analyze most beads and crowns, and consequently our synthetic chemists were not using it to monitor their reactions.

We then replaced this unit with a SpectraTech InspectIR ATR microsampling accessory with video-capture capability. This device demonstrated excellent sensitivity and was used extensively over a 5 year period to monitor solid-phase reactions. During this time, however, limitations of this accessory became apparent. Bead clusters tended to fly apart when pressurized by the silicon ATR crystal, relegating the analysis of every resin sample to "single-bead" work. Although there are some advantages to single-bead work, this type of analysis required us to take spectra on at least two beads per resin sample to ensure bead-to-bead homogeneity. In addition, most crowns could not be analyzed directly because of poor optical contact. Another disadvantage of this accessory was the need to fill the MCT detector with liquid nitrogen two or three times per day; this requirement was not viewed favorably by most chemists.

Because of these limitations, the InspectIR was replaced with a Durascope single-bounce diamond ATR accessory from SensIR Technologies. Like the SplitPea, this device fits directly into the sample compartment of most FTIR instruments, using the room temperature (DTGS) detector of the spectrometer. Like the InspectIR, these units can be purchased with an internally installed video camera, making sample alignment very easy and allowing visual verification of good optical contact prior to data acquisition. Although this accessory can generate spectral artifacts (see Section V), it is remarkably easy to use and is capable of producing very high quality data in a very short time (seconds). In contrast to the InspectIR, which was limited to single-bead analysis, many beads are analyzed simultaneously by the Durascope, providing a better statistical sampling of resin samples. (Note: In principle, this device is capable of single-bead analysis of beads of 100 μm or greater diameter.) Additionally, crowns can be analyzed directly and nondestructively within seconds using this unit.

III. EXPERIMENTAL METHODS

A. Nuclear Magnetic Resonance Methods

All spectra were recorded at ambient temperature on a Varian Unity Plus 500 MHz spectrometer equipped with a Varian Nanoprobe. Forty-microliter glass rotors were used with MAS speeds of 2–3 kHz. All resin samples were recorded in ~40 μL CD_2Cl_2, using ~1–2 mg of resin. The crown samples were prepared according to Sefler and Gerritz (15), with d_5-pyridine and d_7-DMF as the solvents for the PS and MD crowns, respectively. All spectra were referenced to the appropriate solvent signal and processed using ACD SpecManager software v4.0 with a line broadening of 2 Hz.

B. Infrared Methods

All IR spectra were recorded using a Bruker Optics Vector 22 spectrometer equipped with side-port switching optics. A SpectraTech InspectIR ATR microsampling accessory (silicon ATR crystal) was mounted externally to the side port of a Vector 22, while the SensIR Technologies' Durascope ATR microsampling accessory (diamond ATR crystal) was mounted in the sample compartment. All data were acquired and processed using Bruker's OPUS-NT (v2.01) spectroscopy software.

IV. EXAMPLES OF USING NMR AND IR TO MONITOR SOLID-PHASE REACTIONS

Separately, both IR and NMR have limitations, but together the two techniques are capable of characterizing most molecules synthesized on solid supports. In this section, we present examples where both IR and NMR data were used to follow reactions on a variety of solid supports.

A. Monitoring Reaction Completion

The coupling of primary amines to an aromatic aldehyde functionalized linker by reductive amination is a common reaction used in solid-phase sequences. Recently, Fivush and Willson (39) published the use of a novel aldehyde resin for performing this reaction. This acid-sensitive methoxybenzaldehyde resin (AMEBA resin) is typically reacted with a primary aromatic or aliphatic amine and a reducing agent to form a resin-bound secondary amine. IR and/or NMR can easily follow this reaction.

Figure 1 shows NMR spectra for a series of reductive aminations. The top spectrum is of the starting AMEBA resin, and the aldehyde proton c can be seen clearly at 10.28 ppm as well as the aromatic proton b, which is shifted downfield to 7.77 ppm by the aldehyde. The center spectrum shows an incomplete reaction with n-butylamine. The aldehyde has been completely consumed, as indicated by the disappearance of the signal at 10.28 ppm. This spectrum, however, shows signals from both the desired secondary amine product and unreduced imine. The proton signals b and c, at 7.89 and 8.60 ppm, respectively, are diagnostic for the imine. A rough comparison of the integrals of the methyl g protons shows this reaction to be ~65% complete. By contrast, the bottom spectrum shows a completed reaction with p-methoxy benzylamine, and none of the imine signals are present. In addition, the new signals arising from the aromatic protons c and f are present at 7.25 and 6.84 ppm.

Figure 2 shows ATR-IR data (2200–775 cm^{-1}) that were observed at each step of a reductive amination performed with the AMEBA resin. The data shown here have been normalized to the polystyrene out-of-plane ring vibration at 699 cm^{-1}. The IR spectrum of the starting AMEBA resin (top panel) is rich with spectral bands that are diagnostic for the 2,4-dialkoxybenzaldehyde portion of the resin. Bands at 2765 (figure insert), 1678, and 1396 cm^{-1} in this spectrum are assigned to the CH stretching, C=O stretching, and the in-plane CH bending vibrations, respectively, of the aldehyde group on the resin. Because the carbonyl group in DMF absorbs at 1678 cm^{-1}, we recommend examining all of these regions to confirm reduction of the aldehyde. In cases where aldehyde groups remain unreacted after an initial reductive coupling, the extent of reaction is best estimated using peak intensities of either the aldehyde C=O stretching or CH in-plane bending bands. In addition to these diagnostic aldehyde bands, successful reductive coupling can also be supported by changes in the intensities of peaks arising from vibrations of the aromatic ether and portions of the 2,4-dialkoxy-benzaldehyde moiety. These peaks occur near 1600 and 1260 cm^{-1} and are assigned to C=C and C—O stretching vibrations, respectively. As shown in Figure 2, both bands are strongly intensified in the AMEBA spectrum (top) but not in the spectrum of the reduced benzylamine product (bottom). Although we do

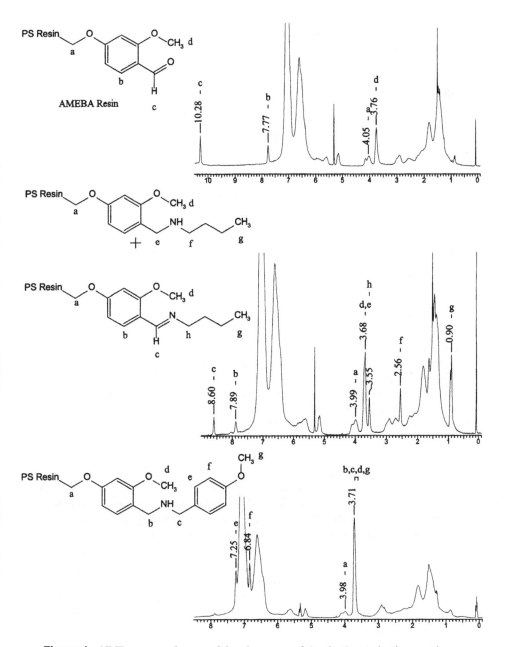

Figure 1 NMR spectra of successful and unsuccessful reductive amination reactions.

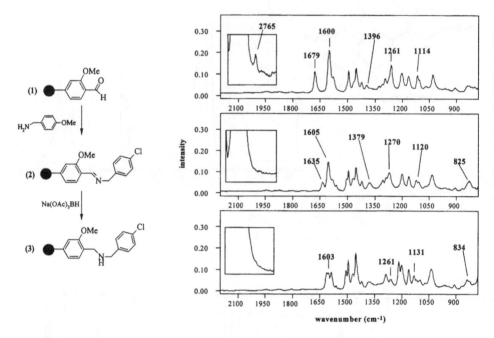

Figure 2   Reductive aminations are one of the most basic reactions used in solid-phase organic syntheses. A typical reaction scheme is shown here, with ATR-IR spectra that were observed for resin samples at each step in this reaction. These data illustrate how to reliably ensure reduction of the aldehyde functionality on the resin and identify a marker band that is diagnostic for the desired secondary amine product. These spectra were taken using a SensIR Technologies Durascope.

not use these bands for quantitative purposes, they do provide additional qualitative evidence of a successful coupling reaction.

Formation of the imine in Figure 2 was achieved by coupling 4-chlorobenzylamine with the AMEBA resin in the absence of reducing agent. The ATR-IR spectrum taken of this intermediate is shown in the middle panel of this figure. Here, the diagnostic bands for the aldehyde group are clearly gone, whereas the bands near 1600 and 1260 cm^{-1} remain intensified, albeit not as greatly as in the original aldehyde spectrum. The imine group is characterized by bands at 1635 and 1379 cm^{-1} that are due to the C=N stretching and CH in-plane bending modes of the newly formed imine. [Note that the imine CH stretching mode is not observed in the spectrum of the imine intermediate, in contrast to the aldehyde (compare top and middle inserts).] Although either of these bands can be used to estimate relative amounts of imine still present after the reduction reaction, we have found that estimates made using the band at 1635 cm^{-1} appear to agree more closely with NMR data than estimates based on the band at 1379 cm^{-1}. Consequently, we have used the normalized intensity of this band to estimate residual imine content as low as 5–10% in spectra of resins where imine reduction was incomplete. In addition to bands diagnostic for the imine, a band at 825 cm^{-1} also appeared that was not observed in the spectrum of the AMEBA resin. This band is due to the synchronous CH out-of-plane bending vibration of the protons on the p-substituted phenyl ring in the amine monomer, providing direct evidence of desired product formation.

In the spectrum of the fully reduced amine (Fig. 2, bottom panel), bands arising from vibrations of the aldehyde and imine groups are no longer observed, and the intensities of the aromatic ring vibration near 1600 cm^{-1} and the ether C—O stretching vibration near 1261 cm^{-1} are no longer intensified. All of these spectral changes can and should be used to confirm complete reduction of both the aldehyde and imine groups. Note again that the aromatic CH mode near 834 cm^{-1} provides direct evidence for the formation of the desired product. Because reductive aminations are performed in one step, IR data are not taken to ensure imine formation prior to reduction to the corresponding secondary amine. Consequently, it is important to identify bands in the product spectrum that signify product formation as well as disappearance of the aldehyde. Identification of the band at 834 cm^{-1} in the bottom spectrum of Figure 2 illustrates how to use a diagnostic band to confirm actual product formation. Spectroscopic evidence supporting formation of the desired product, however, may not be apparent in the spectrum of the product resins, particularly for amines containing no strong IR-absorbing groups. We routinely employ difference spectroscopy to identify diagnostic product bands. In a difference spectrum, bands below the baseline are due to the functional group transformed by the reaction, whereas bands above the baseline are due to the structural fragment of the molecule added to the resin.

In Figure 3, difference spectra are shown for the imine intermediate and fully reduced secondary amine of Figure 2. These spectra were obtained by subtracting the spectrum of the AMEBA resin from each of the product resins (imine and fully reduced amine). In this example, bands due to the PS resin were removed by interactively scaling the starting resin spectrum during the subtraction process until the PS ring band at 699 cm^{-1} was roughly zero; this process is identical to subtracting normalized spectra. In the difference spectrum of the imine (Fig. 3a), bands diagnostic for the aldehyde group on the starting resin lie below the baseline while bands above the baseline at 1635 and 1379 cm^{-1} are due to the newly formed imine group (see Fig. 2 discussion for vibrational mode assignments). Although a number of other bands are observed above the baseline in this figure, those at 1504, 1471, 863, and 821 cm^{-1} are the most important, because they arise directly from ring stretching and proton bending vibrations of the p-substituted phenyl ring of the amine monomer.

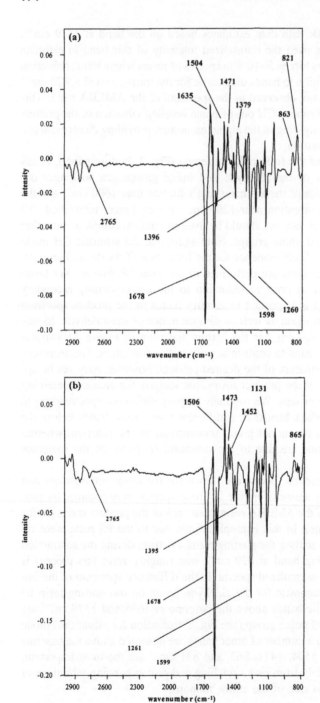

The difference spectrum shown in Figure 3b is typical of those we use to confirm formation of secondary amine products on solid supports. In addition to bands diagnostic for the *p*-substituted phenyl ring in 4-chlorobenzylamine (1506, 1473, and 865 cm^{-1}), bands at 1452 and 1131 cm^{-1} are assigned to the CH$_2$ scissor mode and C$-$N stretching mode of the reduced imine. Both of these difference spectra clearly show formation of the imine intermediate and desired secondary amine. In the absence of contaminants, our experience has been that the bands observed above the baseline (diagnostic for product molecules) compare remarkably well to IR spectra of simple model reference compounds found in IR libraries—for example, 4-chloroethylbenzene or 4-chlorotoluene in this case.

Infrared data can be used not only to show that a solid-phase reaction is incomplete but also to estimate the extent of reaction completion. When data are used in this manner, however, it is necessary to normalize both spectra to a band arising from the support. The support band chosen for normalizing data essentially acts like an internal standard, compensating for differences in depth of penetration that may occur from one solid-phase measurement to the next. Although any number of bands in the spectrum of the starting support could be used to normalize data, we typically use either the PS ring deformation band at 699 cm^{-1} or the PS combination band near 1945 cm$^-$ to perform this operation. In Figure 4, the results of using normalized and raw data to estimate the extent of completion of a reaction are contrasted. In this example, the single-bead ATR-IR spectrum of the starting aldehyde resin is shown in panel (a). The single-bead ATR-IR spectrum in panel (b) was taken of a resin sample that was only partially coupled. Using the integrated peak areas of the aldehyde C$=$O band from these spectra, this reaction was predicted to be roughly 75% complete. When the spectrum of the partially reduced resin (Fig. 4b) was normalized to the spectrum of the original aldehyde spectrum (Fig. 4a), however, using peak areas of the PS ring vibration at 699 cm^{-1} (Fig. 4c), the extent of reaction was predicted to be only 55%.

The NMR and IR spectra shown for these reductive amination reactions demonstrate the importance of checking the products on the solid support. Any unreduced imine would either remain on the resin during cleavage or be hy-

---

Figure 3 Infrared difference spectra are shown for (a) the imine and (b) secondary amine products formed in the second and third steps of the reaction sequence shown in Figure 2. These difference spectra were obtained by subtracting the AMEBA spectrum in Figure 2 from the imine and amine product spectra, yielding the difference spectra displayed here. Bands below the baseline are characteristic of functional groups lost during a reaction, and bands above the baseline are characteristic of functional groups added to the resin.

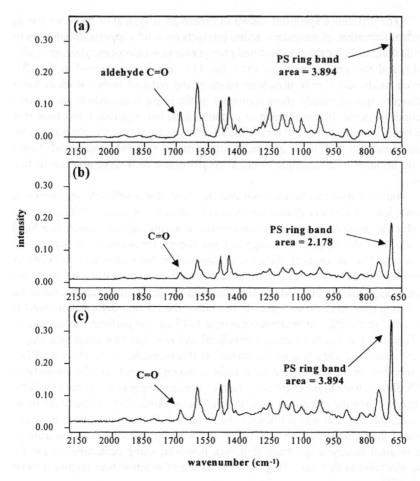

Figure 4 Single-bead ATR-IR spectra are shown for (a) an aldehyde starting resin, (b) a resin following incomplete reductive amination, and (c) the same product resin but with data normalized to the PS ring band at 699 cm^{-1} in the spectrum of the starting resin. Data were acquired using a SpectraTech InspectIR ATR microsampling accessory.

drolyzed during later reactions or washings. The chemist would simply observe a lower-than-expected yield without knowing whether the problem was in the formation or reduction of the imine. The IR and NMR spectra clearly indicate what happens in this type of reaction and can be used to estimate to what extent the reaction is completed.

B. Monitoring Failures of Solid Supports, Spacers, and Linkers

When performing solid-phase reactions or planning sequences, one tends to ignore the structure and nature of the linker and solid support and instead concentrate on the "interesting" portion of the molecule. Indeed, the linker, spacer, and support often become a generic, inert "ball" in a reaction scheme. We must remember, however, that both the linker and the support have chemical and physical properties of their own, and these properties can determine the success or failure of a reaction.

1. Failure or Contamination of the Solid Support

An example of failure of the solid support is shown in Figure 5. The spectra are from a reaction using a polystyrene (PS) crown with a Sasrin-type linker. These crowns are constructed of polypropylene, with a PS graft on the surface. The intent was to produce the primary bromide and displace the bromide with benzylamine (40). The top spectrum has signals consistent with the starting material, but the bottom spectrum of the product contains only the NMR solvent (pyridine), water, and polypropylene signals. Even the PS signals around 7 ppm are missing, which indicates that the PS graft has been removed.

We have also used the ATR-IR microsampling accessory with imaging capability to diagnose problems with the solid support. One benefit of this accessory is that the analyst is given an opportunity to visually inspect the solid support before acquiring data. We used this opportunity to identify two potential problem areas that impact both the quality of the IR data and, more significantly, the outcome of the chemistry being performed on the resin.

The first of these problems involves actual mechanical damage to PEG-grafted PS resins such as TentaGel or ArgoGel. The PS core of these resins is brittle, leaving them prone to mechanical damage when subjected to vigorous handling techniques. When this type of damage occurs, the resin sample is observed in the magnified image as a mixture of "normal" spherical beads and irregularly shaped bead fragments. We have noted this damage in a number of instances, and it is usually not extensive. In these cases, the resin sample appeared to be "contaminated" with a powder that gave the same IR spectrum as the normal spherical beads. In the worst case, however, an entire resin sample was reduced to powder by the stirring action of a magnetic stirring bar, rendering this sample useless for further chemistry.

The second potential problem that we have identified from magnified images of bead samples is one of actual contamination. In this case, solid reagents were observed to be mixed in with and often clinging to the surface of the beads and crowns. This type of contamination does not usually cause chemistry prob-

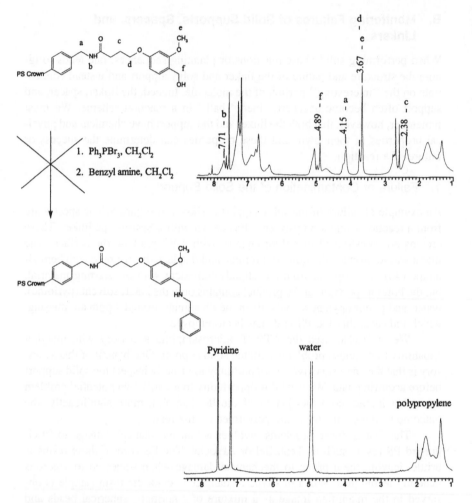

Figure 5 NMR spectra illustrating failure of a solid support.

lems, because the reagents are often washed away in the next step. It can be a serious IR problem though, because the bead spectrum will now contain features of the contaminant, complicating data interpretation. This problem can be avoided by simply rewashing the sample before taking the spectrum. Resin samples are particularly easy to clean up because they are placed on a glass microscope slide for analysis. When these samples are suspected of being contaminated, they can be washed directly on the slide with a series of rinses using methanol, acetone, and chloroform. Contamination does cause difficulties for IR analysis and serves as a reminder that the highest quality IR work is done by analysts who are fully aware of the history of the sample (i.e., reaction conditions) and cognizant of the spectroscopic properties of potential contaminants.

Contaminants are also detected in NMR spectra, but are not as problematic as for IR because their signals can usually be distinguished from those of resin bound molecules. The linewidths of the contaminant signals are typically narrower than the broadened signals arising from molecules attached to a solid support. A comparison of the linewidths of bound signals vs. those free in solution can be seen in the top spectrum of Figure 1. The methylene chloride solvent peak at 5.32 ppm is considerably sharper than any of the labeled signals from the attached aldehyde. One caveat is that if the contaminants are insoluble in the NMR solvent used and/or are trapped inside the solid matrix, their NMR signals will be broadened.

2. Failure of the Linker

Another problem often encountered during reaction optimization involves premature cleavage of the linker on the solid support. Obviously, the overall yield of the sequence will suffer if this occurs. We have chosen the Heck reaction in Scheme 1 to illustrate how readily this problem can be identified using NMR and/or IR data. The Heck reaction was performed on three different types of solid supports: Chiron SynPhase Hyperlabile polystyrene (PS HYP) crowns, Chiron SynPhase Hyperlabile methacrylic acid/dimethylacryl-amide copolymer (MD HYP) crowns, and Sasrin resin (Bachem Biosciences).

Nuclear magnetic resonance spectra of the PS HYP crowns are shown in Figure 6. The spectrum in Figure 6a clearly shows the para-substituted aromatic protons f and g as well as terminal alkene protons h in the downfield region of the spectrum, as expected. These p-substituted aromatic protons, however, are missing in the spectrum of the product (Fig. 6b), indicating that cleavage has occurred at the ester bond, leaving the primary alcohol of the Sasrin linker.

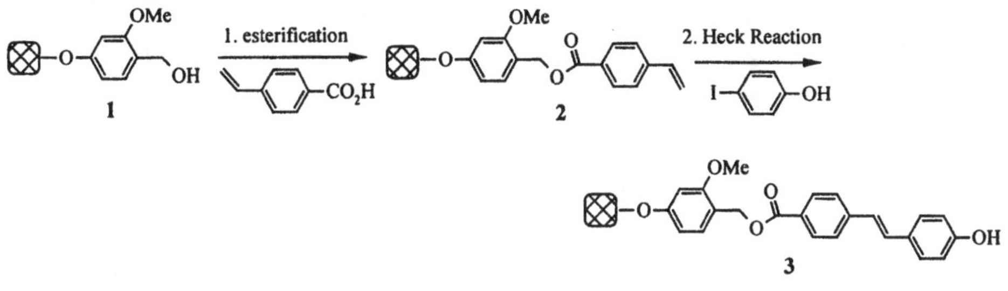

Scheme 1

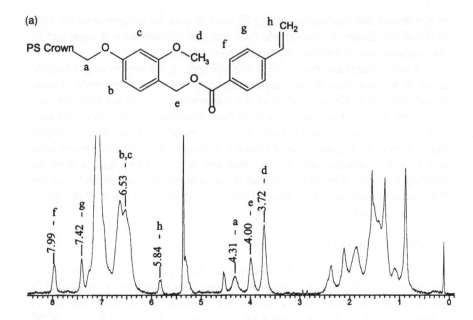

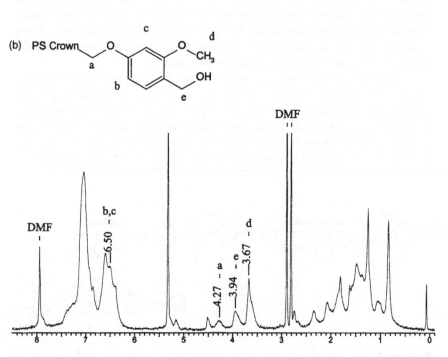

Figure 6 NMR spectra of a Heck reaction performed on a Chiron SynPhase PS HYP crown illustrating cleavage of the linker.

The ATR-IR data taken of the PS crowns, MD crowns, and Sasrin resin are shown in Figures 7–9, respectively. The spectra of the starting supports are shown at the top of each figure. The middle spectrum in each figure was observed after the supports were subjected to the esterification reaction (Scheme 1). In each case, bands that are diagnostic for the formation of the desired ester (**2**) in Scheme 1 were observed: 1715 (C=O stretch), 1265 (antisymmetric C—O—C stretch), and 1100 cm^{-1} (symmetric C—O—C stretch). Although observation of these bands provided adequate evidence that the desired intermediate had formed, bands diagnostic for the terminal =CH$_2$ portion of the vinyl group were also assigned. These bands occurred near 1405 and 915 cm^{-1} and were assigned to the in-plane bending and wagging vibrations of the vinyl =CH$_2$ protons. The band near 915 cm^{-1} is of particular interest because it would not be observed for structure **3** (Scheme 1) and therefore could be used to estimate completion of this Heck coupling. Extensive cleavage of the ester linker under the conditions used for the Heck coupling is shown in all three cases by large decreases in the intensities of bands diagnostic for the aryl ester (i.e., the C=O and C—O stretching bands). Using normalized data, ~75% of the ester bonds on the Sasrin resin and ~85% of the ester bonds on both crowns were estimated to have been cleaved.

The findings from IR and NMR provided a consistent analytical picture for this reaction sequence and clearly identified the Heck reaction as the problematic step. The chemist was then able to alter the reaction conditions to prevent cleavage of the ester bond. Figure 10 shows the NMR spectra of the Heck reaction performed on the MD HYP crowns after modifying the reaction conditions. In the spectrum of Figure 10a, one can again see the expected signals of the starting material. The spectrum of Figure 10b, however, appears consistent with the desired product of the Heck reaction, as not only can the aromatic protons f and g be seen, but also the new aromatic signals k and l are observed, confirming that this coupling reaction has occurred. In addition, the complete disappearance of the terminal alkene signals at 5.93 and 5.39 ppm provide evidence that the reaction has gone to completion.

Without the reaction modifications suggested from the NMR and IR analysis, the chemist would have isolated little or no product once the final compounds were cleaved from the supports. Given the observed analytical data, however, the chemist was able to modify reaction conditions and successfully resolve this cleavage problem. We have found that product loss due to linker cleavage is quite common but can be minimized or even eliminated when appropriately diagnosed with spectroscopic data, as shown by this example. Once the compatibility between a solid support and the desired reaction conditions is known, the chemist can redesign conditions that minimize these problems and thereby maximize yields during library production.

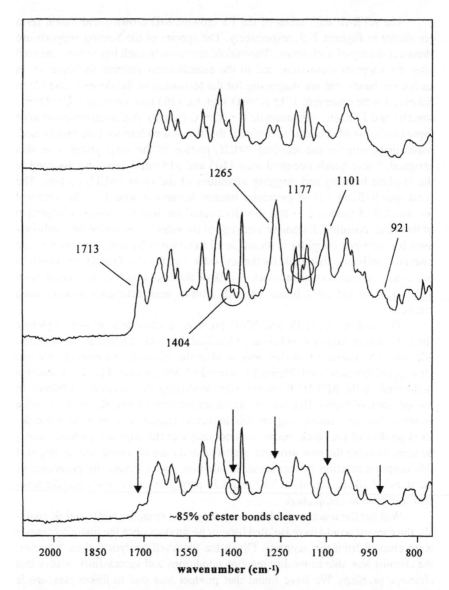

Figure 7 ATR-IR data taken of Chiron SynPhase PS HYP crowns subjected to the reaction sequence shown in Scheme 1. The ATR-IR spectrum of the unreacted crown is shown at the top, and the middle and bottom spectra are those of crowns following the esterification and Heck reactions, respectively. These spectra were taken directly of the crown surface using a SensIR Technologies Durascope.

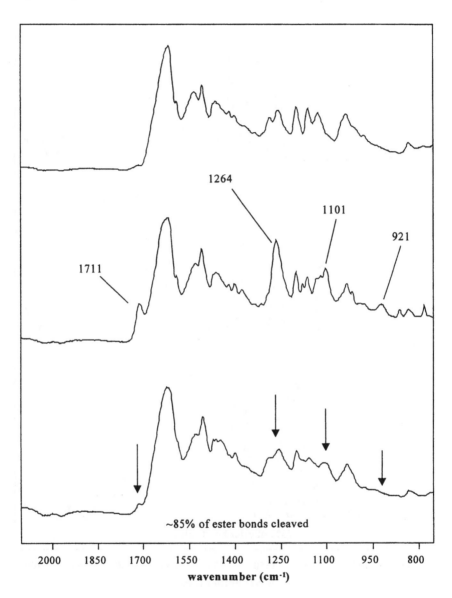

Figure 8 ATR-IR data taken of Chiron SynPhase MD HYP crowns subjected to the reaction sequence shown in Scheme 1. The ATR-IR spectrum of the unreacted crown is shown at the top, and the middle and bottom spectra are those of crowns following the esterification and Heck reactions, respectively. These spectra were taken directly of the crown surface using a SensIR Technologies Durascope.

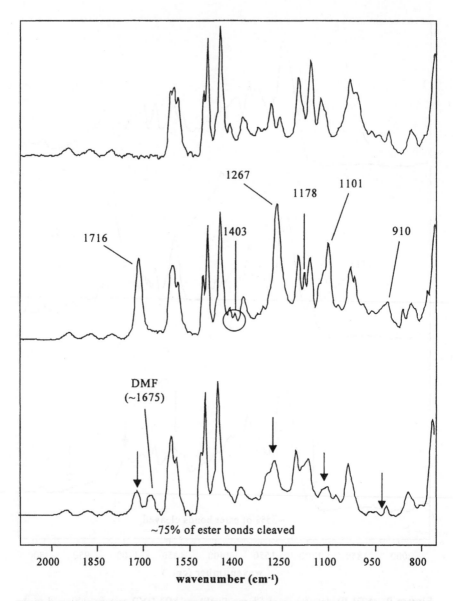

Figure 9 ATR-IR data are shown for samples of Sasrin resin subjected to the reaction sequence in Scheme 1. The ATR-IR spectrum of the unreacted resin is shown at the top; the middle and bottom spectra are those of resin samples following the esterification and Heck reactions, respectively. These spectra were taken using a SensIR Technologies Durascope.

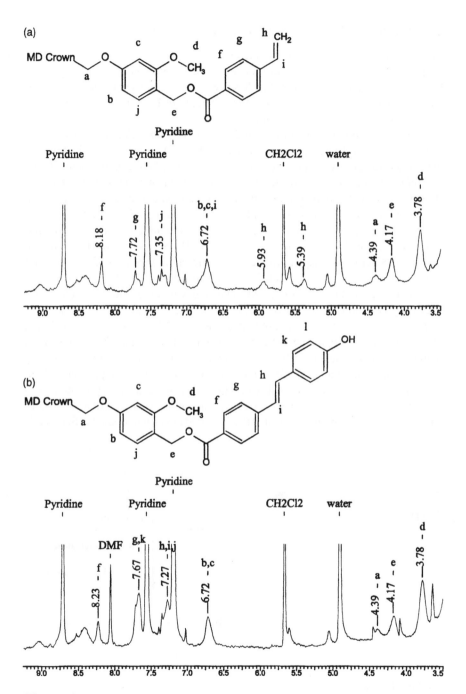

Figure 10 NMR spectra of a successful Heck reaction performed on a Chiron SynPhase MD HYP crown.

C. Monitoring Cleavage from the Solid Support

In addition to providing spectroscopic evidence of product formation on a solid support, IR and NMR data can also be used to determine molecular changes that have occurred on a solid support that has been cleaved. Indeed, it could be argued that the cleavage step is one of the most critical yet least understood and least investigated steps in reaction optimizations.

The reaction shown in Scheme 2 exemplifies how cleavage data can be used to understand and optimize a reaction. Here, 4-aminobenzyl alcohol was reacted with an aromatic thionyl chloride functionalized PS resin (theoretical load ~1 mmol/g) to give a sulfonate ester 2. IR spectra of resin samples 1–3 are shown in Figure 11. Formation of the desired sulfonate ester 2 is clearly shown by the IR data in Figure 11a and 11b. In the spectrum of the starting thionyl chloride resin, the antisymmetric and symmetric stretching vibrations of the SO_2 group are assigned to the bands at 1373 and 1171 cm^{-1}, respectively, while the band at 1082 cm^{-1} is assigned to the C—S stretching vibration of the S-phenyl group. After the resin was reacted with 4-aminobenzyl alcohol, the spectrum shown in Figure 11b was observed. The antisymmetric and symmetric SO_2 stretching vibrations of the newly formed sulfonate ester were observed at 1359 and 1174 cm^{-1}, and the C—S stretching vibration at 963 cm^{-1}. The actual load of sulfonate ester was then estimated by cleaving the sulfonate bond by using a secondary amine at several different concentrations (0.8 and 2 equiv; Scheme 2). The sulfonate ester bands at 1359 and 963 cm^{-1} completely disappeared at both amine concentrations (Fig. 11c). The complete loss of these bands provided spectroscopic evidence that cleavage of the aminoaniline was quantitative at both concentrations. This finding indicated that the yield of desired product 2 (Scheme 2)

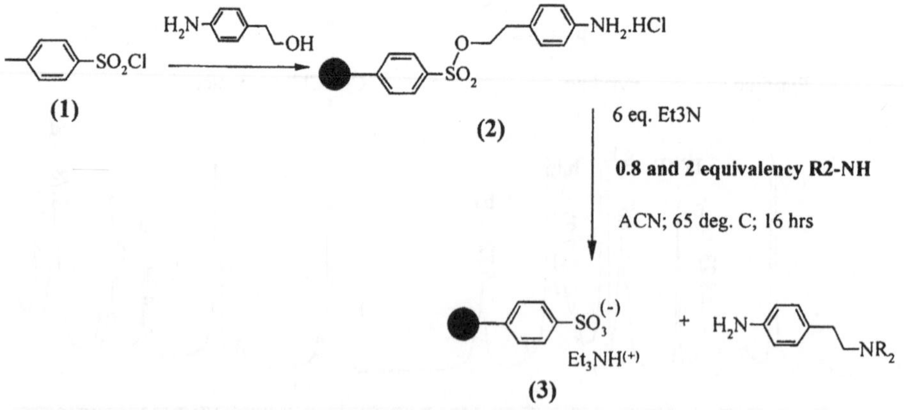

Scheme 2   Assessing load from cleavage data.

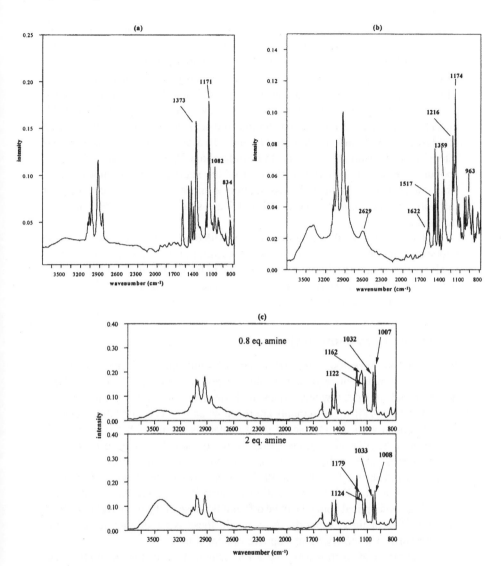

Figure 11 ATR-IR spectra for (a) the thionyl chloride functionalized starting resin **1** in Scheme 2; (b) the sulfonate product resin **2** in Scheme 2; (c) sulfonate resin samples after cleavage of the sulfonate ester at two different concentrations of base (**3** in Scheme 2). These spectra were taken using a SensIR Technologies Durascope.

was probably less than 80% of the theoretical load reported for this resin (1.06 mmol/g). Two more concentrations of amine were then selected, 0.6 and 0.4 equiv, and the cleavage was repeated. Again, quantitative cleavage of the sulfonate ester was observed. (Note: Data for these cleavage reactions are not shown.) This finding was unexpected because it suggested that loading of the sulfonate ester **2** was much lower than expected. A review of the IR data in Figures 11a and 11b supported this finding. The normalized peak heights of the sulfone bands at 1374 and 1359 cm^{-1} in Figures 11a and 11b were compared. We found the intensity of the band at 1359 cm^{-1} to be only about 35% as intense as the corresponding peak at 1374 cm^{-1} in the data of the starting thionyl chloride resin. This comparison is valid if the assumption is made that the group absorption intensities of the sulfone moiety in **1** and **2** are similar, which is reasonable in our opinion.

The IR data in this example thus provided a clear picture of what was happening on the solid support at each concentration of secondary amine used to cleave the resin. In addition, the comparison of band intensities for the antisymmetric SO_2 stretching bands provided a spectroscopic method for estimating the load of amino alcohol in future coupling reactions with this resin.

D. Infrared Technique Notes

In general, we have found that IR spectra of solid-supported materials can be acquired in a matter of seconds with little or no sample manipulation. We have, however, developed a few techniques to obtain the highest quality data possible.

The best ATR measurements are made when the ATR crystal makes good optical contact with the sample. If good contact is not made between the solid support and the ATR crystal, the spectra will have smaller signal-to-noise ratios and greater spectral artifacts. Good optical contact is usually obtained for PEG functionalized resins, because the PEG spacer gives these resins more gel-like properties. For most PS resins and all macrosupports (e.g., crowns and lanterns), achieving good optical contact can be difficult because of the more rigid mechanical properties of these supports. As a result, these types of samples are often overpressurized when trying to improve optical contact.

The effect of overpressurizing a PS resin sample against an ATR crystal is illustrated in Figure 12a. Two negative spectral artifacts can be noted in this figure. The first one involves a nonuniform rise in the baseline that is most severe at the high frequency end of the spectrum. We observed this effect using two different accessories, one equipped with a diamond ATR crystal and one equipped with a silicon crystal, and consequently believe that it is independent of the ATR crystal used. Although the rise is only modest in this spectrum, it can be extreme and make operations like spectral subtractions impossible. This effect is particularly troublesome in resins whose mechanical properties have been noticeably altered by the molecules being synthesized on the resin. In these cases, it becomes almost impossible to obtain a good quality IR spectrum of the sample resin because almost any amount of pressurizing can cause severe baseline distortion.

The second negative artifact shown in Figure 12a is unique to the diamond crystal in the Durascope. This artifact occurs between 2300 and 1900 cm^{-1} and is caused by absorption of most of the IR light transmitted in this region; the photon throughput in this region is less than 10% owing to intense absorption by the diamond crystal. We believe that this artifact arises because we reference sample spectra to a background of air. First, when the background is collected, no sample is touching the ATR crystal. When the sample is then placed in contact with the crystal and pressed against it, the spatial position of the diamond crystal changes. This change in position causes the distortion shown in Figure 12a. As in the case of baseline rise, this distortion becomes more severe as the pressure between the sample and ATR crystal is increased.

Although these artifacts are a nuisance, we have developed techniques for minimizing both of them. For the spectral distortion due to the diamond ATR crystal, the solution is to simply underpressurize the sample. With the IR spectrometer operating in a real-time scan mode, the sample is pressed just enough to bring spectral signals above the baseline. If the artifact is observed, the pressure is decreased until it is either minimized or eliminated. We now use this technique routinely to analyze crowns, which are more prone to this problem than beads. In general, spectral acquisitions are initiated when the intensity of the strongest peak in the real-time scan mode is in the range of 0.04–0.1 absorbance units. Even for these very weak signals, four to eight sample scans are completely adequate for obtaining high quality IR spectra.

The solution to these problems for non-PEG-based PS resins was to make these resins more gel-like. We do this by adding a very small amount (1–2 μL) of $CHCl_3$ to a resin sample prior to analysis, causing the resin to swell and soften. When most of the solvent has evaporated, the softened resin is pressed against the diamond crystal. As long as the resin remains slightly ''wet'' with $CHCl_3$, it will melt against the ATR crystal with even the slightest application of pressure.

Data acquired by ATR-IR using this technique are shown in Figures 12b and 12c. The data in Figure 12b show that the spectral distortions in Figure 12a have been removed by lowering the pressure on the sample. Strong pressurizing of this sample was no longer needed because the resin had first been softened with solvent. The added benefit of softening the resin is that optical contact is greatly enhanced, as shown in Figure 12c. The top spectrum was taken without $CHCl_3$ softening, whereas the bottom spectrum was taken after softening with $CHCl_3$. The band intensities of the $CHCl_3$-swollen resin were approximately seven times as intense as the corresponding bands in the spectrum of the dry

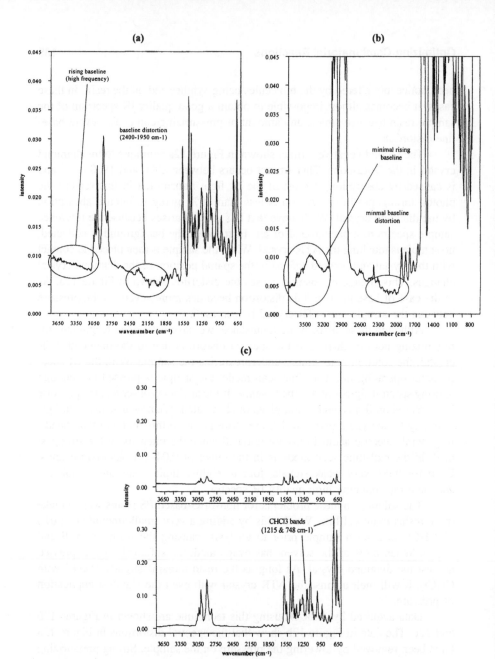

Figure 12 (a) ATR-IR data of a PS resin showing effects of overpressurizing the solid support with an ATR accessory. (b) ATR-IR spectrum of a PS resin pressed correctly against the ATR crystal (diamond). Note the absence of spectral distortions for samples that are not overpressurized. (c) ATR-IR spectrum of PS resin (top) compared to spectrum of the same resin sample softened a priori with $CHCl_3$. Data in (c) demonstrate the advantages of softening a PS resin sample before acquiring data. The six- to sevenfold increase in signal is attributed to superior optical contact when $CHCl_3$ is used. These spectra were taken using a SensIR Technologies Durascope.

resin. This improvement is completely due to the improved optical contact achievable with swollen resins. The only drawback of this technique is that $CHCl_3$ will produce strong absorption bands near 1215 and 750 cm^{-1}.

Using the techniques described above, we have been able to obtain good quality IR spectra for all of the solid supports used by our chemists.

V. COMPARISON OF IR AND NMR ADVANTAGES AND DISADVANTAGES

The examples we have shown were chosen to illustrate how IR and NMR can be used separately or together to optimize solid-phase organic reactions. Next, we would like to point out some of the advantages and disadvantages we have found for the two techniques.

Infrared has a distinct advantage over NMR in the cost and complexity of equipment. SensIR Technologies' Durascope is less expensive than a typical HRMAS probe accessory for the NMR instrument and has the added advantage of portability. The Durascope is easy to set up and should fit nicely into the sample compartment of any benchtop FTIR spectrometer. The NMR experiments require either dedicating an NMR system to MAS experiments or switching an instrument between liquid- and solid-phase samples. The former is expensive, and the latter is time-consuming for the NMR staff.

The expense and complexity of the NMR equipment, however, is compensated for by added experimental flexibility. Newer HRMAS probes are equipped with gradients and are capable of running both proton and heteronuclear experiments. With these probes, the spectroscopist can acquire all the necessary data for complex structure elucidation problems. Also, the gradients allow diffusion experiments such as diffusion-based spectral editing to be performed (41). These techniques can be used to distinguish signals of the compound of interest from both slower or faster diffusing signals, such as those arising from polystyrene resin or solvent and reagent contaminants.

These contaminants, however, can cause major difficulties for the interpretation of the IR spectra. For example, triphenyl phosphine oxide, which has several intense bands in the fingerprint region of the IR spectrum, is a common by-product of Mitsunobu reactions and is very difficult to wash from the support. Even common solvents such as DMF or THF can be difficult to remove completely and can increase the complexity of IR data interpretation. Lack of experimental flexibility has always been a limitation for IR, unlike NMR, where the operator has an ever-expanding repertoire of experiments.

Infrared spectroscopy does have major advantages over NMR in speed, sensitivity, and ease of sample preparation. An IR spectrum of a single bead can be acquired in seconds. In contrast, an NMR spectrum of a single bead can be obtained only with higher-loading resins, and then the acquisition can take hours. Solid-phase NMR samples are typically prepared with at least 20–30 beads to acquire high quality data in minutes or less. Sample preparation is also a major bottleneck for solid-phase NMR. PEG functionalized resins can be difficult to transfer into the small rotors used for HRMAS. IR, on the other hand, requires little or no sample preparation; the sample is simply placed on a microscope slide and analyzed. As a consequence of these advantages, IR has a much greater throughput than NMR for solid-phase samples.

The throughput for solid-phase NMR, however, can be improved a bit with robotics. Varian manufactures an autosampler for use with the Nanoprobe. This equipment is used for robotic sample acquisition and allows for up to 48 samples to be run unattended. Currently, autosamplers are not commercially available for solid-phase IR, but we expect one to be available soon.

The final hurdle for both IR and NMR is data interpretation. NMR has a slight advantage here over IR, because most chemists are well trained and familiar with the interpretation of NMR data. Yet, solid-phase NMR spectra do have different characteristics than solution spectra. Until the chemist is comfortable with these differences, spectral analysis can be difficult. Both techniques would benefit from computational methods that would calculate the solid-phase IR and NMR spectra of a given compound. Until such software becomes available, we recommend building a searchable reference database of solid-phase spectra.

VI. CONCLUSION

As you can see, both NMR and IR have advantages and disadvantages. Either one of these techniques can be used by itself to monitor many types of reactions. Yet, more complex problems will be solved with a higher degree of confidence when data from both techniques are used together. Ideally, we recommend that the well-equipped analytical lab be capable of obtaining NMR and IR data for solid-phase samples. The chemist in pursuit of solid-phase reaction optimization can use these techniques to positively identify products, to gain a better understanding of reaction kinetics, to estimate the extent of completion of a reaction, and even to study the physicochemical properties of resins.

REFERENCES

1. B Yan, L Liu, CA Astor, Q Tang. Determination of the absolute amount of resin bound hydroxyl or carboxyl groups for the optimization of solid-phase combinatorial and parallel organic synthesis. Anal Chem 71:4564–4571, 1999.
2. A Madder, N Farcy, NGC Hosten, H De Muynck, PJ De Clercq, J Barry, AP Davis.

A novel sensitive colorimetric assay for visual detection of solid-phase bound amines. Eur J Org Chem 1999(11):2787–2791, 1999.

3. NJ Haskins, DJ Hunter, AJ Organ, S Rahman, C Thom. Combinatorial chemistry: Direct analysis of bead surface associated materials. Rapid Commun Mass Spectrom 9:1437–1440, 1995.

4. SR McAlpine, S Schreiber. Visualizing functional group distribution in solid-support beads by using optical analysis. Eur J Chem 5(12):3528–3532, 1999.

5. AR Vaino, DB Goodin, KD Janda. Investigating resins for solid phase organic synthesis: The relationship between swelling and microenvironment as probed by EPR and fluorescence spectroscopy. J Comb Chem 2(4):330–336, 2000.

6. Y Luo, X Ouyang, RW Armstrong, MM Murphy. A case study employing spectroscopic tools for monitoring reactions in the developmental stage of a combinatorial chemistry library. J Org Chem 63:8719–8722, 1998.

7. R Riedl, R Tappe, A Berkessel. Probing the scope of the asymmetric dihydroxylation of polymer-bound olefins. Monitoring by HRMAS NMR allows for reaction control and on-bead measurement of enantiomeric excess. J Am Chem Soc 120:8994–9000, 1998.

8. C Dhalluin, C Boutillon, A Tartar, G Lippens. Magic angle spinning nuclear magnetic resonance in solid-phase peptide synthesis. J Am Chem Soc 119:10494–15000, 1997.

9. SK Sarkar, RS Garigipati, JL Adams, PA Keifer. An NMR method to identify nondestructively chemical compounds bound to a single solid-phase-synthesis bead for combinatorial chemistry applications. J Am Chem Soc 118:2305–2306, 1996.

10. WL Fitch, G Detre, CP Holmes. High-resolution ^{1}H NMR in solid-phase organic synthesis. J Org Chem 59:7955–7956, 1994.

11. PA Keifer. High-resolution NMR techniques for solid-phase combinatorial chemistry. Drug Disc Tech 2:468–478, 1997.

12. MA Gallop, WL Fitch. New methods for analyzing compounds on polymeric supports. Curr Opin Chem Biol 1:94–100, 1997.

13. MJ Shapiro, JR Wareing. NMR methods in combinatorial chemistry. Curr Opin Chem Biol 2:372–375, 1998.

14. PA Keifer. Influence of resin structure, tether length, and solvent upon the high-resolution ^{1}H NMR spectra of solid-phase-synthesis resins. J Org Chem 61:1558–1559, 1996.

15. AM Sefler, S Gerritz. Using one- and two-dimensional NMR techniques to characterize reaction products bound to Chiron SynPhase crowns. J Comb Chem 2:127–133, 2000.

16. R Jelinek, AP Valente, KG Valentine, SJ Opella. Two-dimensional NMR spectroscopy of peptides on beads. J Magn Reson 125:185–187, 1997.

17. CH Gotfredsen, M Grotli, M Willert, M Meldal, JO Duus. Single-bead structure elucidation. Requirements for analysis of combinatorial solid-phases libraries by Nanoprobe MAS-NMR spectroscopy. J Chem Soc Perkin Trans I 7:1167–1171, 2000.

18. G Lippens, M Bourdonneau, C Dhalluin, R, Warrass, T Richert, C Seetharaman, C Boutillon, M Piotto. Study of compounds attached to solid supports using high resolution magic angle spinning NMR. Curr Org Chem 3:147–169, 1999.

19. PA Keifer, L Baltusis, DM Rice, AA Tymiak, JN Shoolery. A comparison of NMR spectra obtained for solid-phase-synthesis resins using conventional high-resolution, magic-angle-spinning, and high-resolution magic-angle-spinning probes. J Magn Reson Ser A 119:65–75, 1996.

20. B Yan, G Kumaravel, H Anjaria, A Wu, RC Petter, CF Jewell Jr, JR Wareing. Infrared spectrum of a single resin bead for real-time monitoring of solid-phase reactions. J Org Chem 60:5736–5738, 1995.

21. B Yan. Monitoring the progress and the yield of solid-phase organic reactions directly on resin supports. Acc Chem Res 31:621–630, 1998.

22. B Yan, H-U Gremlich, S Moss, GM Coppola, Q Sun, L Liu. A comparison of various FTIR and FT Raman methods: Applications in the reaction optimization stage of combinatorial chemistry. J Comb Chem I(1):46–54, 1998.

23. B Yan, H-U Gremlich. Role of Fourier transform infrared spectroscopy in the rehearsal phase of combinatorial chemistry: A thin-layer chromatography equivalent for on-support monitoring of solid-phase organic synthesis. J Chromatogr B 725:91–102, 1999.

24. H-U Gremlich. The use of optical spectroscopy in combinatorial chemistry. Biotechnol Bioeng (Comb Chem) 61(3):179–187, 1998/1999.

25. J Hammond, B Kellam, AC Moffat, RD Jee. Non-destructive real-time reaction monitoring in solid-phase synthesis by near-infrared reflectance spectroscopy-esterification of a resin-bound alcohol. Anal Commun 36:127–129, 1999.

26. D Walker, J Anderson, F Tarczynski. Optimization of solid-phase syntheses via online spectroscopic monitoring. SPIE Conf on Optical Online Industrial Process Monitoring 859:90–97, 1999.

27. BD Larsen, DH Christensen, A Holom, R Zillmer, OF Nielsen. The Merrifield peptide synthesis studied by near-infrared Fourier-transform Raman spectroscopy. J Am Chem Soc 115:6247–6253, 1993.

28. J Hochlowski, D Whittern, J Pan, R Swenson. Applications of Raman spectroscopy to combinatorial chemistry. Drugs Future 24(5):539–554, 1999.

29. B Yan, G Kumaravel. Probing solid-phase reactions by monitoring the IR bands of compounds on a single "flattened" resin bead. Tetrahedron 52(3):843–848, 1996.

30. RE Marti, B Yan, MA Jarosinski. Solid-phase synthesis via 5-oxazolidinones. Ring opening reactions with amines and reaction monitoring by single-bead FT-IR microspectroscopy. J Org Chem 62:5615–5618, 1997.

31. Q Sun, B Yan. Single bead IR monitoring of a novel benzimidazole synthesis. Bioorg Med Chem Lett 8:361–364, 1998.

32. DE Pivonka, TR Simpson. Tools for combinatorial chemistry: Real-time single bead infrared analysis of a resin-bound photocleavage reaction. Anal Chem 69(18):3851–3853, 1997.

33. DE Pivonka, DL Palmer. Direct infrared spectroscopic analysis of reagent partitioning in polystyrene bead supported solid-phase reaction chemistry. J Comb Chem 1(4):294–296, 1999.

34. SS Rahman, DJ Busby, DC Lee. Infrared and Raman spectra of a single resin bead for analysis of solid-phase reactions and use in encoding combinatorial libraries. J Org Chem 63:6196–6199, 1998.

35. DE Pivonka. On-bead quantitation of resin bound functional groups using analogue techniques with vibrational spectroscopy. J Comb Chem 2:33–38, 2000.

36. DE Pivonka, K Russell, T Gero. Tools for combinatorial chemistry: In situ infrared analysis of solid-phase organic reactions. Appl Spectrosc 50(12):1471–1478, 1996.

37. W Huber, A Bubendorf, A Grieder, D Obrecht. Monitoring solid phase synthesis by infrared spectroscopic techniques. Anal Chim Acta 393:213–221, 1999.

38. H-U Gremlich, SL Berets. Use of FT-IR internal reflection spectroscopy in combinatorial chemistry. Appl spectrosc 50(4):532–536, 1996.

39. AM Fivush, TM Willson. AMEBA: An acid sensitive aldehyde resin for solid phase synthesis. Tetrahedron Lett 38:7151–7154, 1997.

40. K Ngu, DV Patel. Preparation of acid-labile resins with halide linkers and their utility in solid phase organic synthesis. Tetrahedron Lett 38:973–976, 1997.

41. MJ Shapiro, J Chin, A Chen, JR Wareing, Q Tang, RA Tommasi, HR Marepalli. Covalent or trapped? PFG diffusion MAS NMR for combinatorial chemistry. Tetrahedron Lett 40:6141–6143, 1999.

9
Analytical Methods in Library Feasibility and Rehearsal Studies

Bing Yan, Tushar Kshiragar, Liling Fang, Pascal Fantauzzi, and Jianmin Pan
ChemRx Advanced Technologies, Inc., South San Francisco, California

I. INTRODUCTION

The synthesis of high quality combinatorial libraries depends on the development of optimal synthesis, analysis, and cleanup protocols. In our lead discovery library synthesis, we employ solid-phase and solution-phase parallel synthesis methods to produce libraries containing approximately 5000 compounds each (see Chapter 1). Our approach involves a series of feasibility and rehearsal procedures designed to optimize the final production protocol. This is in contrast to the reaction optimization approach of process chemistry, which seeks the highest yield for a single reaction. The optimization in combinatorial chemistry seeks a set of reaction conditions that provides an overall good yield for a diverse set of reactions. This set of reaction conditions may not be the best for every single reaction because it is unlikely that a single set of conditions exists that would give the highest yield and optimal kinetics for so many reactions. This complex process is referred to as chemistry optimization or validation. It is the most costly and time-consuming step, but it is the essential step in combinatorial or parallel synthesis.

The development of a combinatorial synthesis protocol is a tedious process. It includes the initial concept, determination of the feasibility of the chemistry, the optimization of general reaction conditions, the validation of building blocks, and the testing of rehearsal libraries. Numerous analytical technologies are required in each step of this process. Among the analytical challenges, two are particularly significant and therefore deserve a full discussion in this chapter.

One challenge is the analysis of the synthetic product on solid support in order to gain qualitative and quantitative information regarding the reactions to be optimized. The other is that means must be provided to rapidly analyze a large number of samples in solution to gain quantitative information on diverse reactions. In the following section we discuss applications of several analytical methods in the synthesis optimization process.

II. ANALYTICAL METHODS IN THE FEASIBILITY STUDY PHASE

The initial process of chemistry development involves a systematic study of a variety of reaction schemes and conditions in order to determine which ones are optimal. To effectively select optimal reaction conditions, it is essential to probe organic reactions. Although solution-phase reactions can be monitored easily, there is still a gap between desirable analytical methods for monitoring solid-phase reactions and those actually available. Rapid quantitative analysis directly on solid support is crucial for monitoring chemical synthesis. Indirect analysis after cleavage can provide only indirect evidence about the reaction on solid phase. We generally use one or more of the following five methods to leverage our solid-phase chemistry.

A. Single-Bead FTIR Method

The IR spectrum of every organic compound is unique. Every functional group in the molecule has multiple vibrational modes and therefore provides multiple frequencies for us to monitor. Due to its fingerprinting capability, Fourier transform infrared spectroscopy (FTIR), particularly single-bead FTIR, is an effective tool for monitoring organic reactions on solid phase (1,2). Although many sampling methods have been used to acquire vibrational spectra from resin samples, we have found that transmission spectral measurement from a single bead generates high quality spectra in terms of both reproducibility and signal-to-noise ratio. In our experiments, resin beads are flattened between two salt or diamond windows using the hands or a compression cell. Spectra are then collected in transmission mode on a single flattened bead on one side of such a window. The background is collected using an empty region of the same window. This operation eliminates the saturation effect due to the variation of bead diameters (50–100 μm). Furthermore, we use polystyrene bands as internal standards to correct for signal variations associated with the variable pathlengths in flattened beads. Compared to attenuated total reflection (ATR) measurements, the transmission experiments better monitor the chemical entity in the bead interior because more than 99% of the reaction is occurring inside the bead. Examples describing the single-bead FTIR method and its application in reaction optimization are described in detail in Chapter 7.

B. Color Test

Color testing is rooted in classical organic analysis. Based on the color change induced by a specific reaction between a given reagent and an organic functional group, the presence of this functional group can be qualitatively determined. Specific reagents have also been developed to react with organic functional groups that are attached to solid support. The functionality is indicated by a color or fluorescence change on the bead. During a reaction, a given qualitative color test can be used for quantitative analysis by repeatedly testing a functional group in a starting material until it disappears. Some common reagents for testing solid-bound organic functional groups are listed in Table 1. Additional color testing reagents can be found in Table 1 of Chapter 10.

C. Quantitative Analysis of Organic Functional Groups

Whereas qualitative tests are an essential part of the daily routine in monitoring solid-phase organic synthesis (SPOS), quantitative methods are particularly important for a direct determination of resin loading and the yield of a synthetic product. Quantitative analysis is based on a specific reaction of a reagent with an organic functional group coupled with the quantitative measurement of a chromophoric product produced in the reaction or direct determination of the amount of reagent consumed (6). Table 2 lists some reagents for quantitative analysis of solid-bound organic functional groups.

For example, the method for the analysis of resin-bound hydroxyl or carboxyl groups in 1 and 2 (Scheme 1), respectively, is based on the rapid and efficient reaction of an approximately twofold excess of 9-anthroylnitrile or 1-pyrenyldiazomethane (PDAM) with the resin-bound hydroxyl or carboxyl groups

Table 1 Qualitative Color Test for Resin-Bound Organic Compound

Functional group	Test reagent	Color	Ref.
Primary amine	Ninhydrin	Blue	3
Secondary amine	Chloranil	Blue	4
Phenol	$FeCl_3$	Purple	5
Aldehyde	Dansylhydrazine	Fluorescent	6
Alcohol	9-Anthroylnitrile	Fluorescent	7
Carboxyl	PDAM	Fluorescent	7

Table 2 Quantitative Analysis Method for Organic
Functional Groups on Solid Support

Functional group	Reagent	Ref.
Primary amine	Ninhydrin	8
Aldehyde	Dansylhydrazine	6
Alcohol	9-Anthroylnitrile	7
Carboxyl	PDAM	7
N, S, Cl, Br, I	Elemental analysis	9

in these two resins. A spectroscopic measurement of the reagent concentration
in the supernatant reveals the loading on the resin after subtracting the noncovalently absorbed amount of reagent molecules in the control resin (7). These works
demonstrate that accurate determinations can be achieved at various loading levels of each functional group. The data point corresponding to the lowest amount
of hydroxyl groups was determined to be at 0.05 mmol/g. The flexibility and
dynamic range of these methods are essential for monitoring the synthesis of a
product and the disappearance of a starting material in SPOS.

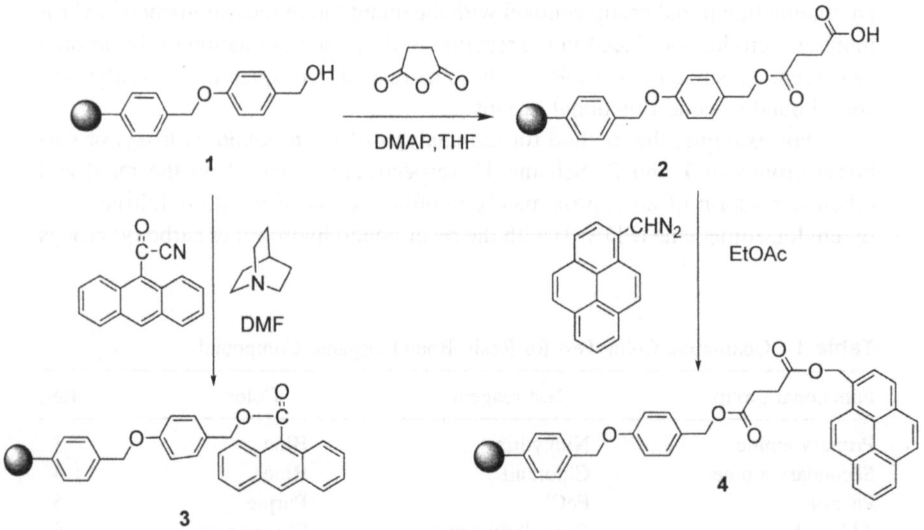

Scheme 1

D. Elemental Analysis

Elemental analysis is the gold standard method in organic chemistry for determining the purity of an organic compound. In SPOS, elemental analysis of resin-bound organic compounds provides a valuable method for monitoring the completion of organic reactions on resin if the reaction involves a change in elements such as N, S, F, Cl, Br, or I. Studies (9) have shown that elemental analysis results are consistent with theoretical values or values derived through spectrophotometric determinations for resin-bound "standard" compounds. For synthesis of resin samples, this method has been shown to be reproducible and accurate in most cases (for example, see Scheme 2). Complete washing and drying, however, are crucial for accurate analysis.

E. Analysis After Cleavage

Cleaving and analyzing the synthetic intermediate at any point of the synthesis is a straightforward way to probe the outcome of a solid-phase reaction. An example of the analysis of the completion of a resin loading reaction is shown in Scheme 3 and Figure 1. If the formation of **8** is complete, the reaction of the product with 4-methylbenzoic chloride produces only one product **10** after amine cleavage, and as a result we should detect only one peak in the HPLC trace. In the case of incomplete loading, two HPLC peaks are detected. In addition to the expected peak for **10**, a second peak attributable to the side product **11** from the reaction of **7** and the reagent is observed (Fig. 1). On the basis of this result, the coupling reaction should be continued until this peak disappears. In general, any synthesis product or side product can be analyzed by HPLC after cleavage. In cases where there is an uncertainty about product assignment, LC/MS is often used.

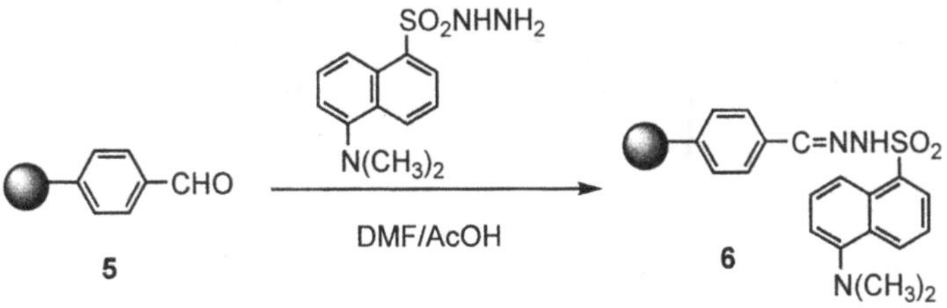

Scheme 2

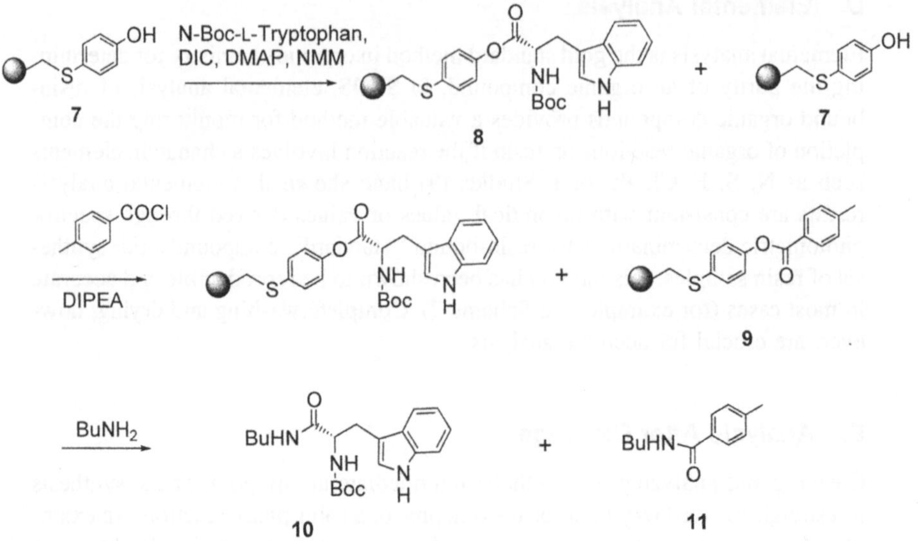

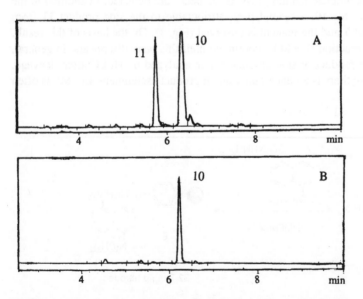

Scheme 3

Figure 1 HPLC analysis of the loading level for an incomplete reaction (A) and a complete reaction (B). For details see Scheme 3.

III. ANALYTICAL METHODS IN THE REHEARSAL PHASE

Rehearsal study is necessary to adapt the initially developed reaction conditions to more reactions involving an increasing number of building blocks. The diversity of building blocks is crucial for building lead discovery libraries. These diverse building blocks intrinsically have variable chemical reactivity. Therefore, the selection of building blocks consists of two parts: a preliminary selection based on diversity computation and an experimental selection based on chemical reactivity.

A. Diversity Selection

Although several computational approaches to calculate diversity exist, the selection of diversity still remains rather subjective. Although procedures for the calculation of diversity differ, there is some agreement on the inclusion of pharmacologically relevant side chains, hydrogen bond donors and acceptors, and when appended on the scaffold these side chains will probe a large diversity space. We have employed a proprietary diversity selection tool called IcePick (10) in our diversity selection.

B. Chemical Selection

Reaction conditions optimized with feasibility studies may not be general enough for hundreds of diverse building blocks. Furthermore, other factors such as solubility start to cause complications when more diverse building blocks are involved. Scale-up also generates other problems such as agitation of the reaction mixture. To validate building blocks for library synthesis, it becomes important to obtain information associated with identity, purity, and rapid quantification for cleaved compounds.

1. Identity Analysis

The desired product can be simply identified by its molecular weight. Flow injection analysis (FIA) MS has been widely used as a routine identity analysis method. However, solvent, contaminant, and excess reagents will all show up in the spectra. If they charge well, their signals may be significant even if they exist only in trace amounts. Molecular competition for protons may also cause a significant ion suppression effect that may partially or completely erase the product signal. Because of the problems associated with FIA MS, LC/MS becomes an indispensable method. Figure 2 shows the identity analysis of 576 compounds synthesized in six 96-well plates. We define that a compound is identified if its relative ion abundance reaches 20% of the base ion intensity in FIA MS

MS, P1

LC/MS, P1

MS, P2

LC/MS, P2

MS, P3

LC/MS, P3

MS, P4

LC/MS, P4

MS, P5

LC/MS, P5

MS, P6

LC/MS, P6

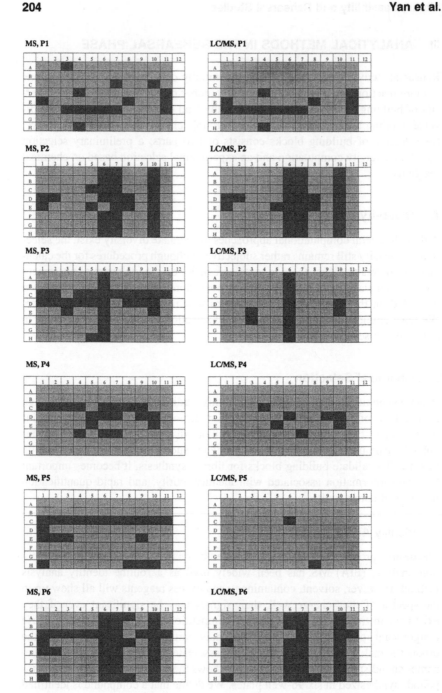

analysis or its LC peak area reaches 20% of the total peak area in the chromatogram as detected by UV_{214} (UV detection at 214 nm) in LC/MS analysis. In the present example, 71% of compounds are identified by FIA MS, but 83% are found by LC/MS. Therefore, 12% of compounds that are present at more than 20% are missed by FIA MS analysis because of the problems just discussed. Among these six plates, plates 2 and 6 are found to be problematic because the synthesis of 34% and 31% of compounds, respectively, was not successful. Several building blocks are identified as unsuitable under the synthesis conditions. These include three building blocks in plates 2 and 6 used to synthesize compounds in columns 6, 7, and 10.

2. Purity Determination

The concept of purity in traditional organic synthesis is that of quantitative purity or absolute purity, defined as the percentage of the absolute amount of desired compound in the total sample by weight. When a large number of samples must be analyzed, the percentage of the HPLC peak area of the product relative to the total peak areas detected by UV_{214} or evaporative light scattering detection (ELSD) is often used to indicate the purity of the compound. This purity is a relative purity. It is expected that the relative purity will not match the quantitative purity. However, a direct comparison is not available. In order to evaluate purity measurements, we routinely carry out three kinds of purity measurements: (a) quantitative purity measurement for six compounds in each rehearsal library; (b) relative purity measurement using UV_{214} detection, and (c) relative purity measurement using ELSD.

The quantitative purity of a reference compound is determined by obtaining its absolute weight by using an HPLC calibration curve. The calibration curve is generated using the same compound synthesized separately, purified to homogeneity, and characterized by ^1H and ^{13}C NMR and elemental analysis. It is made by accurate weighing, dilution, and HPLC analysis of the standard compound. Reference compounds with the same structures as these standards are synthesized with the library, following the generic synthesis procedures in the building block validation and rehearsal libraries. These compounds are then quantitatively analyzed by HPLC/UV using their individual calibration curves. From the individual standard curve, the absolute quantity of each compound is determined. With refer-

---

Figure 2 A plate view of the identity analysis for plates 1–6 from flow-injection analysis (left) and LC/MS analysis (right). A compound is identified if its relative ion abundance reaches 20% of the base ion intensity in FIA MS analysis or its LC peak area reaches 20% of the total peak area in the chromatogram as detected by UV_{214} in LC/MS analysis. Light color means a pass and dark color a failure.

ence to the crude weight measured before HPLC analysis, the quantitative purity of each compound is determined. The quantitative purity is often lower than the relative purity determined by UV_{214} and ELSD detection. This indicates that there are invisible (by UV_{214} and ELSD) impurities in the real sample. These impurities may include solvent, salt, TFA, plastic residues, and other unidentified substances. Examples of reference compound analysis are given in Section IV.

Two of the relative purity measurement methods are based on HPLC separation and on-line UV_{214} or ELSD detection. Purity analysis by UV_{214} is often used to indicate the purity of the product based on the wild assumption that the product and all side products have the same extinction coefficient at 214 nm. This assumption for all practical purposes is wrong. ELSD has been validated as a more general detector (11–13). However, we have found that purity measured by ELSD is consistently higher than that measured by UV_{214}. From studies of compounds with different molecular weights, we have found that an ELSD

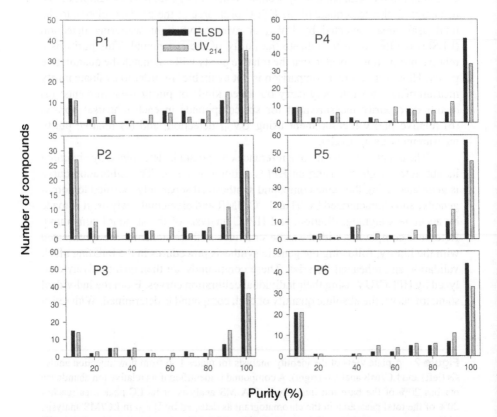

Figure 3 Compound purity distribution of plates 1–6 from HPLC analysis based on UV_{214} or ELSD detection.

detector cannot quantitatively detect compounds that have low molecular weight and are relatively volatile (14). Figure 3 shows the purity analysis of the same 576 compounds as in Figure 2 by UV_{214} and ELSD. As in the identity analysis, plates 2 and 6 are problematic because they contain 31% and 24%, respectively, of compounds with purities less than 10%.

We find that ELSD is a more general detector for purity analysis than UV, but when low molecular weight impurities are present, UV_{214} detection better describes the library quality. We use both detection methods to characterize our rehearsal libraries.

3. Quantity Determination

We determine the absolute weight of six reference compounds in each rehearsal library using standard calibration methods as described above. The relative error

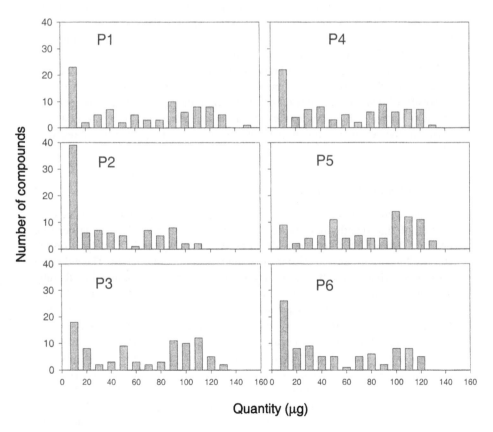

Figure 4   Compound amount distribution in plates 1–6 as determined by LC/MS/ELSD.

of this method is ±5%. We also determine the absolute weight of every compound by on-line LC/MS/UV/ELSD based on a general calibration curve. This general calibration curve is generated from 42 standard compounds from seven diverse libraries. The relative error of this high throughput LC/MS/UV/ELSD method is ±28% (14). The absolute quantities of 576 compounds were determined by LC/MS/ELSD as shown in Figure 4. Consistent with both identity and purity analyses, plates 2 and 6 contain more compounds in very low quantity. The quantitative measurement is important because the experimental data show that even though many samples have a high purity, they are present in very low quantities. The quantity of a compound is an important measure for combinatorial synthesis in our rehearsal validations.

IV. CASE STUDY

The development of a synthesis protocol requires the application of some or all of the analytical methods discussed above to help optimize the synthesis conditions. The development of a *trans*-aminomethylcyclohexane-4-carboxamide library synthesis protocol is described below.

One of the scaffolds for library development is Fmoc-protected unnatural amino acid **15** as shown in Scheme 4. The natural or unnatural amino acids directly afford two obvious sites for diversification: the carboxylic acid portion and the amine portion. In addition, the primary amine functionality can be first converted to the secondary amine and subsequently acylated, providing three sites of diversity.

A. Synthesis Feasibility

Our approach to derivatization of the unnatural amino acid **15** involves the conversion of chloromethylpolystyrene **12** to the aldehyde resin **13**. Single-bead IR detected distinct bands of aldehyde carbonyl stretching, and Cl analysis on the starting Merrifield resin also confirmed the reaction's completion. The aldehyde resin was derivatized to the secondary amine **14** by reductive amination, thereby incorporating the first diversity element. Monitoring methods for this step included the detection of the disappearance of the aldehyde IR bands and a positive chloranil test (blue-purple color, specific for secondary amines) in the product. The addition of the Fmoc-protected amino acid to form **15** introduces two new carbonyl bands at 1680 cm^{-1} (amide) and 1710 cm^{-1} (Fmoc) in the IR spectrum and gives a negative chloranil test. Subsequently the Fmoc amino acid was deprotected by standard deprotection methods (20% piperidine in DMF) to afford primary amine **16**. A clean removal of the Fmoc IR band at 1710 cm^{-1} and a positive ninhydrin color test (blue) indicated the completion of this reaction step. This step was followed by a second reductive amination using aldehydes to establish

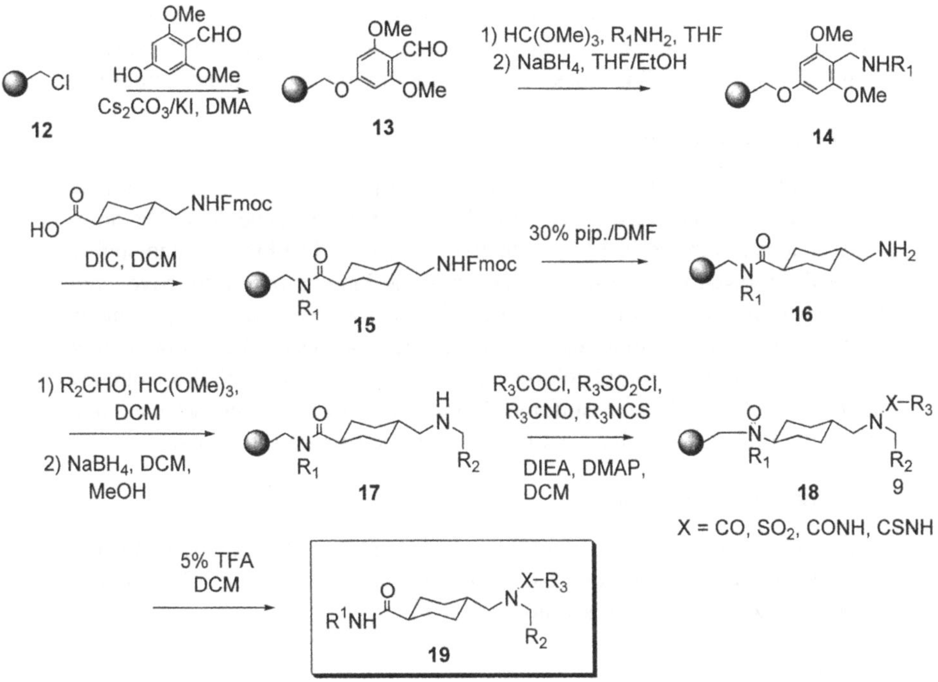

Scheme 4

the second diversity element. The ninhydrin test was used to monitor this step until the reaction was complete. The secondary amine **17** was then treated with a variety of acylators, including acid chlorides, sulfonyl chlorides, isocyanates, and isothiocyanates, to afford the resin-bound products **18**. Newly formed carbonyl IR bands and a negative chloranil test indicated the completion of this reaction step. The desired products **19** were obtained by cleavage off support with 5% TFA in dichloromethane.

This reaction sequence is attractive because it demonstrated that any Fmoc-protected amino acid can potentially be converted to a series of analogs by attaching three types of diversity elements. In addition, the chemistry takes advantage of widely available commercial building blocks including amines as the first diversity element, aldehydes as the second, and acylators as the third.

B. Validation of Building Blocks

Diverse building blocks to be used in this library were selected by our proprietary software IcePick. The chemistry and building block validation is carried out as close to the final production method as possible. In-process reaction monitoring methods developed in the feasibility stage were used at this stage for statistically selected samples. Building block validation for this library was done in three parts.

1. Amine Validation

A range of amines such as aliphatic amines, aromatic amines, and heteroaromatic amines were validated using the procedure described in Scheme 4. The amines were attached to the resin 13 by reductive amination. The Fmoc amino acid was coupled, followed by piperidine cleavage of the Fmoc group. Reductive amination reactions were carried out using isovaleraldehyde as a representative aldehyde, and the acylation reactions were performed using the following representative acylators: 2-furoyl chloride, cyclohexyl isocyanate, ethyl-3-isothiocyanatopropionate, and benzenesulfonyl chloride. The acylators were selected to represent the different reactive functionalities in the diverse acylator set.

2. Aldehyde Validation

Both aliphatic and aromatic aldehydes were validated. Butylamine and benzylamine were used as the representative reducing amines, and ethyl-3-isothiocyanopropionate was used as an acylator.

3. Acylator Validation

Similarly, representatives of all four types of acylators were validated according to Scheme 4. Benzylamine and benzaldehyde were used as the representative reducing amine and aldehyde, respectively.

Table 3 lists some representative building blocks that were used in the validation. A pass result indicates that the product has a relative purity of >80% by LC/MS/UV$_{214}$ analysis, and the passed building blocks were used in the final library production. A fail result indicates that the product may have formed but is <80% by LC/MS/UV$_{214}$ analysis. A few identifiable side products were observed, and these were either due to incomplete reductive amination or due to incomplete acylation.

C. Rehearsal Library Study

A 352-compound rehearsal library was produced according to the layout in Figure 5 using the method developed in the validation studies.

Results of the analysis of the rehearsal library indicated that although reference compounds were >80% pure by LC/MS/UV analysis, their quantitative purity was low (average of six standards = 33.5%). This indicated that the impurities might not be UV-active. However, our results from validation studies indicated that the chemistry was robust. The impurities may be from other sources, such as inappropriate cleavage conditions.

Table 3 Representative Building Blocks Used in the Validation Study

Entry	Name	Structure	Pass/Fail
	Aldehydes		
1	Butyraldehyde		Pass
2	2-Furaldehyde		Pass
3	Benzaldehyde		Pass
4	2-Quinoline carboxaldehyde		Pass
5	Cyclohexane carboxaldehyde		Pass
6	2-Thiophene carboxaldehyde		Fail
7	4-Dimethylamino benzaldehyde		Fail
	Acylators		
8	3-(Diethylamino) propyl isothiocyanate		Pass
9	Phenyl isocyanate		Pass
10	3,4-Dimethoxybenzenesulfonyl chloride		Pass
11	2-Ethoxybenzoyl chloride		Pass
12	3-(Methylthio)phenyl isocyanate		Pass
13	4-n-Butylphenyl isocyanate		Fail

Table 3 Continued

Entry	Name	Structure	Pass/Fail
	Amines		
14	Cyclopentylamine		Pass
15	4-(3-Aminopropyl) morpho-line		Pass
16	3,4-Dimethoxyphenethylamine		Pass

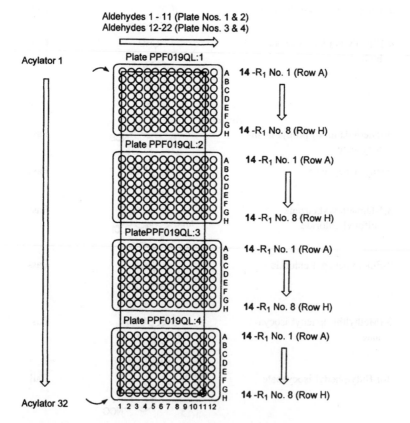

Figure 5 The synthesis layout of the rehearsal library.

D. Optimization of the Cleavage Step

A review of the protocol identified the cleavage conditions as a possible source for the introduction of impurities. The plates containing the resins in the first rehearsal library were treated with 10% TFA in DCM, followed by a wash with the same solution, and the combined solvents were evaporated. The plates were then subjected to two more washes with 10% TFA in DCM (method A). The excessive washing procedure, as well as the increased exposure of the collection plates to DCM, could have extracted impurities from the plates or the resin beads. These impurities most likely do not ionize in the MS, do not elute with compounds, and do not absorb at 214 nm.

To optimize the cleavage conditions, a second rehearsal library was designed such that the cleavage step involved a single treatment with 5% TFA in DCM and a single wash with DCM (method B). In addition, the extraction of the compound from the wells was carried out using three different solvent systems as described in Table 4.

The results from this study (Table 4) indicated that method B gave better purities in the final product. It is also interesting to note that although the solvent system used for extraction of the compounds from the wells does not affect the purity, the overall yield differs. The results from the above study gave circumstantial evidence that the impurities in the first rehearsal library were introduced by the cleavage method and that greater overall quantitative purity was obtained by using method B.

A *trans*-aminomethylcyclohexane-4-carboxamide library for general screening purposes was synthesized using the developed synthetic scheme and validated building blocks.

V. SUMMARY

Using the development of a synthesis protocol for a *trans*-aminomethylcyclohexane-4-carboxamide library as an example, we reviewed an array of analytical

Table 4 Effects of Cleavage Solvents on the Purity and Quantity of Products

	2:1 MeOH/DCM		MeOH		ACN	
Compound	Amt (mg)	Purity(%)	Amt (mg)	Purity(%)	Amt (mg)	Purity(%)
1	21.0	72.5	18.7	74.8	32.3	75.6
2	8.6	26.8	7.0	26.3	10.4	30.4
3	17.6	68.9	9.0	62.0	19.2	67.4
Average	15.7	56.1	11.6	54.4	20.6	57.8

methods used in our laboratories to optimize combinatorial synthesis. These methods focus on two key aspects of synthesis optimization: (a) to quantitatively and qualitatively monitor reactions on solid support and (b) to rapidly determine identity, purity, and quantity simultaneously for a large number of compounds in solution.

REFERENCES

1. B Yan, G Kumaravel, H Anjaria, A Wu, R Petter, CF Jewell Jr, JR Wareing. Infrared spectrum of a single resin bead for real-time monitoring of solid-phase reactions. J Org Chem 60:5736–5738, 1995.
2. B Yan. Monitoring the progress and the yield of solid phase organic reactions directly on resin supports. Acc Chem Res 31:621–630, 1998.
3. E Kaiser, RL Colescott, CD Bossinger, PI Cook. Color test for detection of free terminal amino groups in the solid-phase synthesis of peptides. Anal Biochem 34: 595, 1970.
4. T Vojkovsky. Detection of secondary amines on solid phase. Peptide Res 8:236, 1995.
5. CR Johnson, B Zhang, P Fantauzzi, M Hocker, KM Yager. Tetrahedron 54:4097, 1998.
6. B Yan, W Li. Rapid fluorescence determination of the absolute amount of aldehyde and ketone groups on resin supports. J Org Chem 62:9354–9357, 1997.
7. B Yan, L Liu, C Astor, Q Tang. Determination of the absolute amount of resin-bound hydroxyl or carboxyl groups for the optimization of solid-phase combinatorial and parallel organic synthesis. Anal Chem 71:4564–4571, 1999.
8. VK Sarin, SBH Kent, JP Tam, RB Merrifield. Quantitative monitoring of solid-phase peptide synthesis by the ninhydrin reaction. Anal Biochem 117:147, 1981.
9. B Yan, CF Jewell, SW Myers. Quantitatively monitoring of solid-phase organic synthesis by combustion elemental analysis. Tetrahedron 54:11755, 1998.
10. J Mount, J Ruppert, W Welch, AN Jain. IcePick: A flexible surface-based system for molecular diversity. J Med Chem 42:60, 1999.
11. CE Kibbey. Quantitation of combinatorial libraries of small organic molecules by normal-phase HPLC with evaporative light-scattering detection. Mol Diversity 1: 247, 1995.
12. BH Hsu, E Orton, SY Tang, RA Carlton. Application of evaporative light scattering detection to the characterization of combinatorial and parallel synthesis libraries for pharmaceutical drug discovery. J Chromatogr B 725:103, 1999.
13. L Fang, M Wan, M Pennacchio, J Pan. Evaluation of evaporative light-scattering detector for combinatorial library quantitation by reversed phase HPLC. J Comb Chem 2:254, 2000.
14. L Fang, J Pan, B Yan. High-throughput determination of identity, purity and quantity of combinatorial library members using LC/MS/UV/ELSD. Biotech Bioeng Comb Chem 71:162, 2001.

10

Approaches to Reaction Monitoring by Product Derivatization

Miles S. Congreve, Corinne Kay, and Jan Scicinski
Cambridge University Chemical Laboratory, GlaxoSmithKline, Cambridge, England

I. INTRODUCTION

The synthesis of combinatorial libraries using solid-phase chemistry has become a routine strategy in the practice of drug discovery (1–3). Traditionally, the syntheses of nonpeptide libraries were initially developed in solution, enabling reaction conditions and intermediates to be characterized by using conventional analytical techniques, and the resulting synthetic route was then transferred to the solid support using an appropriate linker (4). However, differences in reaction rates and yields between solid and solution phase (5) and the need to shorten reaction optimization time has led to an adoption of strategies in which the chemistry is developed directly on the solid support. Although this approach depends upon effective monitoring of reactions carried out on resin, the requisite analytical techniques remain limited (6–8).

Synthesis of compound mixture libraries (9) using, for example, the split–mix–pool strategy (10) additionally adds to the complexity of the identification of materials from these pools. This has generally been addressed by introducing a chemical (11,12) or oligonucleotide (13,14) coding sequence onto the bead (15). Structure determination by direct identification of materials cleaved from a single bead, for example by mass spectrometry, is often problematic owing to the very small amount of material produced (often on the order of 100–300 pmol). Indeed, few methods are sufficiently sensitive to give information on the quantities of material available from single beads. These problems in both the development of chemistry in a split–mix–pool format and subsequent analysis of completed libraries are exacerbated by the need for high throughput and cost-effective methods.

This chapter describes some general methods used for following the course of solid-phase chemical reactions and the analysis of reaction products by cleavage and/or derivatization of resin-bound materials. Commonly used color tests employed to indicate the presence or absence of a functional group are outlined. Quantification by cleavage of protecting groups is described. Mass spectrometric techniques applied to resin chemistry and single-bead analysis are briefly reviewed, and examples of approaches used to enhance the detection of products by chemical derivatization are given. Finally, dual linker approaches developed to facilitate the analysis of resin-bound products are examined in some depth.

II. COLORIMETRIC TESTS ON THE SOLID SUPPORT

Spot tests for resins are fast, require little resin material (a few beads), and are capable of detecting less than 0.5% of the total available sites. Although the most commonly detected functionality is the amino group, a broad range of functional groups can be derivatized, giving rise to colored beads or solutions. The Kaiser test (ninhydrin color test) (16,17) and the bromophenol blue test (3′,3″,5′,5″-tetrabromophenolsulfonylphthalein) (18) both give a blue color in the presence of amines, whereas a positive trinitrobenzene sulfonyl chloride (TNBS) test (19) gives a characteristic red color (Table 1, entries 1–3). Other less frequently used amine color tests are the chloranyl (20) and picric acid (trinitrophenol) (21,22) tests. Finally, primary amines can also be detected with the fluorescamine test (23), whereas secondary amines such as proline can be visualized by using the isatin test (24) (2,3-indoline dione).

A convenient test (25) for determining the absolute amount of hydroxyl or carboxyl groups directly on the solid support following reaction of resins with 9-anthroylnitrile or 1-pyrenyldiazomethane (PDAM), respectively, has recently been reported. Other methods for the visualization of hydroxyl groups (26–28) and phenols (29) have also been described. The determination of absolute amounts of aldehyde and ketone groups on polystyrene and TentaGel resins was achieved (30) by derivatization with dansyl hydrazine followed by quantitation of fluorescence in the supernatant. Reaction of free thiols with 5,5′-dithiobis(2-nitrobenzoic acid) (DTNB) gives a characteristic yellow color, which is the basis of the Ellman test (31,32) (Table 1, entry 4). Finally, guanidines can be visualized using the Sakaguchi test (33), which gives a red color with α-naphthol.

Table 1 Principally Employed Color Tests and Chemical Derivatives for Resin Analysis

Entry	Test/color	Species detected	Functionality detected	Ref.
1	The ninhydrin Kaiser test. Blue.	 Ruhemann's Purple Complex	Primary amines (secondary amines to a lesser extent)	16,17
2	The bromophenol blue test. Blue.		Primary and secondary amines	18
3	The TNBS test. Orange.[a]		Primary and secondary amines	19
4	The Ellman test. Yellow.		Free thiols	31,32
5	The DMT and NPIT tests. Orange.		Alcohols, and primary amines (directly). Secondary amines and anilines (indirectly).	38–40
6	The Fmoc quantitation test.		Detection and quantitation amines following cleavage with piperidine	35,36

[a] The chromophore produced in the TNBS test may further react with amines to give an aniline derivative by loss of sulfur dioxide.

III. QUANTIFICATION BY PRODUCT DERIVATIZATION

Generally these methods aid detection by derivatization with a chromophoric group for UV or HPLC detection and are discrete from the chemical encoding of beads, which is not discussed here and has been reviewed elsewhere (15).

Popularized by Atherton and Sheppard (34), the 9-fluorenylmethoxycarbonyl (Fmoc) quantification method is based on the measurement of the UV absorbance at 290 nm (35) or 301 nm (36) of the piperidine-dibenzylfulvene adduct resulting from the deprotection procedure (Table 1, entry 6). The related Bsmoc quantification (37) (benzo[*b*]thiophenesulfone-2-methyloxycarbonyl) has been exploited for peptide synthesis. Similarly, quantification of (dimethoxytrityl) (DMT)-protected amines and alcohols can be achieved by spectrophotometric measurement of the color of the released trityl cation at 498 nm following acid treatment of the resin, a procedure used routinely during on-line monitoring of peptide (38) and oligonucleotide (39) synthesis (Table 1, entry 5). DMT quantification is also used indirectly in the nitrophenyl isothiocyanate-*o*-trityl (NPIT) test (40) for primary and secondary amines as well as anilines (Table 1, entry 5). Finally, *o*-nitrobenzenesulfonyl deprotection of amino acids has also been reported to be a useful color test, because deprotection releases a yellow chromophore (41,42).

IV. MASS SPECTROMETRY TECHNIQUES AND APPROACHES TO PRODUCT IDENTIFICATION

A description of mass spectrometry (MS) techniques applied to the analysis of combinatorial libraries could form a book in itself, and the field has been extensively reviewed (43–48). The principal methods used to analyze combinatorial libraries are briefly outlined here.

Electrospray ionization (ESI) (49,50) and matrix-assisted laser desorption ionization (MALDI) (51,52) techniques for MS have been available as routine methods for the detection of small quantities of material for a number of years. These techniques have proved to be very suitable for detection of products cleaved from solid supports. Such soft ionization methods together with the Fourier transform ion cyclotron resonance (FTICR) analyzer technique (53,54) have revolutionized mass spectral analysis of sensitive compounds such as peptides. It is now generally accepted that the default analytical technique for the vast majority of combinatorial libraries is mass spectrometry or a combination of mass spectrometry and chromatography (44).

A. Electrospray Ionization

Electrospray ionization coupled with mass spectrometry (ESI-MS) remains the most versatile tool for high throughput sample analysis and is the most suitable for combination with separation techniques such as liquid chromatography (LC) and capillary electrophoresis (CE). Early studies focused on the analysis of mixtures of peptides (55–58) and oligonucleotides (59), largely using the data qualitatively as a "fingerprint" for the library. There have been a number of recent reports on the use of ESI-MS for characterization of nonpeptide libraries. For example, Carell and coworkers (60,61) reported analysis of a crude library of 55 xanthene derivatives, with a success rate of 93% in identifying the products using the technique. Importantly, both positive and negative ion scanning as well as tandem experiments were used, giving enhanced structural information on the products.

For the analysis of small-molecule combinatorial libraries, ESI-MS offers many advantages over other mass spectrometric methods. These include soft ionization (reducing fragmentation), rapid analysis, the ability to analyze low molecular weight compounds, automation, and compatibility with separation techniques. Although ESI-MS lends itself to high throughput analysis of combinatorial libraries, the ability to be able to effectively, quickly, and efficiently assess the huge amount of data generated has become a major issue. As high throughput ESI-MS has begun to be implemented in the pharmaceutical industry (62), a number of systems designed to streamline the data analyses have been developed. For example, Hegy et al. (63) reported a method that included computerized techniques such as file-formatted sample input information, automated analysis, data interpretation and processing, and final data reporting. Throughput has been further increased by shortening MS run time while minimizing sample carryover (64) and by developing systems allowing parallel sample processing. Wang et al. (65) described an ESI-coupled multiple-probe autosampler system dedicated to simultaneous injection and analysis of eight samples, thus enabling a 96-well microtiter plate to be processed in 5 min ("ultrahigh throughput"). In addition, Fourier transform ion cyclotron resonance (FTICR) is a new MS technique using ES ionization that shows promise for analysis of combinatorial libraries, allowing exact mass determination and library characterization (66,67). ESI-TOF (QTOF) (time-of-flight) instruments also afford a promising approach to accurate mass determination and are beginning to be used within the industry for library product characterization.

B. Matrix-Assisted Laser Desorption Ionization

Matrix-assisted laser desorption ionization mass spectrometry (MALDI-MS) is a powerful and sensitive technique for the ionization of large and nonvolatile molecules and is particularly useful in the analysis of low concentrations of material, especially peptides and oligonucleotides, that may be detected in the picomole to femtomole concentration range (68–71). The technique is almost always coupled to the TOF analysis method. MALDI-TOF is not generally a high throughput technique, often requiring careful manual intervention to align samples with the integral laser of the instrument. MALDI is much more tolerant than ESI of the high salt concentrations used in biological systems, enabling direct sampling without purification. These techniques have been applied to the field of combinatorial chemistry, including single-bead analysis (72,73); the analysis of peptide libraries (74–76), of analogs of lysobactin (77), and of oligonucleotides (78–81); and even high throughput screening applications (82). In addition, MALDI often does not require chemical cleavage of the compounds from the bead. For example, Geysen and coworkers (83) described an approach using photocleavable linkers that were directly irradiated during MALDI analysis to effect compound cleavage. MALDI coupled with Fourier transform mass spectrometry (MALDI-FTMS) (84), in which accurate mass determination and MS/MS fragmentation to aid structure identification are both possible, has also been applied to library characterization.

C. Time-of-Flight Secondary Ion Mass Spectroscopy

Time-of-flight secondary ion mass spectroscopy (TOF-SIMS) (85) has been used to monitor organic reactions on solid supports. This technique presents several advantages over ESI-MS and MALDI-MS, the most important being that accurate and sensitive mass measurements can be achieved in a spatially resolved format. Additionally, the weak intensity of the incident primary ions in static SIMS means that the bombarded surface per scan is small (100 μm × 100 μm). This can facilitate the use of a single bead for nondestructive MS analysis, which is not possible with MALDI-MS or ESI-MS (86). However, the technique does not yet offer the high throughput possible with ESI methods. The sensitivity of this technique, which is in the low femtomole range, has allowed the use of TOF-SIMS to analyze amino acid residues on polystyrene resin using the vapor-phase clipping technique (87) from the Sasrin linker (88). Similarly, TOF-SIMS has been applied to Fmoc-amino acid peptide synthesis on polystyrene resins with direct irradiation of the beads (89–92). TOF-SIMS has also been applied to the analysis of substrates bound to polypropylene crowns (86).

D. Single-Bead Sensitivity

The ability to synthesize large pooled libraries by split-and-mix or similar techniques requires analytical procedures that are sensitive enough to be able to detect

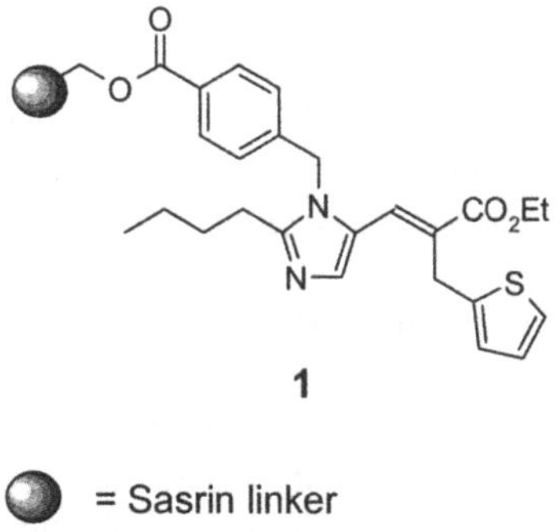

1

= Sasrin linker

Figure 1 Angiotensin II antagonist model compound.

material from single beads of resin. By exploiting high-loading beads prepared using dendrimer chemistry, Bradley and coworkers (93) were able to fully characterize, by NMR and ESI-MS, small peptide fragments released from a single bead. However, the quantity of material released from standard beads can be subnanomole, restricting the analytical techniques available. The use of MALDI, as described in the above examples, has proved efficient in the identification of compounds released from single beads; however, the work in this area has largely focused on the analysis of peptides and oligonucleotides. The analysis of small druglike molecules released from single beads is significantly more difficult, owing to variability in the physical properties between the structurally diverse members of libraries. To address this issue, Benkovic and coworkers (94) performed a direct comparison of ESI, MALDI, and TOF-SIMS, demonstrating the use of all three methods for single-bead analysis. As a model compound, a resin-bound angiotensin II antagonist **1** (Fig. 1) was chosen and synthesized using a Knoevenagel condensation approach on the Sasrin linker. For each of the three methods, fragments were studied to resolve possible structural ambiguities. It was shown that all three MS techniques possess appropriate sensitivity for compounds prepared combinatorially even from a single 40 μm diameter bead. It is important to note, however, that there is still a lack of techniques that facilitate the high throughput analysis of material released from single standard beads.

E. Combined Mass Spectrum and Quantitative Detection Methods

An additional dimension to library analysis is introduced through the coupling of mass spectroscopy with separation methods such as liquid or gas chromatography or capillary electrophoresis (CE). In the absence of a second and quantitative detection technique, misinterpretation of MS results due to variable ionization properties of expected products and by-products is common. However, combined use of chromatographic and MS systems is suitable only for spray ionization techniques such as ESI or atmospheric pressure chemical ionization (APCI) and essentially excludes the use of MALDI-MS.

Gas chromatography coupled with mass spectrometry (GC/MS) has been successfully used for quality control of a β-mercaptoketone library (95), but this method is restricted to molecules that are sufficiently volatile. A number of recent reports have described the use of various techniques for purity assessment and product and by-product identification (96–102), and in addition the combination of high performance liquid chromatography and mass spectrometry (HPLC/MS) has been reviewed (103). An example from Jung and coworkers (104) successfully describes characterization of two aryl ether libraries by LC/ESI-MS, showing the power of a tandem approach. Use of LC/MS, however, for multicomponent mixtures has severe limitations as the number of components increases. In contrast, CE/ESI-MS and MS/MS have been applied to a library mixture of 171 compounds, allowing identification of 160 (94%) of the components (105).

Considerable attention has been paid to evaporative light scattering detection (ELSD) as a more universal alternative to UV detection for HPLC (106). Although initially reported to have a nearly linear mass-dependent response, irrespective of chemical composition (107), it has become clear that the ELSD response can vary considerably between analogs of similar structure, thus limiting its usefulness for quantitative assessment of purity and yield (44). A new development for quantitative detection is the chemiluminescent nitrogen detector (CLND) for HPLC (108). This technique is discussed in more detail elsewhere in this book.

V. CHEMICAL APPROACHES TO AID COMPOUND IDENTIFICATION BY MASS SPECTROMETRY

Despite the many successful applications of the techniques described in this chapter for product detection and library characterization, there are serious limitations in the use of MS for small molecules. Much of the published work focuses on peptide or nucleotide molecules with well-understood properties and reliable fragmentation. When synthesizing new classes of small molecules with diverse substituents, the ease of detection of the library members cannot be ensured. In particular, many small molecules do not possess good ionization properties, and this problem is exacerbated when substrates are cleaved from a single bead because of the very small amounts of materials produced. These difficulties make the task of assessing the degree of success of the library chemistry somewhat subjective, even if extensive monomer rehearsals have been carried out prior to synthesis. The following sections describe approaches to address these issues either by chemical derivatization of the products or solid supports to aid detection and analysis of the materials synthesized.

A. Mass Ladders

Youngquist and coworkers (109) extended a method, earlier reported by Kent and coworkers (110), for rapidly sequencing peptides isolated from support-bound combinatorial libraries as an alternative to detection of the product or to solid-phase library tagging techniques for identification of the compounds. Their strategy allowed preservation of sequence information available during the peptide synthesis and the use of this information to determine the sequence of the final peptidic product. During the synthesis of pentapeptide libraries using either seven or 16 amino acids (7^5 or 16^5 compounds), anchored to the resin through a five amino acid termination sequence, 10% of the growing peptide chains were capped at each step with N-acetyl-D,L-alanine in competition with the next amino acid (90%), generating a series of sequence-specific termination products on each bead. The linking sequence included a methionine residue, allowing cleavage with cyanogen bromide to give terminal homoserine lactones, and an arginine group, which owing to the presence of a fixed positive charge, ensured sensitivity and a reasonably uniform MS response for each member of the termination synthesis family. Three proline residues were also included to increase the molecular mass of all library fragments beyond 500 Da, ensuring that each unit had a mass greater than the "chemical noise" produced by desorption of the UV-absorbing matrix. The beads were screened and active beads were collected and cleaved with cyanogen bromide, and the solutions were analyzed by MALDI-TOF. Using this approach, it was possible to sequence all of the active peptides at a rate of 25–30 residues per hour, and only three of the 80 active beads from the screened libraries could not be sequenced. In addition, the identification of deletions in the sequence as well as by-products was achieved.

A number of related approaches have been described. Rinnova et al. (111) used 2,4-dinitrofluorobenzene to end-cap unreacted resin-bound peptides after each coupling step of a peptide synthesis. The dinitroaniline-substituted truncated peptide by-products were then easily identified and removed after cleavage from the solid support, making it possible to detect where deficiencies in the synthesis arose. Burgess et al. (112) exploited a "mass ladder" using a photolabile linking group. This had the advantage that it was possible to analytically cleave the peptides using much milder conditions, thus extending the versatility of the approach. Vasseur and coworkers (113) further extended the generality of the technique to include oligonucleotide synthesis.

B. Self-Coded Libraries

Hughes (114) described another approach to library design that avoids deconvolution and tagging strategies by monomer selection that avoids MS degeneracy in the library products. The approach allows "split–mix" synthesis without recourse to high resolution accurate mass determination, because all library products have unique nominal masses. These "self-coded" libraries can have between two and four substitutents and can be up to several thousand members in size. This strategy relies on the products having adequate ionization properties for their routine detection and places restrictions on the choice of monomers that can be used. Stankova et al. (115) earlier reported a similar approach in which library hits from single beads were identified by MS and MS/MS analysis. Although the approach was successfully used to identify nonpeptidic inhibitors of streptavidin, very careful data interpretation and time-consuming resynthesis of the compounds to confirm activity were required.

C. Isotopic Labeling

Geysen and coworkers (116) introduced a technique they called "ratio encoding" in which mixtures of isotopically labeled building blocks are incorporated as part of a library synthesis. Peptoid and imidazole libraries were synthesized with various ratios of stable isotopes incorporated into the monomers. For the peptide synthesis ^{13}C-labeled and unlabeled α-bromoacetic acid were used. Ten ratios were employed between 9:91 and 90:10 to encode the first monomer, the second position was encoded by using a unique mass, and the third position was known from the location of the final synthesized pool (spatially encoded). The method allowed identification of the library products from individual beads from the pools by MS with 95% confidence.

A very recent related approach from Nazarpack-Kandlousy et al. (117) introduces the technique of "regiochemical tagging" in which isotopically labeled library building blocks are used to differentiate between what would otherwise be isobaric library products. Using QqTOF MS (single MS and tandem MS/MS), 27 isotopically labeled isomeric oxime ether regioisomers could be distinguished by their accurate mass and fragmentation signatures.

The use of mixtures of isotopes in this way as a signature for compounds derived from the chemistry, rather than extraneous impurities or background noise, has been incorporated as part of other detection strategies described in the following sections.

D. The Dual Linker Approach to Product Identification

Tam et al. (118) first described the concept of using two linkers in series for solid-phase synthesis of peptides in which each linker could be cleaved under

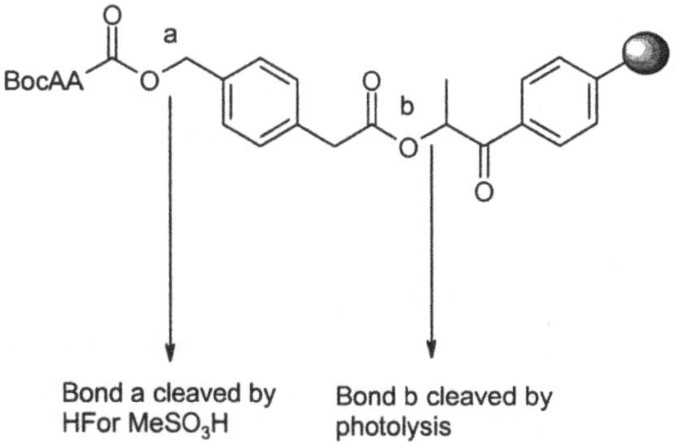

Bond a cleaved by   Bond b cleaved by
HF or MeSO₃H   photolysis

Figure 2 Boc-aminacyl-o-CH₂-Pop-resin.

orthogonal conditions. Figure 2 shows one such system, the Boc-aminacyl-O-CH₂-Pop-resin. Bond a could be cleaved with strong acid (HF or trifluoromethyl-sulfonic acid) without cleavage of bond b to release the peptide. Bond b could be cleaved by photolysis to release the side-chain-protected peptide substrate still substituted with 4-methylphenylacetic acid. Both bonds a and b could be cleaved by nucleophiles. The "multidetachable" resin supports were developed as a flexible approach to solid-phase chemistry and, for example, allowed peptide fragments to be built up, cleaved from the support (via bond b), and purified before reattachment to a base resin for completion of the synthesis. There have been a number of other reports of multidetachable resins for similar applications (119–132). The systems described, however, were not used to aid analysis of the chemical reactions, and it was not until recently that the potential of dual linker systems for this application was realized.

Analytical dual linker approaches were first explored for combinatorial libraries as an approach to encoding of resin bead samples (15). In one-compound-per-bead solid-phase synthesis involving an encoding strategy, the active sites on the resin are differentiated to allow a small proportion of tags, which are coded for the building blocks used in the synthesis, to be incorporated onto the support. After conventional cleavage and biological evaluation of the desired library product, the tag code is removed from the bead and then "read." Because the relationship between the cleaved material and the bead is retained, the structure of the library product can be deduced.

The first dual linker approach to split–mix library encoding was developed by Geysen et al. (133), who relied on the mass differentiation of tags to overcome some of the limitations of conventional encoding strategies. Conventional techniques rely on the use of orthogonal chemistry to incorporate tags in separate synthesis steps, often complicating the synthesis of combinatorial libraries. By using a mass encoding strategy based on the introduction of isotopically labeled amino acids onto the resin, Geysen et al. postulated that due to the high throughput and sensitivity of MS techniques it would be possible to read all of the tags in a library and thus gain valuable SAR information. A number of encoding strategies were proposed (Fig. 3).

In single-peak positional encoding (Fig. 3a), after library synthesis, linker 2 was cleaved to release the compound A—B—C for biological screening. If the compound tested positive, linker 1 was cleaved and the code analyzed by MS. Up to 10 different codes, made up of multiple variants of unlabeled and ^{13}C-labeled glycine and alanine, were designed to appear in a convenient part of the mass spectrum. The codes were used to encode the first position (A) of the synthesis. This approach was further developed by incorporating a second reference mass (Fig. 3b). The encoded information is now determined by the difference in mass of the coding unit and the reference mass unit. In a similar way a mixture of tags can be incorporated together to produce a "bar code" unequivocally determining the structure of the compound produced (Fig. 3c). The concept of "peak splitting" was also described in which tags that were labeled with 1:1 ratios of stable isotopes were incorporated into the code (Fig. 3d), giving rise to a clear ion-pair fingerprint in the mass spectrum and allowing faster identification of the fragments. The approaches outlined above rely on reliable detection from single beads by MS, and Geysen et al. overcame this issue by the introduction of a basic amine (lysine) into the coding region to ensure ionization. This pivotal

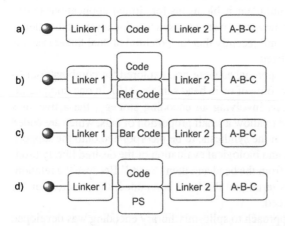

Figure 3 Isotope or mass encoding strategies. (a) Single-peak positional encoding; (b) double-peak positional encoding; (c) the bar code uses two or more isotopically mixed components; (d) peak splitting (PS) to aid fragment identification. (Adapted from Ref. 133, with permission.)

paper, therefore, outlined many of the concepts that were embraced in the development of dual linker analytical construct systems that have been developed subsequently and are described below.

E. Dual Linker Analytical Constructs

The mass spectroscopic techniques outlined earlier have been very useful for the detection and characterization of solid-phase chemistry products, especially peptidic compounds, even from single beads of resin. However, there can be difficulties in the interpretation of the results when side reaction lead to mixtures of products. Although this is not an issue in cases where confirmation of the presence or absence of the desired product is required, it becomes very important during reaction development, when information on side reactions and by-products is crucial to the optimization of the chemistry. Furthermore, as discussed earlier, many of the nonpeptidic solid-phase chemistries now being exploited in library synthesis afford diverse products that do not necessarily have common or reliable ionization characteristics for confident mass spectral analysis nor do they possess uniform UV extinction coefficients aiding detection. This can limit the applicability of MS- and LC/MS-based analytical systems for analysis (134).

Geysen and coworkers have sought to address these issues by building on the encoding strategies described earlier. They proposed the concept of analytical constructs (135–135b), in which the solid-supported substrate and linker is anchored to the resin through an analytically enhancing group and a second orthogonally cleaved linker (Fig. 3d). The analytical enhancer contained a mass spectroscopic "sensitizer" (110,136,137) (t-Boc-ε-lysine) ensuring uniform ionization of all products in the mass spectrum and a "peak splitter" based on mixtures of ^{13}C-labeled and unlabeled glycine residues, to give a distinctive signal in the mass spectrum for all materials previously bound to the resin. Cleavage at linker 1 (Rink amide) also deprotected the t-Boc-ε-lysine residue to produce a charged amine that served as the MS sensitizer. The construct system, therefore, allowed either cleavage of the products of the chemistry being carried out attached to the analytical enhancer for detection and analysis by MS or conventional cleavage at the orthogonal linker for isolation and biological testing.

Independently, Carrasco et al. (138) reported a similar approach using an orthogonal double linker with an incorporated MS sensitizing group for direct analysis by MALDI-MS (Fig. 4). A photocleavable linker was used to attach the ionization tag sequence to the resin support. A chemically cleavable linker was then added to allow both synthesis and release of the target molecules. The combination of the photocleavable linker and the ionization tag enabled the direct analysis of the resin beads by MALDI-MS without a precleavage step. The chemically labile Rink linker again permitted the orthogonal recovery of the desired substrates prepared on the solid support. The approach was applied successfully to

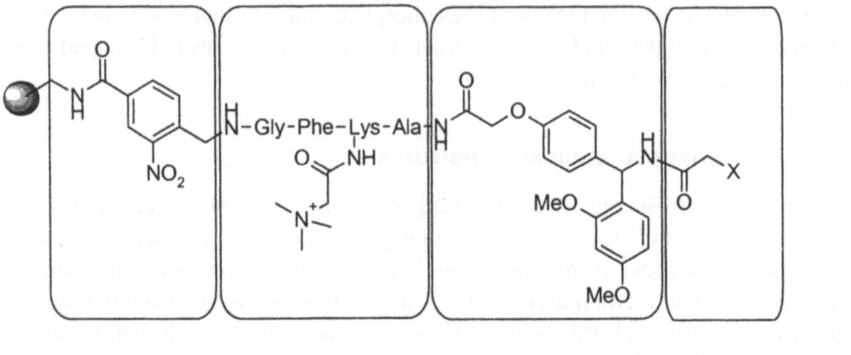

| Photocleavable | Ionisation | Chemically | Desired substrate |
| linker | sequence | cleavable linker | |

Figure 4 Carrasco and Kent's dual linker analytical construct. X = Br or CN. (Adapted from Ref. 138, with permission.)

the displacement of resin-bound bromoacetamide with potassium cyanide. A possible limitation of the construct described was the presence of the quaternary ammonium group, which may limit the general application of the methodology.

McKeown et al. (139) further developed Geysen et al.'s proposal by designing dual linker analytical constructs based on a photolabile carbamate linker and orthogonal acid-cleavable linkers. The system had the advantage that photolysis released the MS sensitizing group directly, but the amine functionality was blocked from chemical reactivity during library chemistry prior to cleavage from the resin. Again, incorporation of a peak splitter (1:1 $^{14}N_2/^{15}N_2$) ensured that the MS peaks were "hallmarked" as characteristic doublets, allowing facile interpretation of the results (Fig. 5).

The power and utility of this approach were demonstrated by preparing N-acylated dipeptides, which contained no strongly chromophoric groups, on the analytical constructs. The dipeptides were cleaved both analytically by photolysis and conventionally from the orthogonal Rink linker by acidolysis. A comparison of the resultant mass spectra was then made (Fig. 6). The spectra obtained for the compounds generated by conventional synthesis (Fig. 6a) were largely uninformative. These compounds, despite their excellent purity (NMR, >95%), had poor ionization properties, and it was difficult to distinguish the molecular ions from background noise and extraneous peaks. In contrast, in the spectrum obtained from the molecules attached to the analytical enhancer, the molecular ions were readily identified, the sensitizing group was highly effective, and the isotopic peak splitter distinguished all construct-derived signals as doublets. The sensi-

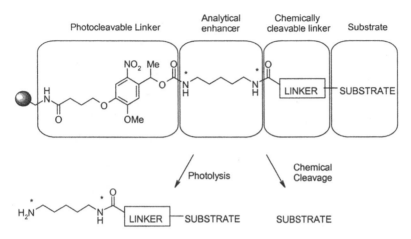

Figure 5 Analytical construct with peak splitter and MS sensitizer. * = 50% labeled with 2 × ^{15}N. Resin = ArgoGel. (Adapted from Ref. 139, with permission.)

tivity obtained by including an amine in the construct fragment is evident from Figure 6b, which represents material cleaved from a single bead.

Extending this approach, analytical construct methodology has been successfully applied to single-bead product detection for a library of products (140). Using the pyrimidine safety catch linker (141) based construct in combination with an acid-cleavable second linker, amide products were built up by split–mix–pool library synthesis (Fig. 7). The synthesis of a 4^3 library in radiofrequency-encoded IRORI Kans (microreactors) (142,143) gave 64 products that could be analyzed by cleavage under acidic conditions to release the library products (route B, Fig. 7), or by construct cleavage by oxidation and displacement with *N*-methylpiperazine (route A, Fig. 7). The use of this amine introduces the amine sensitizer group. Comparison of the ESI-MS data collected from single beads for the 64 products indicated by analysis of the construct fragments that all of the products had been successfully synthesized. However, only 25 of the library products could be unambiguously identified after cleavage from the acid linker from a single bead. A further 11 compounds appeared to be present but with high background levels of noise giving uncertainty of the quality of the products, and 28 compounds could not be detected. An example of comparison of data generated by cleavage at both the first and second linkers is given in Figure 8. The peak split construct peak clearly indicates the presence of the expected product, whereas cleavage at the second linker does not give the expected ion. Cleavage of all of the resin samples (25 mg from each Kan) and ^{1}H NMR

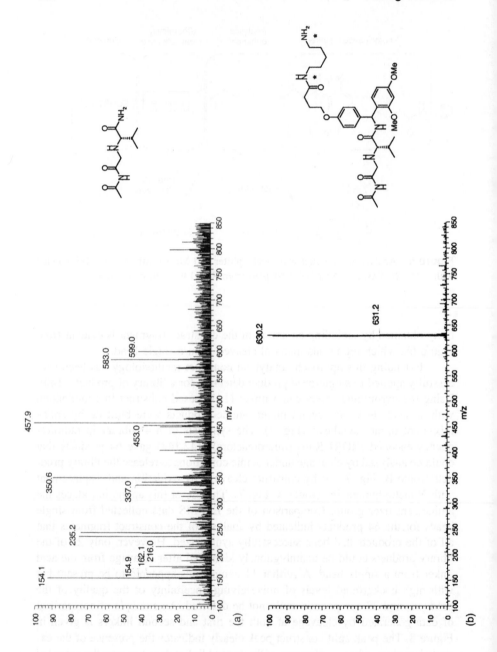

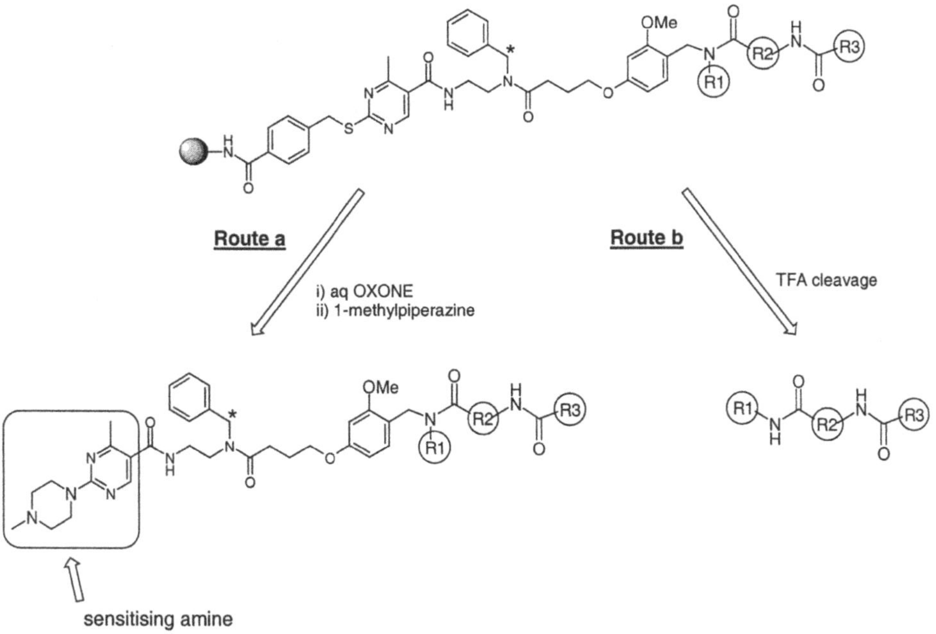

Figure 7 Route a: Analytical fragment release. Route b: Classical cleavage for biological screening. (Adapted from Ref. 140.)

analysis of the library products confirmed that each compound had been successfully synthesized. These data indicate that for these small-molecule products, single-bead cleavage and analysis by ESI-MS is not the optimal technique and that the use of an analytical construct gave significantly more reliable results for detection of the library products.

Murray et al. (144) applied an analytical construct approach to the evaluation of linker stability. Both the 1-(4,4-Dimethyl-2,6-dioxocyclohexylidene)ethyl (Dde) (145) and the 2-nitrobenzenesulfonamide (146) linkers were incorporated "downstream" of the analytical fragment as linker 2 (Scheme 1). Initially, acidolysis of resins **2** and **3** released the capped analytical fragments **4** and **5**, respectively, which showed strong signals for the expected ion pairs in the mass spectra.

Figure 6 Electrospray MS of Ac-Gly-Val prepared on the analytical construct. (a) Conventional synthesis and cleavage ($MH^+ = 216$). (b) With construct resin and photolytic cleavage ($MH^+ = 628/630$). This was the spectrum obtained from a single bead. (Adapted from Ref. 139, with permission.)

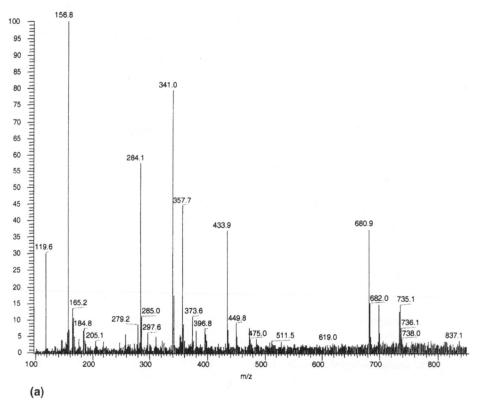

Figure 8 Electrospray MS of the library component corresponding to monomer combination BFK, (a) using conventional cleavage (MH$^+$ = 341) and (b) using analytical construct fragment release (MH$^+$ = 969.5). (Adapted from Ref. 140.)

Incubation of bead aliquots to a range of commonly used reagents was performed in parallel, followed by subsequent acid cleavage and analysis by LC/MS. The results of these studies (Table 2) showed that the Dde and sulfonamide groups possess a high degree of orthogonality with conventional, acid-cleavable resin linkers, but, as might be expected, the Dde group was attacked by sodium borohydride whereas tin(II) chloride reduced the aromatic nitro group in the sulfonamide. The sulfonamide linker proved to be quite labile under strongly basic conditions and, to a lesser degree, toward fluoride anion, suggesting possible alternative cleavage conditions for the removal of Fukuyama et al.'s sulfonamide protective group (147). The data demonstrated the power of the construct technology employed in both solid-phase reaction analysis and evaluation of linker stability.

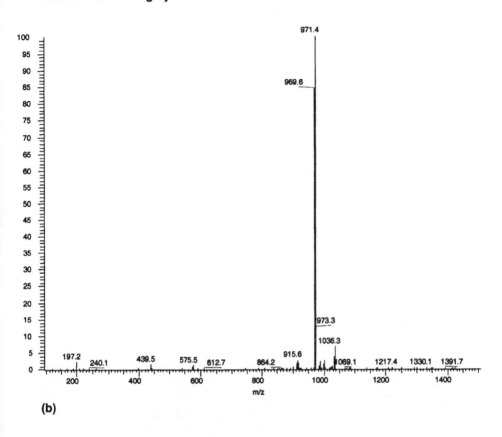

(b)

Incorporation of a UV chromophore into resin-bound products specifically to aid reaction monitoring is an attractive concept. For example, Hartwig and coworkers (148) investigated conditions for Heck cross-coupling reactions by coupling a series of aryl halides on Wang resin to acrylates tethered by an alkyl chain to 4-methyl-7-hydroxycoumarin **6** (Scheme 2). Fluorescence analysis of the beads gave the quantity of product **7** present on the resin, thus giving a measure of reaction success. This technique was then used to rapidly scan a large number of ligands under different conditions in a high throughput format, allowing identification of the optimum ligand for this transformation. A further enhanced analytical construct that additionally incorporated a UV chromophore to similarly enable quantification of the products formed on solid support has recently been described (149). In addition to the MS splitter and sensitizer groups, a chromophoric label was also introduced (Fig. 9). For useful quantification it was recognized that the absorption maxima of the UV spectrum must occur in a region generally free

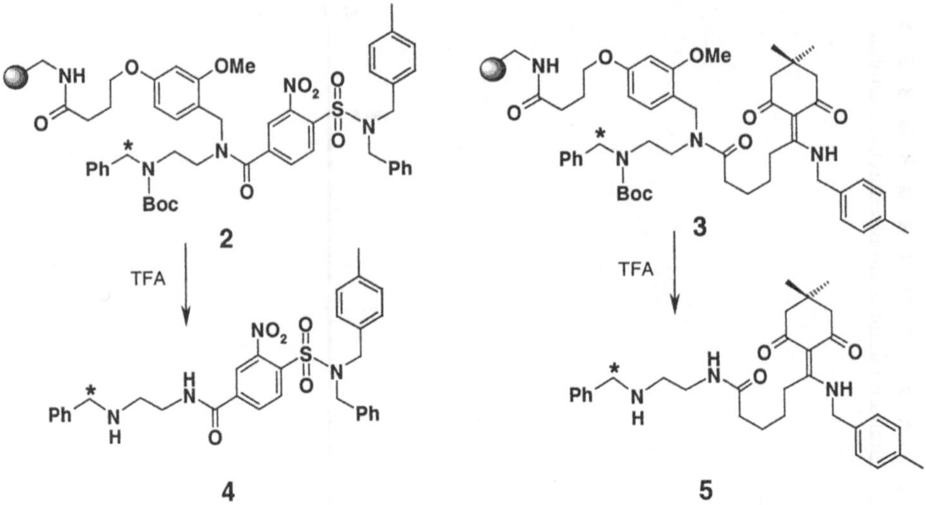

Scheme 1 Analytical constructs used for reaction scanning. (Adapted from Ref. 144, with permission.)

from absorbances due to other species and be strong enough to enable detection of material released from a single resin bead. The anthracene moiety, possessing distinctive and strong absorbances at 254 nm (ε 180,000, CH_3CN) and at 386 nm (ε 9000, CH_3CN) was selected. The system provided a means whereby components present in a product mixture could be identified by MS but also be quantified quickly and accurately by measurement at the key UV wavelengths.

The effectiveness of the chromophore as a tool for quantification was determined by studying products synthesized as a mixture. A construct-containing resin with the Rink linker at the second position was functionalized with a mixture of three acids, and the ratio of the three components was determined by cleaving a resin sample at linker 1 to release a mixture of **8a**, **8b**, and **8c** and measuring the peak area due to each component by HPLC at 254 nm and 386 nm (Fig. 10). It was also established that the chromophore was suitable for detection of the mixture of products from a single bead of resin. Cleavage at linker 2 using TFA released a mixture of the three corresponding primary amides, and comparison of ¹H NMR with HPLC data showed that the anthracene-containing construct could be used to accurately determine relative quantities of materials in the product mixture.

The practical utility of this system was then demonstrated in the synthesis of the fibrinogen receptor antagonist **13** (Scheme 3) (150). The target was chosen because the compound in itself lacks a diagnostic chromophore. An acid-labile

Table 2 Reaction Scans for the Sulfonamide and Dde Linkers in Analytical Construct Resins 2 and 3

Entry	Reaction conditions[a]	4	5	Entry	Reaction conditions[a]	4	5
1	Mercaptoethanol, DBU	×	✓	12	PhCH$_2$Br, Cs$_2$CO$_3$, DMF	✓	✓
2	20% Piperidine, DMF	✓	✓	13	NaBH$_4$, THF	✓	×
3	2% Hydrazine, DMF	✓	×	14	PhCHO, Na(OAc)$_3$BH, 1%	✓	✓
4	Phenyltetrazole, KOtBu	✓	✓	15	SnCl$_2$·2H$_2$O, DMF	×	✓
5	10% MeOH, THF	×	✓	16	m-CPBA, CH$_2$Cl$_2$	×	×
6	NaOMe, MeOH, THF	✓	✓	17	H$_2$O$_2$ (10%), 30 vol, THF	✓	✓
7	Ph$_3$ P=CH(C=O)Me, THF	✓	✓	18	PPh$_3$, PhCH$_2$OH, DEAD, CH$_2$ Cl$_2$	✓	✓
8	PyBOP, HOBt, DIPEA	✓	✓	19	Pd(PPh$_3$)$_4$, HOBt, THF	✓	✓
9	Ac$_2$O, DIPEA, CH$_2$Cl$_2$	✓	✓	20	KOTMS, THF	×	✓
10	PhCOCl, pyridine	✓	✓	21	TBAF (1 M), THF	×	✓
11	MeSO$_2$Cl, DIPEA	✓	✓	22[b]	hv, DMSO	✓	✓

[a]Resin aliquots, ~20 mg, were incubated with reagents (5 equiv) in a HiTop Block (Whatman, Catalogue No PolyWhat 004) at room temperature for 3 h and then cleaved with 50% TFA, CH$_2$Cl$_2$. LC/MS analyses were performed on a Hewlett-Packard HP 1050 instrument (diode array detection) and a Micromass Platform I (8084) mass spectrometer using electrospray ionization in +ve mode; ✓ 85% **4** or **5** by peak area.

[b]Conducted as a separate experiment: 20 mg of resin in 0.5 mL solvent was irradiated at 365 nM for 3 h.

Source: Adapted from Ref. 144, with permission.

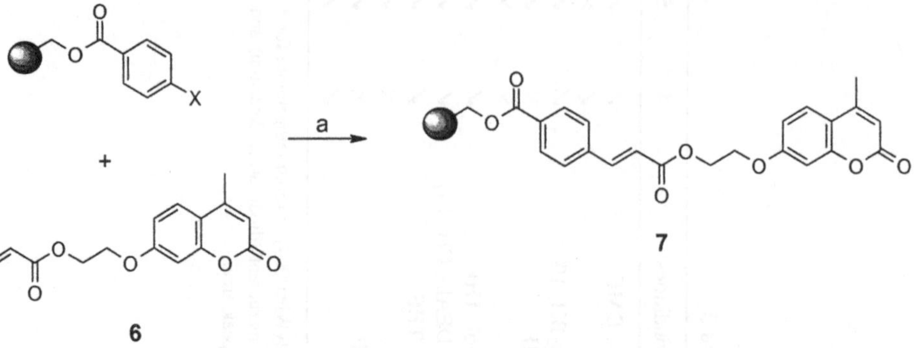

Scheme 2 Reagents and conditions. Catalyst, ligand, DMF, NaOAc, 100°C.

modified Sasrin linker was introduced as linker 2, affording analytical construct **9**. The synthesis was carried out using peptide coupling and deprotection steps, with each step being monitored by cleaving a sample of each resin at linker 1 and analyzing the analytical fragments produced by LC/MS. The chromatograms are reproduced in Figure 11 together with those of parent ions in the mass spectrum. Using the construct it was possible to easily monitor each chemical step.

The ability to accurately quantify products arising from solid-phase synthesis opens a number of new avenues for research. In particular, the quantification of reaction outcomes using only small resin aliquots allows the study of the reac-

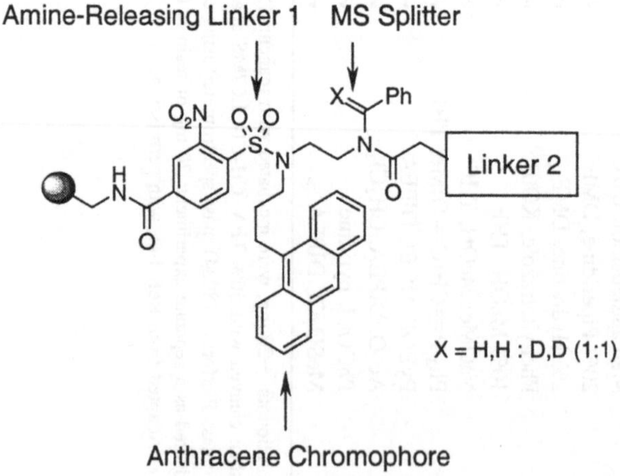

Figure 9 Ultraviolet chromophore containing analytical construct design.

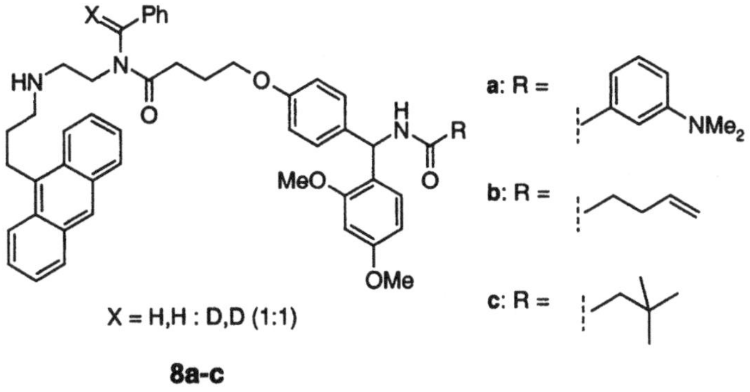

8a-c

Figure 10 Released analytical fragments from a product mixture by cleavage of linker 1. (Adapted from Ref. 149, with permission.)

tion kinetics of immobilized species without appreciably perturbing the system being studied. As an example, the kinetics of the reactions of the 1-hydroxybenzotriazole linker immobilized on a chlorotrityl-based analytical construct (151) were investigated. The analytical construct **14** was chosen to investigate the chemistry. It contained an isotopic mixture of the 9-anthrylpropyl group to give both a clear fingerprint in the mass spectrum and UV quantification as outlined above (Fig. 12). Analytical cleavage of the trityl linker (linker 1) revealed a carboxylic acid

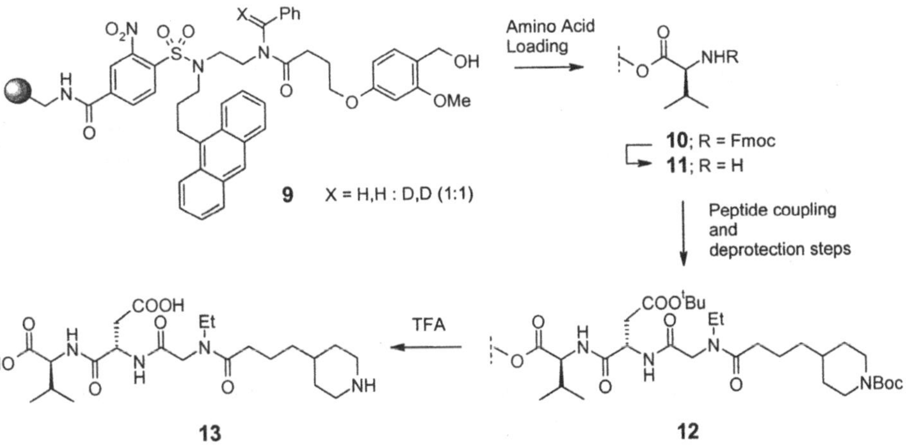

Scheme 3 Synthesis of an analytically problematical molecule using an analytical construct to monitor the chemical steps. (Adapted from Ref. 149, with permission.)

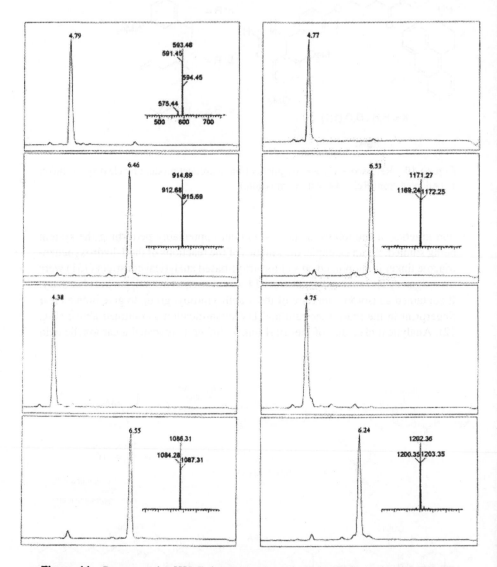

Figure 11 Representative HPLC chromatograms recorded at 386 nm of each intermediate analytical fragment with selected MS ions. (Adapted from Ref. 149, with permission.)

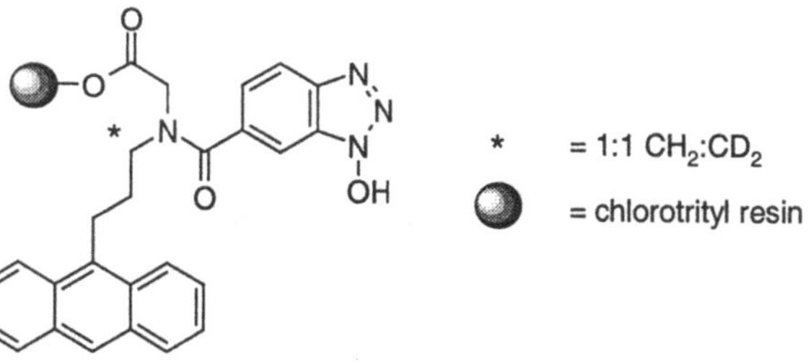

14

Figure 12 Analytical construct employed for kinetic studies, incorporating a 1-hydroxy-benzotriazole linker.

that sensitized the mass spectrum in the ES ionization mode. Automated sampling, cleavage, and HPLC analysis at 386 nm of the loading of a substituted chloropurine (152) onto **14** showed that the reaction was 95% complete after 8 h at 25°C. Further reaction time did not alter the composition of the reaction mixture. By plotting peak areas of starting material versus those of product, it was possible to generate a time-course graph for the reaction (Fig. 13a). In contrast, the reaction of **15** with 2% benzylamine in dichloromethane proceeded smoothly to completion in 10 h (Fig. 13b). The use of analytical constructs for these studies simplified experimental procedures, because it was found that resin washing was not required prior to analytical cleavage. The full potential of this technique would become evident in the study of reactions in which an intermediate or a number of different products are formed, and this is the subject of current work.

VII. SUMMARY

Analytical techniques, until recently, did not keep pace with chemical developments in solid-phase synthesis. It is suggested here that there is an increasingly wide choice of methodologies available to the combinatorial chemist for rapid and high throughput analysis of reactions occurring on resins. In addition to the IR microscopy and MAS NMR techniques dealt with elsewhere in this book and the increasing number of qualitative color tests for functional group identification, mass spectroscopic techniques, especially those employing analytical constructs as analytical enhancers, enable high throughput analysis yielding high quality, information-rich spectra. The bottleneck for analysis in the future appears to be moving from the analytical technique itself to the processing and critical appraisal

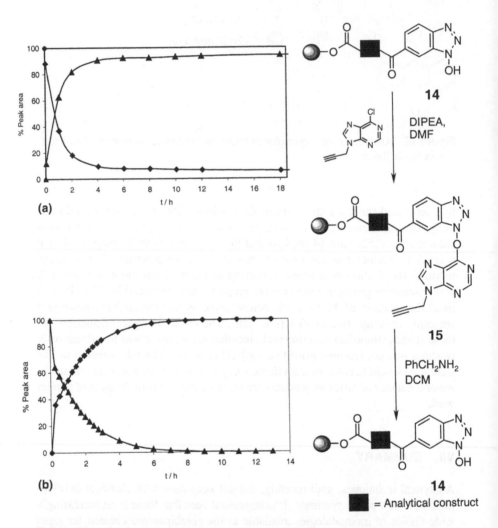

Figure 13 Kinetic data obtained by automated cleavage of the analytical construct showing relative peak areas at 386 nm of (a) loading of the chloropurine onto the 1-hydroxybenzotriazole construct; (b) the displacement of the purine with 2% benzylamine in dichloromethane. Both reactions were performed at 25°C.

of the data generated. Approaches to automated data interpretation for the quality control of synthesized products and experimental design for reaction optimization (153) will be the next steps to impact on this field of chemistry.

REFERENCES

1. S Booth, PHH Hermkens, HCJ Ottenheijm, DC Rees. Solid phase reactions III. A review of the literature Nov 96–Dec 97. Tetrahedron 54:15385–15443, 1998.
2. SE Hall. Recent advances in solid phase synthesis. Mol Diversity 4:131–142, 1999.
3. M Lebl. Parallel personal comments on "classical" papers in combinatorial chemistry. J Comb Chem 1:3–24, 1999.
4. MA Marx, A-L Grillot, CT Louer, KA Beaver, PA Bartlett. Synthetic design for combinatorial chemistry. Solution and polymer-supported synthesis of polycyclic lactams by intramolecular cyclization of azomethineylides. J Am Chem Soc 119: 6153–6167, 1997.
5. P Wipf, A Cunningham. A solid phase protocol of the Biginelli dihydropyrimidine synthesis suitable for combinatorial chemistry. Tetrahedron Lett 36:7819–7822, 1995.
6. B Yan, H-U Gremlich, S Moss, GM Coppola, Q Sun, L Liu. A comparison of various FTIR and FT Raman methods: Applications in the reaction optimization stage of combinatorial chemistry. J Comb Chem 1:46–54, 1999.
7. WL Fitch. Analytical methods for the quality control of combinatorial libraries. Annu Rep Comb Chem Mol Diversity 1:59–68, 1997.
8. ME Szwartz. Analytical Techniques in Combinatorial Chemistry. New York: Marcel Dekker, 2000.
9. RA Houghten, C Pinilla, JR Appel, SE Blondelle, CT Dooley, J Eichler, A Nefzi, JM Ostresh. Mixture-based synthetic combinatorial libraries. J Med Chem 42: 3743–3778, 1999.
10. A Furka, F Sebestyen, M Asgedom, G Dibo. General method for rapid synthesis of multicomponent peptide mixtures. Int J Peptide Protein Res 37:487–493, 1991.
11. MHJ Ohlmeyer, RN Swanson, LW Dillard, JC Reader, G Assouline, R Kobayashi, M Wigler, WC Still. Complex synthetic chemical libraries indexed with molecular tags. Proc Natl Acad Sci USA 90:10922–10926, 1993.
12. JJ Baldwin, JJ Burbaum, I Henderson, MHJ Ohlmeyer. Synthesis of a small molecule combinatorial library encoded with molecular tags. J Am Chem Soc 117: 5588–5589, 1995.
13. S Brenner, RA Lerner. Encoded combinatorial chemistry. Proc Natl Acad Sci USA 89:5381–5383, 1992.
14. MC Needels, DG Jones, EH Tate, GL Heinkel, LM Kochersperger, WJ Dower, RW Barret, MA Gallop. Generation and screening of an oligonucleotide-encoded synthetic peptide library. Proc Natl Acad Sci USA 90:10700–10704, 1993.
15. P Seneci. Combinatorial Chemistry and Technology: Principles, Methods and Applications. New York: Marcel Dekker, 1999, pp 127–167.
16. E Kaiser, RL Colescott, CD Bossinger, PI Cook. Color test for detection of free

terminal amino groups in the solid-phase synthesis of peptides. Anal Biochem 34: 595–598, 1970.

17. VK Sarin, SBH Kent, JP Tam, RB Merrifield. Quantitative monitoring of solid-phase peptide synthesis by the ninhydrin reaction. Proc 7th Am Peptide Soc Symp. Rockford, IL: Pierce Chem Co, 1981, pp 221–224.

18. V Krchňák, J Vágner, M Lebl. Noninvasive continuous monitoring of solid-phase peptide synthesis by acid-base indicator. Int J Peptide Protein Res 32:415–416, 1988.

19. WS Hancock, JE Battersby. A new micro-test for the detection of incomplete coupling reactions in solid-phase peptide synthesis using 2,4,6-trinitrobenzenesulfonic acid. Anal Biochem 71:260–264, 1976.

20. T Christensen. Peptides: Structure and Biological Function. Rockford, IL: Pierce Chemical Co, 1979, pp 385–388.

21. B Gisin. Monitoring of reactions in solid-phase peptide synthesis with picric acid. Anal Chim Acta 58:248–249, 1972.

22. WS Hancock, JE Battersby, DRK Harding. Use of picric acid as a simple method monitoring procedure for automated peptide synthesis. Anal Biochem 69:497–503, 1973.

23. AM Felix, MH Jimenez. Rapid fluorometric detection for completeness in solid phase coupling reactions. Anal Biochem 52:377–381, 1973.

24. JM Stewart, JD Young. Solid Phase Peptide Synthesis. 2nd ed. Rockford, IL: Pierce Chemical Co, 1984, p 121.

25. B Yan, L Liu, CA Astor, Q Tang. Determination of the absolute amount of resin-bound hydroxyl or carboxyl groups of the optimization of solid phase combinatorial and parallel organic synthesis. Anal Chem 71:4564–4571, 1999.

26. O Kuisle, E Quinoa, R Riguera. A general methodology for automated solid-phase synthesis of depsides and depsipeptides. Preparation of a valinomycin analogue. J Org Chem 64:8063–8075, 1999.

27. O Kuisle, M Lolo, E Quinoa, R Riguera. Monitoring the solid phase synthesis of depsides and depsipeptides. A colour test for hydroxyl groups linked to resin. Tetrahedron 55:14807–14812, 1999.

28. A Berkessel, R Riedl. Combinatorial de novo synthesis of catalysts: How much of a hit-structure is needed for activity? J Comb Chem 2:215–219, 2000.

29. JG Breitenbucher, CR Johnson, M Haight, JC Phelan. Generation of a piperazine-2-carboxamide library: A practical application of the phenol-sulfide react and release linker. Tetrahedron Lett 39:1295–1298, 1998.

30.   B Yan, W Li. Rapid fluorescence determination of the absolute amount of aldehyde and ketone groups on resin supports. J Org Chem 62:9354–9357, 1997.

31. GL Ellman. Tissue sulfhydryl groups. Arch Biochem Biophys 82:70–71, 1959.

32. AA Virgilio, JA Ellman. Simultaneous solid-phase synthesis of β-turn mimetics incorporating side-chain functionality. J Am Chem Soc 116:11580–11581, 1994.

33. JM Stewart, JD Young. Solid Phase Peptide Synthesis. 2nd ed. Rockford, IL: Pierce Chemical Co, 1984, p 114.

34. E Atherton, RC Sheppard. Solid Phase Peptide Synthesis: A Practical Approach. Oxford, UK: IRL Press, 1989.

35. NovaBiochem Catalog and Peptide Synthesis Handbook, 1999, Technical Notes 3.10. Laufelfingen, Switzerland.
36. NovaBiochem Catalog and Peptide Synthesis Handbook, 1997–1998. Method 16. Laufelfingen, Switzerland.
37. LA Carpino, M Ismail, GA Truran, EME Mansour, S Iguchi, D Ionescu, A El-Faham, C Riemer, R Warrass. The 1,1-dioxobenzo[b]thiophene-2-ylmethyloxycarbonyl (Bsmoc) amino-protecting group. J Org Chem 64:4324–4338, 1999.
38. MP Reddy, PJ Voelker. Novel method for monitoring the coupling efficiency in solid phase peptide synthesis. Int J Peptide Prot Res 31:345–348, 1988.
39. Applied Biosystems Model 391 DNA Synthesizer Manual. Section 6: Chemistry for Automated DNA Synthesis. Applied Biosystems.
40. SS Chu, SH Reich. NPIT: A new reagent for quantitatively monitoring reactions of amines in combinatorial synthesis. Bioorg Med Chem Lett 5:1053–1058, 1995.
41. SC Miller, TS Scanlan. oNBS-SPPS: A new method for solid-phase peptide synthesis. J Am Chem Soc 120:2690–2691, 1998.
42. AD Piscopio, JF Miller, K Koch. A second generation solid phase approach to Freidinger lactams: Application of Fukuyama's amine synthesis and cyclative release via ring closing metathesis. Tetrahedron Lett 39:2667–2670, 1998.
43. C Enjalbal, J Martinez, J-L Aubagnac. Mass spectrometry in combinatorial chemistry. Mass Spectrom Rev 19:139–161, 2000.
44. WL Fitch. Analytical methods for quality control of combinatorial libraries. Mol Diversity 4:39–45, 1999.
45. LA Loo. Mass spectrometry in the combinatorial chemistry revolution. Eur Mass Spectrom 3:93–104, 1997.
46. JN Kyranos, JC Hogan. High-throughput characterization of combinatorial libraries generated by parallel synthesis. Anal Chem 70:A389–A395, 1998.
47. RD Sussmuth, GJ Jung. Impact of mass spectrometry on combinatorial chemistry. J Chromatogr B 725:49–65, 1999.
48. V Swali, GJ Langley, M Bradley. Mass spectrometric analysis in combinatorial chemistry. Curr Opin Chem Biol 3:337–341, 1999.
49. EC Huang, T Wachs, JJ Conboy, JD Henion. Atmospheric pressure ionization mass spectrometry. Detection for the separation sciences. Anal Chem 62:713A–725A, 1990.
50. JB Fenn, M Mann, CK Meng, SF Wong, CM Whitehouse. Electrospray ionization Principles and practice. Mass Spectrom Rev 9:37–70, 1990.
51. RJ Cotter. Time-of-flight mass spectrometry for the structural analysis of biological molecules. Anal Chem 64:1027A–1039A, 1992.
52. M Karas, F Hillenkamp. Laser desorption ionization of proteins with molecular masses exceeding 10,000 daltons. Anal Chem 60:2299–2301, 1988.
53. MV Buchanan, RL Hettich. Fourier transform mass spectrometry of high-mass biomolecules. Anal Chem 65:245A–259A, 1993.
54. AS Fang, P Vouros, CC Stacey, GH Kruppa, FH Laukien, EA Wintner, T Carell, T Rebek Jr. Rapid characterization of combinatorial libraries using electrospray ionization Fourier transform ion cyclotron resonance mass spectrometry. Comb Chem High Throughput Screening 1:23–33, 1998.
55. S Stevanovic, K-H Wiesmüller, J Metzger, AG Beck-Sickinger, G Jung. Natural

and synthetic peptide pools: Characterization by sequencing and electrospray mass spectrometry. Bioorg Med Chem Lett 3:431–436, 1993.

56. JW Metzger, K-H Wiesmüller, V Gnau, J Brünjes, G Jung. Ion-spray mass spectrometry and high-performance liquid chromatography: Mass spectrometry of synthetic peptide libraries. Angew Chem Int Ed Engl 32:894–896, 1993.

57. S Kienle, K-H Wiesmüller, J Brünjes, JW Metzger, G Jung, J Fresenius. MS-Pep. A computer program for the interpretation of mass spectra of peptide libraries. Anal Chem 359:10–14, 1997.

58. JA Boutin, PH Lambert, S Bertin, JP Volland, JL Fauchère. Physico-chemical and biological analysis of true combinatorial libraries. J Chromatogr B 725:17–37, 1999.

59. SC Pomerantz, JA McCloskey, TM Tarasow, BE Eaton. Deconvolution of combinatorial oligonucleotide libraries by electrospray ionization tandem mass spectrometry. J Am Chem Soc 119:3861–3867, 1997.

60. T Carell, EA Wintner, AJ Sutherland, J Rebek Jr, YM Dunayevskiy, P Vouros. New promise in combinatorial chemistry: Synthesis, characterization, and screening of small-molecule libraries in solution. Chem Biol 2:171–183, 1995.

61. Y Dunayevskiy, P Vouros, T Carell, EA Wintner, J Rebek Jr. Characterization of the complexity of small-molecule libraries by electrospray ionization mass spectrometry. Anal Chem 67:2906–2915, 1995.

62. KC Lewis, WL Fitch, D Maclean. Characterization of a split-pool combinatorial library. LC-GC 16:644–649, 1998.

63. G Hegy, E Gorlach, R Richmond, F Bitsch. High throughput electrospray mass spectrometry of combinatorial chemistry racks with automated contamination surveillance and results reporting. Rapid Commun Mass Spectrom 10:1894–1900, 1996.

64. R Richmond, E Görlach. Sorting measurement queues to speed up the flow injection analysis mass spectrometry of combinatorial chemistry syntheses. Anal Chim Acta 394:33–42, 1999.

65. T Wang, L Zeng, T Strader, L Burton, DB Kassel. A new ultra-high throughput method for characterizing combinatorial libraries incorporating a multiple probe autosampler coupled with flow injection mass spectrometry analysis. Rapid Commun Mass Spectrom 12:1123–1129, 1998.

66. BE Winger, JE Campana. Characterization of combinatorial peptide libraries by electrospray ionization Fourier transform mass spectrometry. Rapid Commun Mass Spectrom 10:1811–1813, 1996.

67. JP Nawrocki, M Wigger, CH Watson, TW Hayes, MW Senko, SA Benner, JR Eyler. Analysis of combinatorial libraries using electrospray Fourier transform ion cyclotron resonance mass spectrometry. Rapid Commun Mass Spectrom 10:1860–1864, 1996.

68. F Hillenkamp, M Karas, RC Beavis, BT Chait. Matrix-assisted laser desorption/ionization mass spectrometry of biopolymers. Anal Chem 63:1193A–1203A, 1991.

69. BT Chait, SBH Kent. Weighing naked proteins: Practical, high-accuracy mass measurement of peptides and proteins. Science 257:1885–1894, 1992.

70. RS Youngquist, GR Fuentes, MP Lacey, T Keough. Matrix-assisted laser desorption ionization for rapid determination of the sequences of biologically active pep-

tides isolated from support-bound combinatorial peptide libraries. Rapid Commun Mass Spectrom 8:77–81, 1994.

71. JR Chapman. Practical Mass Spectrometry: A Guide for Chemical and Biochemical Analysis. 2nd ed. Chichester: Wiley, 1993, Chap. 5.

72. RA Zambias, DA Boulton, PR Griffin. Microchemical structure determination of a peptoid covalently bound to a polymeric bead by matrix-assisted laser desorption ionization time-of-flight mass spectrometry. Tetrahedron Lett 35:4283, 1994.

73. RN Zuckermann, JM Kerr, SBH Kent, WH Moos. Efficient method for the preparation of peptoids [oligo(N-substituted glycines)] by submonomer solid-phase synthesis. J Am Chem Soc 114:10646–10647, 1992.

74. BJ Egner, GJ Langley, M Bradley. Solid phase chemistry: Direct monitoring by matrix-assisted laser desorption/ionization time of flight mass spectrometry. A tool for combinatorial chemistry. J Org Chem 60:2652–2653, 1995.

75. BJ Egner, M Cardno, M Bradley. Linkers for combinatorial chemistry and reaction analysis using solid phase in situ mass spectrometry. Chem Commun 21:2163–2164, 1995.

76. MC Fitzgerald, K Harris, CG Shelvin, G Siuzdak. Direct characterization of solid phase resin-bound molecules by mass spectrometry. Bioorg Med Chem Lett 6:979–982, 1996.

77. BJ Egner, M Bradley. Monitoring the solid phase synthesis of analogues of lysobactin and the katanosins using in situ MALDI-TOF MS. Tetrahedron 53:14021–14030, 1997.

78. D Sarracino, C Richert. Quantitative MALDI-TOF MS of oligonucleotides and a nuclease assay. Bioorg Med Chem Lett 6:2543–2548, 1996.

79. CN Tetzlaff, I Schwope, CF Bleczinski, JA Steinberg, C Richert. A convenient synthesis of 5′-amino-5′-deoxythymidine and preparation of peptide-DNA hybrids. Tetrahedron Lett 39:4215–4218, 1998.

80. I Schwope, CF Bleczinski, C Richert. Synthesis of 3′,5′-dipeptidyl oligonucleotides. J Org Chem 64:4749–4761, 1999.

81. RK Altman, I Schwope, DA Sarracino, CN Tetzlaff, CF Bleczinski, C Richert. Selection of modified oligonucleotides with increased target affinity via MALDI-monitored nuclease survival assays. J Comb Chem 1:493–508, 1999.

82. F Hsieh, H Keshishian, C Muir. Automated high throughput multiple target screening of molecular libraries by microfluidic MALDI-TOF MS. J Biomol Screening 3:189–198, 1998.

83. BB Brown, DS Wagner, HM Geysen. A single-bead decode strategy using electrospray ionization mass spectrometry and a new photolabile linker: 3-amino-3-(2-nitrophenyl)propionic acid. Mol Diversity 1:4–12, 1995.

84. DC Tutko, KD Henry, BE Winger, H Stout, M Hemling. Sequential mass spectrometry and MS_n analyses of combinatorial libraries by using automated matrix-assisted laser desorption/ionization Fourier transform mass spectrometry. Rapid Commun Mass Spectrom 12:335–338, 1998.

85. A Benninghoven, FG Rudenauer, MW Werner. Secondary Ion Mass Spectroscopy: Basic Concepts, Instrumental Aspects, Applications and Trends. New York: Wiley, 1987.

86. JL Aubagnac, C Enjalbal, G Subra, AM Bray, R Combarieu, J Martinez. Applica-

tion of time-of-flight secondary ion mass spectrometry to in situ monitoring of solid-phase peptide synthesis on the Multipin™ system. J Mass Spectrom 33:1094–1103, 1988.

87. CL Brummel, IN Lee, Y Zhou, SJ Benkovic, N Winograd. A mass spectrometric solution to the address problem of combinatorial libraries. Science 264:399–402, 1994.

88. M Mergler, R Tanner, J Gosteli, P Grogg. Peptide synthesis by a combination of solid-phase and solution methods. I. A new very acid-labile anchor group for the solid phase synthesis of fully protected fragments. Tetrahedron Lett 29:4005–4008, 1988.

89. C Drouot, C Enjalbal, P Fulcrand, J Martinez, J-L Aubagnac, R Combarieu, Y De Puydt. Tof-SIMS analysis of polymer bound Fmoc-protected peptides. Tetrahedron Lett 38:2455–2458, 1997.

90. C Drouot, C Enjalbal, P Fulcrand, J Martinez, J-L Aubagnac, R Combarieu, Y De Puydt. Step-by-step control by time-of-flight secondary ion mass spectrometry of a peptide synthesis carried out on polymer beads. Rapid Commun Mass Spectrom 10:1509–1511, 1996.

91. JL Aubagnac, C Enjalbal, C Drouot, R Combarieu, J Martinez. Imaging time-of-flight secondary ion mass spectrometry of solid-phase peptide syntheses. J Mass Spectrom 34:749–754, 1999.

92. C Enjalbal, D Maux, G Subra, J Martinez, R Combarieu, J-L Aubagnac. Monitoring and quantification on solid support of by-product formation during peptide synthesis by Tof-SIMS. Tetrahedron Lett 40:6217–6220, 1999.

93. NJ Wells, M Davies, M Bradley. Cleavage and analysis of material from single resin beads. J Org Chem 63:6430–6431, 1998.

94. CL Brummel, JC Vickerman, SA Carr, ME Hemling, GD Roberts, W Johnson, W Weinstock, D Gaitanopoulos, SJ Benkovic, N Winograd. Evaluation of mass spectrometric methods applicable to the direct analysis of non-peptide bead-bound combinatorial libraries. Anal Chem 68:237–242, 1996.

95. C Chen, LA Ahlberg Randall, RB Miller, AD Jones, MJ Kurth. The solid-phase combinatorial synthesis of β-thioketones. Tetrahedron 53:6595–6609, 1997.

96. M Grieg. Use of automated HPLC-MS analysis and improving the purity of combinatorial libraries. Am Lab 31:28–32, 1999.

97. J-L Aubagnac, M Amblard, C Enjalbal, G Subra, J Martinez, P Durand, P Renaut. Identification of synthetic by-products in combinatorial libraries using high performance liquid chromatography-electrospray ionization mass spectroscopy. Comb Chem High Throughput Screening 2:289–296, 1999.

98. SJ Lane, A Pipe. A single generic microbore liquid chromatography/time-of-flight mass spectrometry solution for the simultaneous accurate mass determination of compounds on single beads, the decoding of dansylated orthogonal tags pertaining to compounds and accurate isotopic difference target analysis. Rapid Commun Mass Spectrom 13:798–814, 1999.

99. PS Marshall. Development and applications of a rapid back-flush microseparation system coupled to a mass spectrometer for the quality control of combinatorial libraries. Rapid Commun Mass Spectrom 13:778–781, 1999.

100. L Zeng, L Burton, K Yung, B Shushan, DB Kassel. Automated analytical/prepara-

tive high-performance liquid chromatography-mass spectrometry system for the rapid characterization and purification of compound libraries. J Chromatogr 794: 3–13, 1998.

101. PH Lambert, JA Boutin, S Bertin, J-L Fauchere, JP Volland. Evaluation of high performance liquid chromatography/electrospray mass spectrometry with selected ion monitoring for the analysis of large synthetic combinatorial peptide libraries. Rapid Commun Mass Spectrom 11:1971–1976, 1997.

102. BD Duléry, J Verne-Mismer, E Wolf, C Kugel, L Van Hijfte. Analyses of compound libraries obtained by high-throughput parallel synthesis: Strategy of quality control by high-performance liquid chromatography, mass spectrometry and nuclear magnetic resonance techniques. J Chromatogr B 725:39–47, 1999.

103. MS Lee, EH Kerns. LCMS applications in drug development. Mass Spectrom Rev 18:187–279, 1999.

104. WJ Haap, JW Metzger, C Kempter, G Jung. Composition and purity of combinatorial aryl ether collections analyzed by electrospray mass spectrometry. Mol Diversity 3:29–41, 1997.

105. YM Dunayevskiy, P Vouros, EA Wintner, GW Shipps, T Carell, J Rebek Jr. Application of capillary electrophoresis-electrospray ionization mass spectrometry in the determination of molecular diversity. Proc Natl Acad Sci USA 93:6152–6157, 1996.

106. BH Hsu, E Orton, S-Y Tang, RA Carlton. Application of evaporative light scattering detection to the characterization of combinatorial and parallel synthesis libraries for pharmaceutical drug discovery. J Chromatogr B 725:103–112, 1999.

107. CE Kibbey. Quantitation of combinatorial libraries of small organic molecules by normal-phase HPLC with evaporative light-scattering detection. Mol Diversity 1: 247–258, 1996.

108. WL Fitch, AK Szardenings, EM Fujinari. Chemiluminescent nitrogen detection for HPLC: An important new tool in organic analytical chemistry. Tetrahedron Lett 38:1689–1692, 1997.

109. RS Youngquist, GR Fuentes, MP Lacey, T Keough. Generation and screening of combinatorial peptide libraries designed for rapid sequencing by mass spectrometry. J Am Chem Soc 117:3900–3906, 1995.

110. BT Chait, R Wong, RC Beavis, SBH Kent. Protein ladder sequencing. Science 262:89–92, 1993.

111. M Rinnova, M Lebl, M Soucek. Solid-phase peptide synthesis by fragment condensation: Coupling in swelling volume. Lett Peptide Sci 6:15–22, 1999.

112. K Burgess, CI Martinez, DH Russell, H Shin, AJ Zhang. Photolytic mass laddering for fast characterization of oligomers on single resin beads. J Org Chem 62:5662–5663, 1997.

113. A Meyer, N Spinelli, J-L Imbach, J-J Vasseur. Analysis of solid-supported oligonucleotides by matrix-assisted laser desorption/ionisation time-of-light mass spectrometry. Rapid Commun Mass Spectrom 14:234–242, 2000.

114. I Hughes. Design of self-coded combinatorial libraries to facilitate direct analysis of ligands by mass spectrometry. J Med Chem 41:3804–3811, 1998.

115. M Stankova, O Issakova, NF Sepetov, V Krchnak, KS Lam, M Lebl, Application

of one-bead one-structure approach to identification of nonpeptidic ligands. Drug Dev Res 33:146–156, 1994.

116. DS Wagner, CJ Markworth, CD Wagner, FJ Schoenen, CE Rewerts, BK Kay, HM Geysen. Ratio encoding combinatorial libraries with stable isotopes and their utility in pharmaceutical research. Comb Chem High Throughput Screening 1:143–153, 1998.

117. N Nazarpack-Kandlousy, IV Chernushevich, L Meng, Y Yang, AV Eliseev. Regio-chemical tagging: A new tool for structural characterization of isomeric components in combinatorial mixtures. J Am Chem Soc 122:3358–3366, 2000.

118. JP Tam, FS Tjoeng, RB Merrifield. Design and synthesis of multidetachable resin supports for solid-phase synthesis. J Am Chem Soc 102:6117–6127, 1980.

119. JP Tam, FS Tjoeng, RB Merrifield. Multi-detachable resin supports for solid phase fragment synthesis. Tetrahedron Lett 51:4935–4938, 1979.

120. JP Tam, FS Tjoeng, RB Merrifield. Multi-detachable resin supports. Proc 6th Am Peptide Soc Symp. Rockford, IL: Pierce Chem Co, 1979, pp 341–344.

121. JP Tam, RD Dimarchi, RB Merrifield. Photolabile multi-detachable p-alkoxyben-zyl alcohol resin supports for peptide fragment or semi-synthesis. Int J Peptide Protein Res 16:412–425, 1980.

122. JP Tam, RD DiMarchi, RB Merrifield. Design and synthesis of a multidetachable benzhydrylamine resin for solid phase peptide synthesis. Tetrahedron Lett 22: 2851–2854, 1981.

123. JP Tam. Multidetachable resin supports. Proc 7th Am Peptide Soc Symp. Rockford, IL: Pierce Chem Co, 1981, pp 153–162.

124. E Giralt, F Albericio, E Pedroso, C Granier, J Van Rietschoten. Convergent solid phase peptide synthesis. II. Synthesis of the 1–6 apamin protected segment on a NBB-resin. Synthesis of apamin. Tetrahedron 38:1193–1208, 1982.

125. C Voss, R Dimarchi, DB Whitney, FS Tjoeng, RB Merrifield, JP Tam. Synthesis of the protected tridecapeptide (56–68) of the VH domain of mouse myeloma im-munoglobulin M603 and its reattachment to resin supports. Int J Peptide Protein Res 22:204–213, 1983.

126. DB Whitney, JP Tam, RB Merrifield. Studies on solid phase fragment synthesis: Synthesis of a 42-residue peptide using a multidetachable resin support and im-proved deprotection conditions. Proc 8th Am Peptide Soc Symp. Rockford, IL: Pierce Chem Co, 1983, pp 167–170.

127. JP Tam. A gradative deprotection strategy for the solid-phase synthesis of peptide amides using p-(acyloxy)benzhydrylamine resin and the S_N2 deprotection method. J Org Chem 50:5291–5298, 1985.

128. P Athanassopoulos, K Barlos, O Hatzi, D Gatos, C Tzavara. Phase change during peptide synthesis. Innovation Perspect Solid Phase Synth Comb Libr Collected Papers Int Symp. 4th ed. Birmingham, UK: Mayflower Scientific, 1996, pp 243–246.

129. S Berteina, S Wendeborn, A De Mesmaeker. Radical and palladium-mediated cyclisations of ortho-iodo benzyl enamines: Application to solid phase synthesis. Synlett 1998:1231–1233.

130. KC Nicolaou, N Watanabe, J Li, J Pastor, N Winssinger. Solid-phase synthesis of oligosaccharides: Construction of a dodecasaccharide. Angew Chem Int Ed 37: 1559–1561, 1998.

131. K Witte, O Seitz, C-H Wong. Solution- and solid-phase synthesis of N-protected glycopeptide esters of the benzyl type as substrates for subtilisin-catalysed glycopeptide couplings. J Am Chem Soc 120:1979–1989, 1998.

132. S Kobayashi, Y Aoki. p-Benzyloxybenzylamine (BOBA) resin. A new polymer-supported amine used in solid-phase organic synthesis. Tetrahedron Lett 39:7345–7348, 1998.

133. HM Geysen, CD Wagner, WM Bodnar, CJ Markworth, GJ Parke, FJ Schoenen, DS Wagner, DS Kinder. Isotope or mass encoding of combinatorial libraries. Chem Biol 3:679–688, 1996.

134. AW Czarnik. Combinatorial chemistry. What's in it for analytical chemists? Anal Chem 70:378A–386A, 1998.

135. HM Geysen. Library Technologies Conference, Washington DC, Apr 27–28, 1998; 5th International Workshop on Molecular and Cell Biology—UNC, Chapel Hill, NC, Sept 18–21, 1997.

135a. DS Wagner, LJ Shampine, H Patel, CD Wagner, JJ Wild, HM Geysen. Proc 46th ASMS, 1029.

135b. FJ Schoenen, HM Geysen, WM Bodnar, L Craddock, I Jeganathan, D Kinder, KE Kinsey, JA Markworth, JC Nelson, G Parke, H Patel, JM Salovich, L Shampine, CD Wagner, DS Wagner, JJ Wild, RL Wilgns. Proc 46th ASMS, 1139.

136. M Bartlet-Jones, WA Jeffery, HF Hansen, DJ Pappin. Peptide ladder sequencing by mass spectrometry using a novel, volatile degradation reagent. Rapid Commun Mass Spectrom 8:737–742, 1994.

137. PC Liao, J Allison. Enhanced detection of peptides in matrix-assisted laser desorption/ionization mass spectrometry through the use of charge-localized derivatives. J Mass Spectrom 30:511–512, 1995.

138. MR Carrasco, MC Fitzgerald, Y Oda, SBH Kent. Direct monitoring of organic reactions on polymeric supports. Tetrahedron Lett 38:6331–6334, 1997.

139. SC McKeown, SP Watson, RAE Carr, P Marshall. A photolabile carbamate based dual linker analytical construct for facile monitoring of solid phase chemistry: "TLC" for solid phase? Tetrahedron Lett 40:2407–2410, 1999.

140. O Lorthioir, SC McKeown, NJ Parr, SP Watson, MS Congreve, JJ Scicinski, C Kay, P Marshall, RAE Carr, MH Geysen. Single bead characterisation using analytical constructs: Application to quality control of libraries. Anal Chem 73:963–970, 2001.

141. O Lorthioir, SC McKeown, NJ Parr, M Washington, SP Watson. Practical synthesis of a new analytical construct: Thiopyrimidine safety-catch linker for facile monitoring of solid phase chemistry. Tetrahedron Lett 41:8609–8613, 2000.

142. KC Nicolaou, X-Y Xiao, Z Parandoosh, A Senyei, MP Nova. Radiofrequency encoded combinatorial chemistry. Angew Chem Int Ed Engl 34:2289–2291, 1995.

143. EJ Moran, S Sarshar, JF Cargill, MH Shahbaz, A Lio, AMM Mjalli, RW Armstrong. Radio frequency tag encoded combinatorial library method for the discovery of tripeptide-substituted cinnamic acid inhibitors of the protein tyrosine phosphatase ptp1b. J Am Chem Soc 117:10787–10788, 1995.

144. PJ Murray, C Kay, JJ Scicinski, SC McKeown, SP Watson, RAE, Carr. Rapid reaction scanning of solid phase chemistry using resins incorporating analytical constructs. Tetrahedron Lett 40:5609–5612, 1999.

145. SR Chhabra, AN Khan, BW Bycroft. Versatile Dde-based primary amine linkers for solid-phase synthesis. Tetrahedron Lett 39:3585–3588, 1998.
146. C Kay, PJ Murray, L Sandow, AB Holmes. A novel, chemically robust, amine releasing linker. Tetrahedron Lett 38:6941–6944, 1997.
147. T Fukuyama, C-K Jow, M Cheung. 2- and 4-Nitrobenzenesulfonamides: Exceptionally versatile means for preparation of secondary amines and protection of amines. Tetrahedron Lett 36:6373–6374, 1995.
148. KH Shaughnessy, PH Kim, JF Hartwig. A fluorescence-based assay for high-throughput screening of coupling reactions. Application to Heck chemistry. J Am Chem Soc 121:2123–2132, 1999.
149. GM Williams, RAE Carr, MS Congreve, C Kay, SC McKeown, PJ Murray, JJ Scicinski, SP Watson. Analysis of solid phase reactions: Product identification and quantification by use of UV chromophore-containing dual linker analytical constructs. Angew Chem Int Ed, Engl 39:3293–3296, 2000.
150. SI Klein, BF Molino, M Czekaj, CJ Gardner, V Chu, K Brown, RD Sabatino, JS Bostwick, C Kasiewski, R Bentley, V Windisch, M Perrone, CT Dunwiddie, RJ Leadley. Design of a new class of orally active fibrinogen receptor antagonists. J Med Chem 41:2492–2502, 1998.
151. JJ Scicinski, MS Congreve, C Jamieson, SV Ley, ES Newman, VM Vinader, RAE Carr. Solid-phase development of a 1-hydroxybenzotriazole linker for heterocycle synthesis using analytical constructs. J Comb Chem 3:387–396, 2001.
152. DE Bays, RPC Cousins, HJ Dyke, CD Eldred, BD Judkins, M Pass, AMK Pennell. Preparation of adenosine derivatives as antiinflammatory agents. PCT Int Appl WO 9967262, 1999.
153. DF Emiabata-Smith, DL Crookes, MR Owen. A practical approach to accelerated process screening and optimization. Org Process Res Dev 3:281–288, 1999.

11

Methods for Quantitative Analysis of Compound Cleavage from Solid-Phase Resins

Nikhil Shah, George Detre, Stephen Raillard, and William L. Fitch
Affymax Research Institute, Palo Alto, California

I. INTRODUCTION

The success of combinatorial chemistry relies heavily on the automation potential derived from solid-phase synthesis. Because property evaluation is usually done in solution, the final step of a solid-phase synthesis is to cleave the material from the solid phase. Critical to these experiments is an arsenal of analytical tools. We have reviewed this area (1–4) and now describe some recent experiences with cleavages and analyses of compounds from single resin beads and from bulk samples.

In large-scale synthesis, quantitation is done by purifying and weighing the product. In small-scale parallel synthesis, there is often a submilligram yield that cannot be quantified by weighing. The currently most popular application of parallel synthesis in large pharmaceutical firms is to do moderate-scale synthesis followed by automated purification to obtain ''IC50 quality'' material. The throughput of this approach can be limited by the quantitation step, even if weighable quantities are obtained.

Libraries prepared on beaded resins by the split/pool approach are characterized by the ''one bead, one compound'' rule. Such libraries, in principle, combine the advantages of split/pool synthesis with those associated with the screening of discrete compounds, because single compounds from individual beads may be assayed for biological activity. The difficulty attending this is that of ensuring the presence of the correct ''one compound'' at the expected concentration after distribution (and cleavage) of the one bead into a bioassay format. Affymax makes encoded, split/pool libraries for high throughput drug discovery. We now describe efforts to ensure the concentration of ligand presented by these libraries.

II. MEASUREMENT OF BULK RELEASE OF SINGLE COMPOUNDS FROM SOLID SUPPORT

Bulk multimilligram samples are readily weighed to an accuracy of better than 10%. But this quantitation is meaningful only after a slow and laborious process of sample purification and drying. Those familiar with preparative organic synthesis know that extreme rigor is required to remove the last traces of solvents, salts, or water after a purification process. The gold standard for this process is to submit the sample to CHN elemental analysis to prove purity. The gold standard is difficult to achieve; submilligram samples cannot be weighed, and demands of throughput will make rigorous purification difficult to implement. Methods are needed for routine high throughput quantitation of solutions of active test molecules. This concept has been termed "compound-independent calibration" and has many applications in the burgeoning fields of combinatorial material sciences (5).

   Internal standard proton NMR (6,7) has submilligram sensitivity and is used in our laboratories for quantitative analysis of single bulk level samples. Figure 1 illustrates that fairly low level samples can be quantified with internal standard NMR. This is a sample of compound I, photocleaved from resin directly into DMSO-d_6. The large resonances at 6.9 and 8.1 ppm are due to the internal standard, 4-nitrophenol, present at 3.3 mM. The sample concentration is calculated to be 0.9 mM.

I

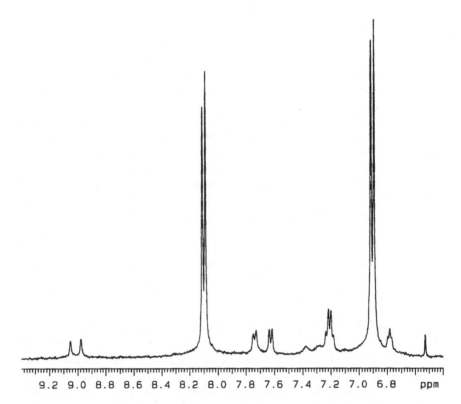

9.2 9.0 8.8 8.6 8.4 8.2 8.0 7.8 7.6 7.4 7.2 7.0 6.8   ppm

Figure 1 The proton NMR spectrum of a cleavage quantitation of compound I. I on TentaGel was photocleaved (4.5 mg into 200 μL of 10 mM 4-nitrophenol in DMSO-d_6 for 2 h). The resultant solution was filtered into a standard NMR tube, the volume was adjusted to 0.6 mL with DMSO-d_6, and 32 scans were collected on a 400 MHz Varian Inova instrument.

High throughput, automated qualitative NMR using flow injection is another NMR technique that has become useful for characterizing libraries (8). Its application for quantitation will be limited until NMR prediction becomes more facile (9).

Recent research (10,11) introduced the idea of using the evaporative light scattering detector as a universal detector for quantitation of combinatorial synthesis products. This method is based on the assumption that all molecules give equal response to this detector. Although promising for crude quantitation, this approach will not be appropriate for high accuracy property measurement.

We and others have been exploring chemiluminescent nitrogen detection (CLND) as an approach to this problem (12–14). CLND is based on the principle of combusting organic molecules to a common product, which can be conventionality measured. This is not a new idea in drug discovery (15), but it is gaining in popularity with the introduction of the Antek technology (16).

The usefulness of CLND is based on the hypothesis that all nitrogen compounds give equal combustion response. We, and others, have confirmed this with most molecules tested. The exception is molecular nitrogen, which requires higher combustion temperatures to oxidize. Molecules that contain $N-N$ bonds may have thermal decomposition pathways that yield N_2 and thus will not give equimolar CLND response (17). This limitation of the CLND detector is important for certain classes of molecules (tetrazoles) but is not a huge problem.

Another limitation of the LC/CLND detector and the reason for its slow acceptance in the industry is that it is not as trivial to use as is a UV or ELSD detector; it is still a specialist's instrument. We have recently introduced an alternative, direct inject CLND (DI-CLND) that *is* easy to use. This detector measures the combustible nitrogen in a sample with two limitations relative to the LC/CLND: The samples are not chromatographically separated, and on-column concentration is not possible, so sensitivity is compromised. We find that these limitations are more than balanced by the ease of use for certain applications. In particular, very fast 100% quantitation of parallel synthesis libraries is achievable (13).

To illustrate these principles, compound II was prepared on TentaGel resin with a photolinkage (18). The resin was characterized by quantitative solid-phase NMR, traditional elemental analysis, and solids CLND (N Shah, unpublished, 2000). By NMR the loading was 0.38 mmol/g, whereas the CHN analysis showed 0.34 mmol/g and the solids CLND showed 0.40 mmol/g. (Although not strictly relevant to the cleavage question, the solids CLND is a convenient, high throughput way to characterize solid-phase synthesis samples.)

II

The photolytic cleavage of compound II was measured using direct inject CLND. A weighed sample of the resin (3–5 mg) was placed in a small glass vial and slurried with 200 µL of solvent. The sealed vial was then exposed to 10 mW/cm^2 light energy, measured at 360 nm (18). After an appropriate time, the vial was removed from the photolysis and diluted for analysis. We showed earlier that the DI-CLND is sensitive to solvent and must be calibrated with the standards dissolved in the same solvent as the samples are dissolved in. The time course for cleavage of II in four solvents is shown in Figure 2.

Figure 3 shows the solid-phase magic angle spinning (MAS) NMR of resin-

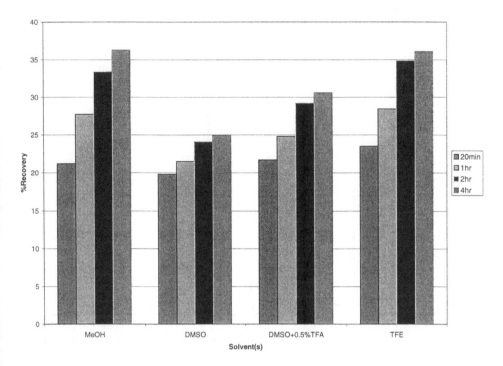

Figure 2   The time course for cleavage of compound II from resin in four solvents.

bound compound II before (a) and after (b) photolysis. MAS NMR is a very useful tool in solid-phase chemistry development, and it has been routine in our laboratories for over 5 years (19). Figure 3b shows that the intact molecule is still present on resin after extended photolysis. We continue research on optimizing photocleavage conditions and have found these NMR and CLND tools of critical importance.

III.   MEASUREMENT OF RELEASE FROM SINGLE BEADS OF SINGLE-COMPOUND SAMPLES

Because of the difficulties in analyzing the compound release from single beads of split/pool libraries, it is often useful to study the release of known compounds from single-compound samples. The CLND, ELSD, and NMR methods are not applicable for the subnanomole amounts of sample released from single beads, so these measurements will need the more sensitive UV or MS detectors. For single compounds, an LC/MS run in selected ion mode is an extremely sensitive

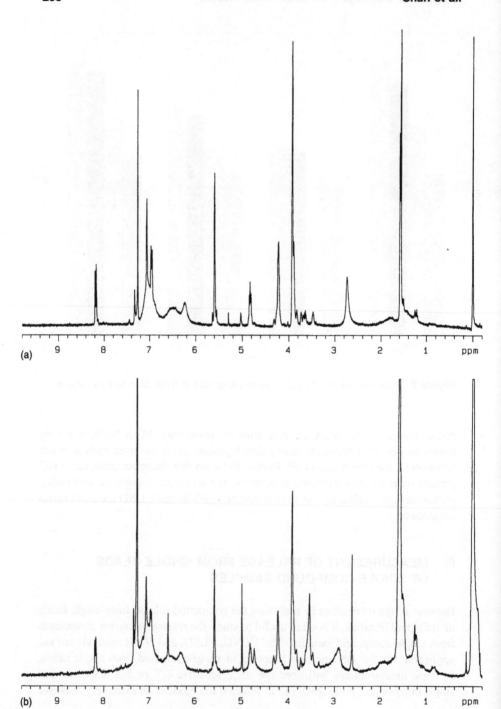

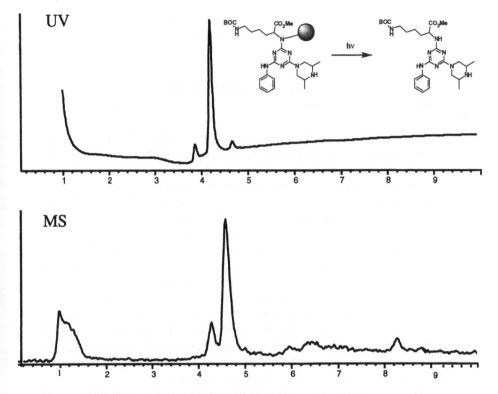

Figure 4 The LC/UV and MS chromatograms for the analysis of compound I cleaved from a single bead.

and specific analytical method. Figure 4 is such an LC/UV/MS analysis for a single bead of compound I.

If adequate sample is available, the material can be bulk cleaved and purified to give a reference standard, which can then be used to make a calibration curve. Typically it is difficult to obtain this much pure compound. An alternative we have employed is to cleave a few hundred beads into a small volume and

Figure 3 Solid-phase magic angle spinning NMR of resin-bound II (a) before and (b) after 4 h photolysis in DMSO. The resin was washed with methanol to remove unbound ligand, filtered, and then swelled in 40 μL of CDCl₃. The spectra were obtained in a Varian Nanoprobe at 400 MHz. The PEG chains were attenuated by presaturation, and 128 transients were collected in each case.

use the CLND detector to "qualify" this reference standard, which can then be used to prepare the calibration curve for the single-bead samples.

IV. MEASUREMENT OF BULK RELEASE FROM SPLIT/POOL LIBRARIES

Encoded split/pool libraries are prepared for high throughput screening at Affymax. These libraries offer the possibility of a large diversity of chemical structures with a minimum of synthetic effort. We prepare and store these libraries attached to their support. They are distributed for screening at a level of one to a few beads per assay when needed. We recently found that these encoded libraries are stable for at least a few years and so offer stability and ease-of-handling advantages over libraries of traditional compounds dissolved in DMSO.

Our method for quality control of this process (20) is to compare the cleaved ligand to the structure predicted by the encoding. The ligand analysis uses LC with UV and MS detection to confirm the structure and assess the purity of the cleaved ligand. Figure 4 illustrates an example of this. However, neither of these techniques has a currently predictable response, so no information is obtained on the amount of ligand released. As discussed below, the development of methods for measuring the compound from a single bead is in its infancy. So we have chosen to measure the bulk release of an encoded library as an estimate of the concentrations that will be released in screening.

The ideal tool for measuring bulk release from a split/pool library is the DI-CLND described previously. A typical encoded library sample contains 500–50,000 members, so chromatographic separation of the individual components is of little value. Rather we choose to measure the release of total nitrogen from a multimilligram sample of the pooled beads. After correction for the average number of nitrogens per member of the library, this will tell us the average release of compound per milligram or per bead.

This is illustrated with a set of triazines with the generic structure III.

III

This one bead, one compound library was prepared as a set of 38 pools of 616 members each. The resin was TentaGel, and PL is the photolinker. The fully encoded split/pool synthesis takes over 10 steps. The standard quality control method is to compare the structure, indicated by encoding, with the LC/MS data found for the photocleaved ligand. In this case the two matched in 92 out of 114 beads checked. This qualitative reliability does not tell us anything about the quantity of compound on the resin.

To measure the quantity of compound, a combined pool (4 mg), which represents the entire library of 23,408 structures, was weighed and photocleaved as described above for 4 h in DMSO/MeOH/TFA (1/3/0.01 v/v/v). The resultant solution was diluted to a final volume of 2.0 mL and found to have a nitrogen content of 0.96 mM. This result can be corrected for the average moles of nitrogen per structure in the library (5.56) to yield a result that the cleavage of this library yields, on average, 80 µmols/g. By measuring the numbers of beads per milligram for an aliquot, the biologists can convert this number to the average picomoles per bead that should be released for single-bead biological screening. The chemists can compare this loading to that seen in the starting resin to evaluate the overall yield of their multistep process.

V. MEASUREMENT OF RELEASE FROM SINGLE BEADS OF SPLIT/POOL LIBRARIES

As described above, it is routine to obtain the UV and MS "peak areas" from the ligand released from a single bead. The decode from that bead then indicates the chemical structure of that ligand, and correlation of the MS and structure confirms the success of the chemical synthesis. But technology for predicting response in UV and MS is not available. One approach that has been described is to add an "analytical construct" to the solid-phase linker that will stay with the ligand on cleavage and enhance and "level" the response of the MS or the UV detector and yield quantitation (21). This approach has been valuable for library chemistry development but puts extra limitations on the chemistry diversity achievable.

We introduce a new concept, the prediction of UV spectral intensity. Software to predict UV has not progressed as swiftly as has software for NMR or IR prediction. And yet modern computational chemistry tools are bringing this into the realm of possibility. We will introduce the idea of UV prediction as a tool for single-bead quantitative analysis.

Normally HPLC detection is done at a single wavelength, 220 or 254 nm being the most common. The slope of the absorbance spectrum is often very steep at these single wavelengths. And a single wavelength poorly captures the size of an absorbance feature that is structure-related. For these reasons we integrate the entire spectrum to increase predictability. The lower limit to this integration will be 220 nm owing to solvent absorbance at lower wavelengths. The upper limit will be 360 nm, because drugs rarely have absorbance in the visible. This integrated absorbance is simple to measure with a diode array UV detector. It is equivalent to a broadband absorbance centered at 290 nm with a bandwidth of 140 nm.

Beer's law states that the light absorption A at a single wavelength is proportional to the sample concentration c:

$$A = \log \frac{I_0}{I} = \varepsilon bc$$

where ε is extinction at that wavelength and b is the light pathlength. From this it is easy to derive the expression:

LC peak area $= EIc$

for any method of measuring the UV signal. The value E is based solely on the properties of the molecule in question and is the value we will predict from structure. The value I is based solely on instrumental values, specifically the detector pathlength, the flow rate, and the injection volume. I will be measured with a known compound for each analytical situation.

To calculate E we are acquiring a database of UV spectra from which a quantitative structure–property relationship can be calculated (22). Preliminary results indicate that this relationship exists but that our desired accuracy of $\pm 20\%$ for 80% of the structures will require quantum-mechanical calculation of spectral intensity descriptors.

VI. CONCLUSIONS

New, high throughput drug discovery protocols demand high throughput methods for quickly assessing the quality and quantity of test molecules presented for assay. We have briefly outlined several methods that are in use in our laboratory or are being actively developed.

ACKNOWLEDGMENTS

We extend our thanks to Chris Holmes, Steven Ferla, and Toño Estrada for preparing the test articles. Thanks also to Mark Gao and Zohra Taiby for help with the CLND work and to Ken Lewis for helpful discussions.

REFERENCES

1. WL Fitch. Analytical methods for quality control of combinatorial libraries. Mol Diversity 4:39–45, 1999.
2. WL Fitch, G Detre, GC Look. Analytical chemistry issues in combinatorial drug discovery. In: JF Kerwin, EM Gordon, eds. Combinatorial Chemistry and Molecular Diversity in Drug Discovery. New York: Wiley, 1998, pp 349–368.
3. WL Fitch. Analytical methods for quality control of combinatorial libraries. In: WH Moos, MR Pavia, AD Ellington, BK Kay, eds. Annual Reports in Combinatorial Chemistry and Molecular Diversity, Vol. 1. Leiden: ESCOM Science, 1997, pp 59–68.
4. MA Gallop, WL Fitch. New methods for analyzing compounds on polymeric supports. Curr Opin Chem Biol 1:94–100, 1997.
5. NA Stevens, MF Borgerding. Effect of column flow rate and sample injection mode on compound-independent calibration using gas chromatography with atomic emission detection. Anal Chem 70:4223–4227, 1998.
6. CK Larive, D Jayawickrama, L Orfi. Quantitative analysis of peptides with NMR spectroscopy. Appl Spectrosc 10:1531–1536, 1997.
7. SW Gerritz, AM Sefler. 2,5-Dimethylfuran (DMFu): An internal standard for the "traceless" quantitation of unknown samples via ^{1}H NMR. J Comb Chem 2:39–41, 2000.
8. PA Keifer, SH Smallcombe, EH Williams, KE Salomon, G Mendez, JL Belletire, CD Moore. Direct-injection NMR (DI-NMR): A flow NMR technique for the analysis of combinatorial chemistry libraries. J Comb Chem 2:151–171, 2000.
9. A Williams. Recent advances in NMR prediction and automated structure elucidation software. Curr Opin Drug Discovery Dev 3:298–305, 2000.
10. BH Hsu, E Orton, S Tang, RA Carlton. Application of evaporative light scattering detection to the characterization of combinatorial and parallel synthesis libraries for pharmaceutical drug discovery. J Chromatogr B 725:103–112, 1999.
11. L Fang, M Wan, M Pennacchio, J Pan. Evaluation of evaporative light-scattering detector for combinatorial library quantitation by reversed phase HPLC. J Comb Chem 2:254–257, 2000.
12. WL Fitch, AK Szardenings, EM Fujinari. Chemiluminescent nitrogen detection for HPLC: An important new tool in organic analytical chemistry. Tetrahedron Lett 38:1689–1692, 1997.
13. N Shah, K Tsutsui, A Lu, J Davis, R Scheuerman, WL Fitch. A novel approach to high-throughput quality control of parallel synthesis libraries. J Comb Chem 2:456–459, 2000.
14. EW Taylor, MG Qian, GD Dollinger. Simultaneous on-line characterization of small organic molecules derived from combinatorial libraries for identity, quantity and purity by reversed phase HPLC with chemiluminescent nitrogen, UV and mass spectrometric detection. Anal Chem 70:3339–3347, 1998.
15. JR Leaton. Automated quantitative determination of selected nitrogen-containing drugs. J Assoc Offic Anal Chem 52:283–286, 1969.
16. JA Borny, ME Homan. Efficient characterization of combinatorial libraries using

mass spectrometry and chemiluminescent nitrogen detection. LC/GC 18:S14–S19, 2000.

17. KC Lewis, D Phelps, A Sefler. Automated high-throughput quantification of combinatorial arrays. Am Pharm Rev 63–68, 2000.

18. CP Holmes, DG Jones. Reagents for combinatorial organic synthesis: Development of a new o-nitrobenzyl photolabile linker for solid phase synthesis. J Org Chem 60: 2318–2319, 1995.

19. WL Fitch, G Detre, CP Holmes, JN Shoolery, PA Keifer. High-resolution ^{1}H NMR in solid-phase organic synthesis. J Org Chem 59:7955–7956, 1994.

20. KC Lewis, WL Fitch, D Maclean. Characterization of a split/pool combinatorial library. LC/GC 16:644–649, 1997.

21. SC McKeown, SP Watson, RAE Carr, P Marshall. A photolabile carbamate based dual linker analytical construct for facile monitoring of solid phase chemistry: "TLC" for solid phase? Tetrahedron Lett 40:2407–2410, 1999.

22. AR Katritzky, U Maran, VS Lobanov, M Karelson. Structurally diverse quantitative structure-property relationship correlations of technologically relevant physical properties. J Chem Inf Comput Sci 40:1–18, 2000.

12

Hammett Relationships and Solid-Phase Synthesis

The Application of a Classical Solution to a Modern Problem

Samuel W. Gerritz
Bristol-Myers Squibb, Wallingford, Connecticut

I. INTRODUCTION

Following the introduction of chloromethylpolystyrene resin by Merrifield in 1963 (1), solid-phase synthesis was rapidly adopted as the method of choice for the synthesis of peptides and oligonucleotides (2–5). Surprisingly, solid-phase synthesis did not receive much attention from the wider synthetic community, and thus the ensuing 25 years of solid-phase research focused largely on the development of "biopolymer-friendly" solid supports (2–7), improved protecting group strategies (8,9), and new synthetic methods for the construction of amide and phosphate bonds (2–5,10,11). Consequently, the solid-phase synthesis of complex peptides and oligonucleotides has become routine and, in many cases, automated. Over the last 10 years the power of solid-phase synthesis has been recognized by medicinal chemists (12–17). The attention of this new audience has shifted the focus of solid-phase research away from the synthesis of large biopolymers, which uses a small set of established chemistries and building blocks. Instead, solid-phase research has broadened to the application of a wide range of chemistries for the synthesis of small molecules that bear little resemblance to peptides or oligonucleotides. These sea changes have been accompanied by a number of new challenges, two of which are particularly relevant to this chapter.

First, the successful adaptation of solution-phase chemistries to the solid phase is highly dependent upon choosing an appropriate solid support, yet the role the solid support plays in a solid-phase reaction is not well understood (18–24). This situation is further complicated by the plethora of solid supports available to the chemist, many of which were originally developed for peptide synthesis, with little comparative information to aid in the selection of the optimal support for a specific chemistry (6,7). In contrast, the selection of solvent for a given synthetic solution-phase transformation is a fairly straightforward process, because a chemist can draw on many references describing the effects of different solvents on a given reaction (25,26).

Second, after a chemical reaction has been successfully validated on a solid support, the paucity of "real-time" analytical techniques to monitor the progress of a solid-phase reaction is a major limitation. With a diverse range of reagents acting on an equally diverse range of solid-supported substrates, ensuring that a reaction has gone to completion is technically arduous. Reaction monitoring via thin-layer chromatography (TLC) is a routine operation in solution-phase synthesis, but no "TLC equivalent" has been developed for reactions that occur on a solid support. As a consequence, reaction times are usually empirically derived (allowing a reaction to proceed overnight seems to be a popular solution), and reagents are carefully screened in a production "rehearsal" (or monomer scan) to remove those that are not sufficiently reactive under the standard set of conditions. The ability to predict reaction times based on the chemical properties of each reagent would be a valuable tool, because it would obviate the rehearsal step and enable a chemist to go directly from chemistry development to library production.

It is interesting to note that these two issues can be viewed as solid-phase variants of research topics that were carefully studied for solution-phase processes by chemists in the first half of the twentieth century. The role of the solid support in a solid-phase reaction is a near neighbor of the solvent effect, a research topic that has been the subject of numerous reviews (25,26). The determination of the appropriate reaction time for a variety of reagents is related to the study of substituent effects on the rate of a given reaction. Historically, one experimental tool has been used to address both of these issues: the Hammett equation. This chapter describes how we have used Hammett relationships to probe the reactivity of solid supports and gain an understanding of the sensitivity of a given solid-phase reaction to changes in reaction conditions. Moreover, the throughput advantages inherent to solid-phase synthesis can be used for the rapid development of new Hammett relationships and corresponding substituent effects. Both approaches highlight the utility of the Hammett equation in the optimization of solid-phase reactions.

II. THE HAMMETT EQUATION AS A QUANTITATIVE PROBE OF CHEMICAL REACTIVITY

The Hammett equation,

$$\log\left(\frac{k}{k^0}\right) = \sigma\rho \tag{1}$$

is an extremely versatile tool for the elucidation of structure–activity relationships between substituents and many types of chemical reactions, including equilibria, rates, and ligand binding (27–31). The power of the Hammett equation lies on the right-hand side of Eq. (1): The effect of a specific substituent on *any* chemical reaction is represented by the substituent constant σ, and the sensitivity of a *specific* reaction to steric and electronic effects is represented by the reaction constant ρ. These two parameters convey a tremendous amount of chemical information, and half a century of research into Hammett relationships has afforded specific values for σ and ρ for many different substituents and reactions, respectively (32,33). For a given chemical reaction, a linear relationship between $\log(k/k^0)$ and σ allows one not only to determine ρ but also to use it in a predictive sense for substituents with known σ values.

In the case of solid-phase synthesis, the value of the reaction constant ρ is particularly useful information because the synthesis of a chemical library generally necessitates that a given set of reaction conditions be repeated with a series of functionally related reactants bearing a sterically and electronically diverse set of substituents. In this case, the ideal reaction would have $\rho = 0$, signifying that individual reaction rates would be insensitive to the range of electronic properties inherent to the different reactants (or "monomer set"). Unfortunately, there are few reactions for which the absolute value of ρ is less than 1.0, and the exponential relationship between the ρ and the rate constant k effectively guarantees a wide range of rates for a series of substituents. Accordingly, it is in the combinatorial chemist's best interest to minimize the magnitude of ρ. It has been demonstrated that ρ is influenced by various reaction conditions such as reagent choice, temperature, and solvent (34–38). In solid-phase reactions, it has been postulated that the solid support serves as a cosolvent (18), and our studies support this supposition (39). Thus, through a series of appropriately designed experiments, one can evaluate virtually any reaction variable and its effect on ρ and thereby optimize reaction conditions using quantitative information.

The sensitivity of a given reaction to substituent effects is independent of the absolute reaction rate. In other words, a change in ρ does not tell us anything about k^0, it tells us only that the ratio k/k^0 has changed. A chemist could thus optimize the ρ of a given reaction to a minimal absolute value, only to find that the rate of the reaction has decreased tenfold. As a result, any changes in ρ must be evaluated in the context of the absolute reaction rate.

Not every chemical reaction yields a linear free energy relationship when its substituents are varied. Rather than being a limitation, the observation of a nonlinear relationship between $\log(k/k^0)$ and σ for a specific reaction provides useful information about the course of the reaction; in general, the nonlinearity is indicative of either a change in mechanism or a change in the rate-determining step. Although nonlinear Hammett plots do not yield quantitative ρ or σ values, a significant amount of qualitative information can be derived from these plots. An excellent discussion of nonlinear Hammett plots is contained in Ref. 31.

III. SOLID SUPPORTS EXPEDITE THE DETERMINATION OF HAMMETT RELATIONSHIPS

The experimental measurement of the ratio k/k^0 is one of the most technically challenging aspects in the determination of Hammett relationships. Obtaining the rate constant for a single reaction is not a trivial task, and the construction of rate-based Hammett plots (versus those based on equilibria) requires that k be determined for every substituent of interest. These individual rate experiments can be eliminated by carrying out competition experiments (40–43) with a large excess of a binary mixture of equimolar reactants, thereby ensuring that the reaction in question obeys pseudo-first-order kinetics.

In solution, separating the products from the excess reagents presents a large technical challenge. When the substrate is attached to a solid support, the excess reagents can simply be washed away, leaving behind a binary mixture of solid-supported products that directly reflect the ratio k/k^0. This approach is described generically in Figure 1. In theory, the ratio of products can be measured on the solid support (using techniques described elsewhere in this book), but we have found it more expedient to cleave the products from the solid support and measure the ratio using conventional ''solution-phase'' analytical methods such as ^1H NMR. These competition experiments allow the ratio k/k^0 to be determined

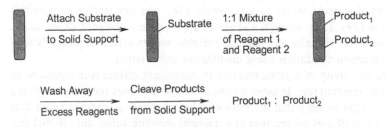

Figure 1   A generic description of solid-phase competition experiments.

by simply measuring the ratio of products without discretely measuring either k or k^0, resulting in a dramatic decrease in the time required to construct a Hammett plot. Furthermore, multiple solid supports can be combined in a single reaction vessel, assuming that the solid supports can be differentiated from one another upon their removal from the vessel.

The combination of these two techniques makes possible the construction of dozens of Hammett plots in the time previously required to construct just one (39).

The experimental design we employed for our competition experiments merits additional discussion. Traditionally, a Hammett plot is constructed from a dataset in which the reaction rate for each substituted analog is compared to the reaction rate for the unsubstituted analog (k/k^0), and the log of this ratio is plotted against the σ for the substituent. The result of this plot is a single linear free energy relationship with a slope equal to ρ. Because competition experiments enable the rapid determination of log (k/k^0), we have developed an experimental design in which every possible combination of analogs is compared. This design leads to the construction of a "combinatorial" Hammett plot, from which is obtained a linear free energy relationship (and the associated ρ value) *for every analog in the set*. In general, the number of experiments required for the fully "combinatorial" set of competition experiments is $n(n - 1)/2$, where n is the number of analogs to be studied. By comparison, the number of experiments required for a "traditional" Hammett plot is $n - 1$. Obviously, the combinatorial design requires additional experiments to be conducted, but the generation of multiple linear free energy relationships is highly advantageous, because an average ρ value and its associated error can be determined from a single set of experiments. In order to generate the same number of linear free energy relationships using a traditional design, $n(n - 1)$ experiments must be run. The twofold increase in efficiency provided by the combinatorial Hammett plots is achieved by using the product ratio from each competition experiment twice in each Hammett plot. For example, a competition experiment that compares p-fluoroaniline to p-anisidine provides two log ratios for the Hammett plot: (a) The log of (p-fluoroaniline/p-anisidine) is plotted against the σ for the p-MeO substituent. (b) The log of (p-anisidine/p-fluoroaniline) is plotted against the σ for the p-F substituent. This experimental design is exemplified in the following two case studies.

IV. SOLID-PHASE HAMMETT CASE STUDY 1: REDUCTIVE AMINATION

Our initial foray into determining solid-phase Hammett relationships was stimulated by the introduction of the backbone amide linker (BAL) (44,45). As described retrosynthetically in Eq. (2), the BAL linker greatly simplifies the solid-phase synthesis of secondary amides. Utilizing a BAL-equipped solid support, one can perform a reductive amination with a primary amine (R^1—NH_2) to obtain a solid-supported secondary amine. This amine can then be acylated with a carboxylic acid (R^2—CO_2H) under a variety of conditions to afford the corresponding solid-supported tertiary amide, which upon treatment with trifluoroacetic acid cleaves the secondary amide from the solid support. This chemistry is very popular with medicinal chemists, owing not only to the generality and robustness of the chemistry but also to the widespread availability of primary amines and carboxylic acids. However, this chemistry is not foolproof. In particular, the reductive amination step can be quite capricious, particularly when substituted anilines are used. Conveniently, substituted anilines are ideal reagents for elucidating Hammett relationships, and this prompted us to study the reductive amination of BAL-equipped polystyrene (PS) crowns (15) via solid-phase competition experiments.

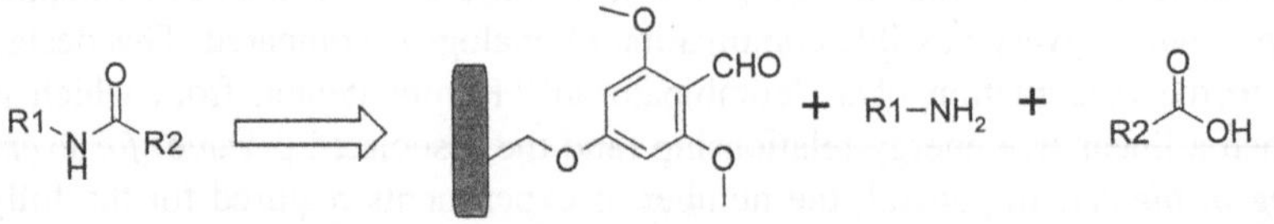

$$R1\underset{H}{\overset{O}{N}}R2 \implies \quad + \quad R1-NH_2 \quad + \quad R2\overset{O}{\underset{}{}}OH \qquad (2)$$

As outlined in Scheme 1, the reductive amination of a BAL-equipped PS crown (loading: 20 μmol/crown) with a 1:1 mixture of para-substituted anilines (using a 15-fold excess of each aniline) provided a binary mixture of secondary amines (**1**). We then acylated **1** with 3,4,5-trifluorobenzoyl chloride to afford a

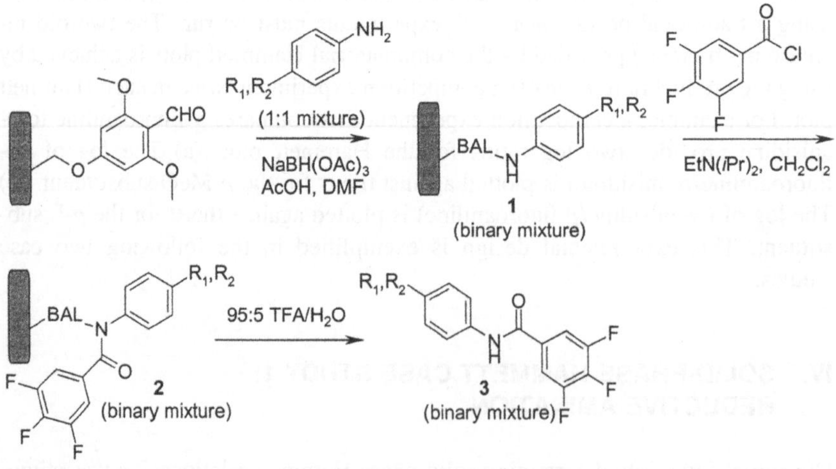

Scheme 1

binary mixture of tertiary amides (**2**), which were subsequently cleaved from the solid support to yield a mixture of two secondary amides (**3**). The ratio of the two products reflects the relative reactivities of the two anilines in the reductive amination step. We selected 3,4,5-trifluorobenzoyl chloride as the acylating agent for two reasons: (a) Benzoyl chlorides are very reactive, thereby reducing the possibility that one of the two anilines would be acylated preferentially. (b) The three fluorine atoms are ^{1}H NMR "invisible," and the protons at the 2 and 6 positions would be shifted downfield (around 7.9 ppm) relative to the aniline protons, facilitating the measurement of product ratios by ^{1}H NMR. We elected to use ^{1}H NMR for the determination of product ratios because it allowed us to avoid the construction of standard curves, which would be required for the spectrophotometric measurement of product ratios. Furthermore, we wanted to measure both the product ratio and the combined yield for each competition experiment, and this can be accomplished in a straightforward fashion using 2,5-dimethylfuran (DMFu) as an NMR internal standard (48).

We selected four anilines to study in our initial set of competition experiments: *p*-anisidine, *p*-toluidine, aniline, and *p*-chloroaniline. As listed in Table 1, the four anilines require six competition experiments in order to satisfy all possible binary combinations. The uniformity of the combined yields provided a level of comfort that the acylation step proceeded unselectively.

Table 2 summarizes the product ratios obtained from the six competition experiments. These data were used to generate a Hammett plot (log ratio vs. σ) for each aniline as shown in Figure 2. Initially, we had hoped that a linear relationship would exist between σ and the log of the product ratio, but our attempts to fit a straight line to each data set were suspect ($R^2 < .95$). However, when a "smoothed" line is drawn through each data series as shown in Figure 3, the nonlinearity of the data becomes readily apparent. Under these circumstances, the value of conducting the combinatorial set of competition experiments is evident,

Table 1 Experimental Design and Combined Yields for Six Competition Experiments Using Four Substituted Anilines

Experiment	Aniline A	Aniline B	Combined yield
1	*p*-Anisidine	*p*-Toluidine	85%
2	*p*-Anisidine	Aniline	80%
3	*p*-Anisidine	*p*-Chloroaniline	84%
4	*p*-Toluidine	Aniline	88%
5	*p*-Toluidine	*p*-Chloroaniline	82%
6	Aniline	*p*-Chloroaniline	80%

Table 2 Product Ratios Obtained from Six Competition Experiments

Aniline (X)	σ	log(p-OMe/X)	log(p-Me/X)	log(H/X)	log(p-Cl/X)
p-OMe	−0.27	0	−0.39	−1.01	−1.35
p-Me	−0.17	0.39	0	−0.64	−0.83
H	0	1.01	0.64	0	−0.22
p-Cl	0.23	1.35	0.83	0.22	0

because one can use the data from four Hammett plots (as opposed to one) in deciding whether a linear relationship exists or not. In the case of Figure 3, the similarity of the four Hammett plots provides compelling evidence that an errant data point is not the source of the nonlinearity.

From a chemical perspective, the nonlinearity of the Hammett plots for the reductive amination of the BAL linker with substituted anilines indicates either a change in mechanism or a change in the rate-determining step. We believe that the latter explanation is more likely, because the reductive amination of an alde-

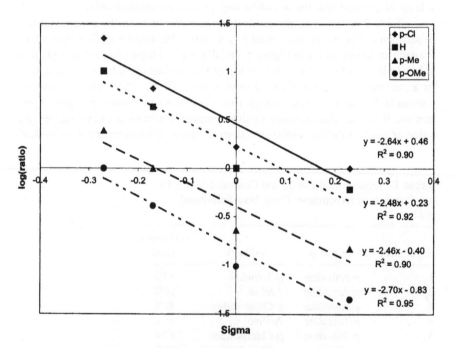

Figure 2 The Hammett plot of the data in Table 2, fit to a straight line.

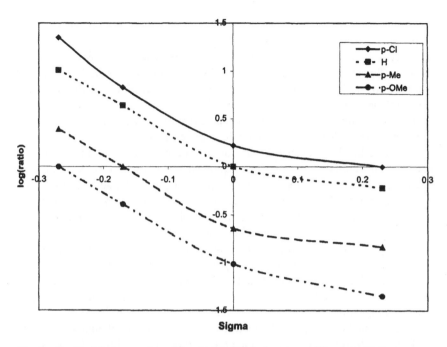

Figure 3 The Hammett plot of the data in Table 2, using a "smoothed" line.

hyde with an aniline involves multiple chemical steps, each of which would be expected to display differential sensitivities to changes in the electronic environment of the aniline. Whereas this information is valuable in a qualitative sense, the lack of a linear relationship between the log of the product ratio and σ renders the determination of ρ unfeasible. Lacking quantitative information (i.e., ρ) to guide our efforts, we abandoned our Hammett study of reductive amination and turned our attention to the formation of amide bonds via the displacement of a pentafluorophenyl ester, a reaction we hoped would not exhibit the nonlinear effects that we observed with reductive amination.

V. SOLID-PHASE HAMMETT CASE STUDY 2: AMIDE FORMATION

As mentioned in the introduction, the formation of amide bonds on solid support has a long and rich history (10,11). Typically, amide bonds are formed on the solid support via reaction of a solid-supported amine with a solution-phase activated carboxylic acid. The reverse approach (i.e., reaction of a solid-supported activated carboxylic acid with a solution-phase amine) is less common and, in our experience, more capricious. We have found that activation of solid-supported carboxylic acids via formation of the corresponding pentafluorophenyl (pfp) ester is very efficient and robust (49,50). The stability of pfp esters is well documented, yet they are sufficiently reactive to be displaced by substituted anilines (51,52). As in the previous example, substituted anilines are ideal reagents for generating Hammett relationships, so we turned our attention to the synthesis of a solid-supported pfp ester for use in competition experiments.

As outlined in Scheme 2, our competition experiments commenced with the attachment of adipic acid to Rink-equipped (53) polystyrene (PS) resin under standard conditions to afford the solid-supported carboxylic acid. The acid was treated with pfp-trifluoroacetate to provide the corresponding pfp ester, which was subsequently reacted with a 1:1 mixture of two anilines to afford a binary mixture of solid-supported secondary amides. The products were cleaved from the solid support using TFA to provide a binary mixture of secondary amides (5).

We selected five anilines for the competition experiments: p-anisidine, p-toluidine, aniline, p-fluoroaniline, and p-chloroaniline. In keeping with the combinatorial design, 10 competition experiments were required to compare every possible combination of anilines. We conducted each competition experiment in duplicate, and the two ratios obtained for each combination of anilines are listed in Table 3. The reproducibility of the product ratios was within 5.0% for the two sets of competition experiments, which is within the accuracy limits of ^{1}H NMR. The combinatorial Hammett plot for the first set of competition experiments is shown in Figure 4 and provides a much better linear correlation ($R^2 > 0.95$ in all five cases) than the previous case study. In addition, the slopes of the five lines are very consistent, ranging between -3.40 and -3.45. A similar Hammett

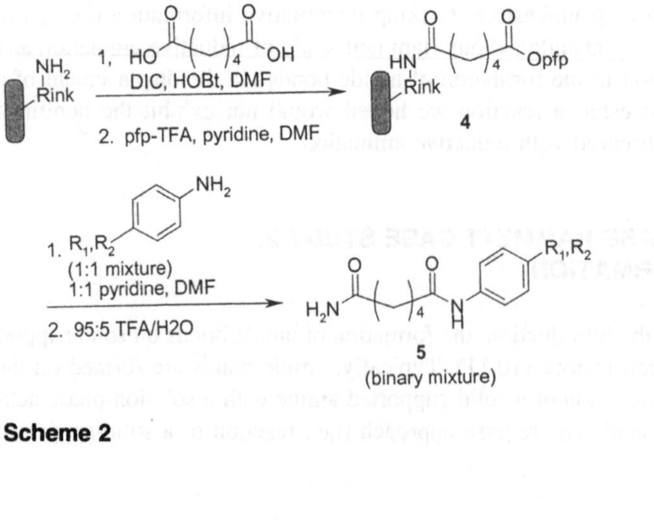

Scheme 2

Table 3 Product Ratios Obtained from PS Resin Following Two Sets of 10 Competition Experiments

Aniline (X)	Sigma (σ_p)	log(ratio) p-OMe/X		log(ratio) p-Me/X		log(ratio) H/X		log(ratio) p-F/X		log(ratio) p-Cl/X	
		1st set	2nd set	1st set	2nd set	1st set	2nd set	1st set	2nd set	1st set	2nd set
p-OMe	-0.27	0	0	-0.55	-0.52	-1.14	-1.14	-1.08	-1.14	-1.82	-1.82
p-Me	-0.17	0.55	0.52	0	0	-0.52	-0.58	-0.51	-0.50	-1.26	-1.32
H	0.00	1.14	1.14	0.52	0.58	0	0	0.03	0.09	-0.70	-0.73
p-F	0.06	1.08	1.14	0.51	0.50	-0.03	-0.09	0	0	-0.77	-0.75
p-Cl	0.23	1.82	1.82	1.26	1.32	0.70	0.73	0.77	0.75	0	0

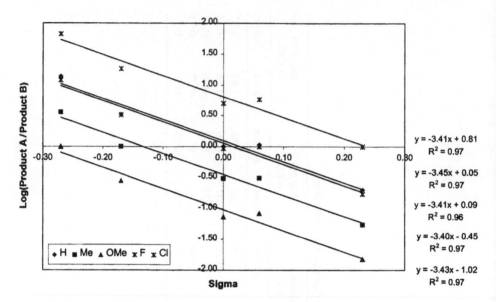

Figure 4   The "combinatorial" Hammett plot for the first set of competition experiments with PS resin.

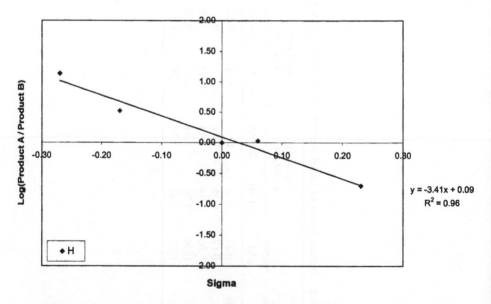

Figure 5   The "traditional" Hammett plot for the first set of competition experiments with PS resin.

plot was constructed from the second set of competition experiments, and the resulting 10 ρ values afforded an average ρ of -3.45 and a standard error of 0.01.

In this context, we can derive the full benefit of the combinatorial Hammett plot. For the sake of comparison, two traditional Hammett plots (substituted anilines versus aniline) were generated from the same data set. The first of these is presented in Figure 5, and although it looks like a good linear free energy relationship, it lacks the visual impact provided by the multiple parallel lines in the combinatorial Hammett plot of Figure 4. More significantly, even though the average ρ value for the two traditional Hammett plots is identical to the value obtained from the two combinatorial Hammett plots (-3.45 in both cases), the standard error is fourfold greater (0.04 vs. 0.01). The difference in error is significant, particularly if one is interested in comparing ρ values across a variety of reaction conditions.

IV. APPLICATION OF SOLID-PHASE HAMMETT EXPERIMENTS TO REACTION OPTIMIZATION

The quality of the Hammett plots generated from the pfp ester-based competition experiments prompted us to change some reaction variables and measure the effect of these changes on ρ. Some of the most significant (and most obvious) reaction variables to examine for a solid-phase reaction are the chemical and physical properties of the solid support. Perhaps the biggest advantage of solid-phase competition experiments is that multiple solid supports can be present in a single reaction flask, provided that they can be separated from one another after the competition experiment. This advantage not only increases the efficiency of conducting the competition experiments but also eliminates the possibility of flask-to-flask variability, which is a significant source of experimental error.

The role of the solid support in solid-phase synthesis has been the subject of much speculation in the literature. Clear differences have been observed when different solid supports have been subjected to identical reaction conditions, but the sources of these differences are anything but clear (18–24). The Hammett equation has been used to quantify solvent effects in solution-phase reactions, and in many cases changing the dielectric constant of the reaction solvent has had a measurable impact on ρ (34–38). By analogy, we felt that we could use competition experiments with a number of different solid supports to determine the effect of the solid support on ρ. Utilizing the same amide bond-forming chemistry as that described in Scheme 2, we selected seven solid supports for our comparison studies. The chemical structures of the seven supports are listed in Table 4. Our experimental procedures for the "multiple supports in one flask" competition experiments were identical to those used for a single support, and

Table 4 Chemical Descriptions for Solid Supports Used in Competition Experiments

Resin	Base polymer[a]	Graft[a]	Loading[b]
MA/DMA crowns	PE	MA/DMA	7.6 μmol/crown
LLPS crowns	PE	PS	8.0 μmol/crown
HLPS crowns	PE	PS	26.0 μmol/crown
PS lanterns	PE	PS	35.0 μmol/lantern
PTFE tubes	PTFE	PS	35.0 μmol/tube
PS resin	PS	None	0.53 mmol/g
PS-PEG resin	PS	PEG	0.20 mmol/g

[a] MA/DMA = methacrylic acid/dimethyl acrylamide copolymer; PE = polyethylene; PS = polystyrene; PTFE = polytetrafluoroethylene; PEG = 3000–4000 MW poly(ethylene glycol).
[b] As reported by the manufacturer.

we maintained a 15-fold excess of reagents relative to the total amount of solid-supported pfp ester present in each reaction. Upon completion of the competition experiments, the solid supports were separated and combinatorial Hammett plots were constructed for each support. As before, the competition experiments were conducted in duplicate, and the resulting ρ values were averaged.

A solution-phase ρ value has not been reported for the displacement of pfp ester with nucleophiles, so we devised a set of solution-phase competition experiments to determine this ρ value. As described in Scheme 3, our solution-phase pfp ester **5** was synthesized from its solid-phase precursor **4** via TFA cleavage from the solid support. Following removal of excess TFA, **5** was reacted in competition experiments under the same conditions as in the solid-phase competition experiments. Product isolation was much more difficult in solution, and as a result we did not run the solution-phase experiments in duplicate.

Table 5 summarizes all of the solid-phase Hammett experiments conducted with seven solid supports and in solution. The average ρ values for the various supports are quite consistent, and the observed variability is low enough that significant differences can be seen for a number of supports. The time required to conduct the 10 competition experiments with the seven solid supports (which yielded 70 data points) was substantially less than that required for the 10 solu-

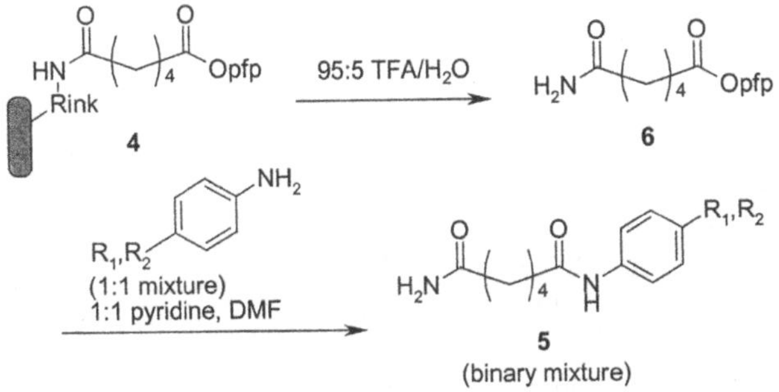

Scheme 3

tion-phase competition experiments (which yielded 10 data points), due entirely to the difficulties encountered in isolating the solution-phase products.

The range of ρ values obtained for the pfp ester displacement reactions carried out on seven solid supports and in solution is most apparent when viewed graphically, as in Figure 6. In particular, the correlation between the solution phase and PS-PEG is striking, as it provides a piece of quantitative evidence that PEG provides a "solution-like" environment for this particular reaction. Of the remaining supports, PS resin, PS lanterns, and HLPS crowns afford roughly the same ρ value at a loading level that is synthetically attractive. The PTFE tubes, MA/DMA crowns, and LLPS crowns are less attractive, mainly because of low loading levels and/or lack of commercial availability. Because the solid support is one reaction variable of many, we selected three solid supports to study in subsequent experiments. The three supports we selected were PS-PEG resin, HLPS crowns, and MA/DMA crowns, and they were chosen to represent a wide variety of ρ values and chemical grafts.

Using these three solid supports, we examined the effect on ρ of solvent (DMF vs. CH_2Cl_2), temperature (25°C vs. 40°C), and the presence or absence of dimethylaminopyridine (DMAP). Rather than examine every variable individually, we conducted three experiments and compared the resulting ρ values with the value obtained under standard conditions (DMF, 25°C, no DMAP). Our findings are summarized in Figure 7. In general, we observed that ρ is strongly influenced by the reaction solvent, with CH_2Cl_2 providing a smaller absolute value of ρ relative to DMF. In the case of MA/DMA crowns, the magnitude of this change highlights the highly synergistic relationship that exists between the reaction solvent and the solid support. The presence of DMAP and addition of heat had a much smaller effect on ρ across the board, but they may serve to increase the overall reaction rate and thus their negligible effect on ρ is still noteworthy.

Table 5 ρ Values Obtained from Hammett Plots for Seven Solid Supports and in Solution

Resin	p-OMe		p-Me		H		p-F		p-Cl		Average ρ	Std error[a]
	1st set	2nd set	1st set	2nd set	1st set	2nd set	1st set	2nd set	1st set	2nd set		
LLPS crowns[b]	−3.78	−3.81	−3.57	−3.46	−3.40	−3.63	−3.54	c	−3.79	−3.70	−3.52	0.03
MA/DMA crowns[b]	−3.50	−3.81	−3.24	−3.66	−3.65	−3.30	−3.52	−3.40	−3.34	−3.94	−3.49	0.05
PS resin	−3.50	−3.43	−3.45	−3.40	−3.49	−3.41	−3.53	−3.45	−3.46	−3.41	−3.45	0.01
PTFE tubes	−3.10	−3.44	−3.51	−3.41	−3.21	−3.61	−3.45	−3.44	−3.19	−3.40	−3.38	0.05
PS lanterns[d]	−3.34		−3.19		−3.70		−3.38		−3.21		−3.36	0.09
HLPS crowns	−3.11	−3.21	−3.45	−3.30	−3.54	−3.48	−3.40	−3.24	−3.14	−3.21	−3.31	0.05
Solution phase[d]	−3.24		−2.80		−3.12		−3.31		−2.99		−3.09	0.09
PS-PEG resin	−3.07	−2.84	−3.11	−3.31	−3.21	−3.00	−2.94	−3.08	−3.12	−2.99	−3.09	0.03

[a] Standard error = $\sigma/\sqrt{N}$, where σ = standard deviation and N = number of observations.
[b] A third set of competition experiments were carried out and included in the average (see supplementary material for data).
[c] Hammett plot not constructed due to missing data.
[d] Only one set of competition experiments were conducted.

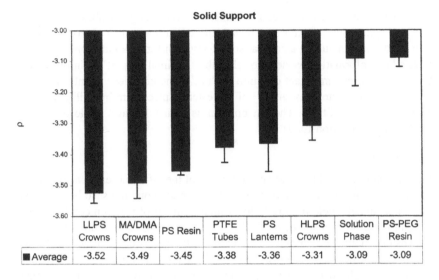

Figure 6   Average ρ values obtained for seven solid supports and in solution.

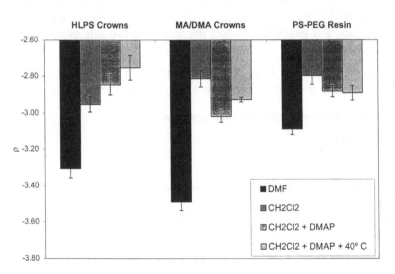

Figure 7   The effect of solvent, temperature, and the addition of DMAP on ρ for HLPS crowns, MA/DMA crowns, and PS-PEG resin.

All of the aforementioned experiments afford a large database of ρ values for the displacement of pfp esters by substituted anilines under a variety of reaction conditions, including a comparison of seven different solid supports. Do we have all the information we need to select the "optimal support"? Unfortunately, the answer to this important question is "no." One key piece of information is still missing: the rate constant (k^0) of pfp ester displacement by aniline for each set of reaction conditions. The rate constant is critical because it allows us solve the Hammett equation for k:

$$k = 10^{\sigma\rho}k^0 \qquad (3)$$

Ideally, a single k^0 value could be used for all of the different reaction conditions, but just as each ρ value is unique to a set of reaction conditions, so is each value of k^0. Fortunately, determining k^0 for a small number of reaction conditions is a straightforward proposition, particularly if macroscopic solid supports such as crowns or lanterns are employed. In addition, the precision of these rate studies is less critical than that of those conducted for the determination of ρ, because k is linearly related to k^0 but exponentially related to ρ. As a result, in many cases one can simply determine the half-life for a given reaction and derive k^0 from the half-life using the equation

$$t_{1/2} = \frac{0.693}{k} \qquad (4)$$

With this information in hand, the "optimal" set of conditions can be selected by calculating the half-lives for the range of substituents to be reacted under each set of conditions. For example, consider the following two sets of conditions using MA/DMA crowns:

1. In DMF, $\rho = -3.49$ and $k^0 = 2.5 \times 10^{-3}$ M^{-1} s^{-1}
2. in CH_2Cl_2, $\rho = -2.81$ and $k^0 = 9.2 \times 10^{-4}$ M^{-1} s^{-1}.*

Combining Eqs. (3) and (4) provides the equation

$$t_{1/2} = \frac{0.693}{10^{\sigma\rho}k^0} \qquad (5)$$

which facilitates the direct comparison of calculated half-lives for each set of conditions, as shown in Table 6.

Table 6 illustrates the complex relationship between k^0 and ρ, because the calculated half-lives of the reactions conducted in CH_2Cl_2 are significantly longer

---

* These k^0 values are supplied for the purpose of calculating half-lives only and were not determined under appropriately rigorous conditions.

Table 6 Comparison of Calculated Rate Constants and Half-Lives for the Displacement of pfp Esters on MA/DMA Crowns

Substituent	σ	DMF (ρ = −3.49)		CH$_2$Cl$_2$ (ρ = −2.81)	
		k	t$_{1/2}$ (min)	k	t$_{1/2}$ (min)
p-MeO	−0.28	2.37E-02	0.49	5.63E-03	2.05
p-Me	−0.14	7.70E-03	1.50	2.28E-03	5.07
H	0.00	2.50E-03	4.62	9.20E-04	12.55
p-Cl	0.22	4.27E-04	27.07	2.22E-04	52.12
p-CO$_2$Me	0.44	7.28E-05	158.58	5.34E-05	216.37
p-NO$_2$	0.81	3.72E-06	3101.29	4.87E-06	2370.80

than those in DMF until substituents with large (>0.5) σ values are examined. As a result, the "optimal" conditions for a given chemical reaction depend upon the range of σ values to be studied. For a narrow range of σ values (e.g., the p-OMe, p-Me, and p-Cl substituents in Table 6), k^0 is the most important variable to optimize, and as the range of σ values widens, ρ becomes increasingly important. An alternative way to view this issue is to consider whether the σ values for some substituents are so large as to be untenable from a library production perspective. For example, the p-NO$_2$ entry in Table 6 has a calculated half-life of nearly 40 h. The inclusion of this monomer in a library synthesis would have a significant impact on the time required to ensure complete reaction. In this respect, one can utilize k^0 and ρ for monomer selection based on calculated half-lives, and this approach is described in the next section.

VII. APPLICATION OF SOLID-PHASE HAMMETT RELATIONSHIPS TO MONOMER SELECTION

The selection of monomers for inclusion in a combinatorial library is a highly iterative process, because generally the selected monomers must satisfy a minimum of three criteria:

1. Each monomer must be sufficiently reactive to undergo the desired transformation in a reproducible manner.
2. Each monomer must not contain functional groups that are incompatible with subsequent chemical steps.
3. Each monomer must fit the chemical property requirements as defined by the library design hypothesis.

Monomer selection is complex because the first two criteria are "reactivity-based," whereas the third criterion is "chemical property based," and frequently the two sets of criteria are mutually exclusive. To make matters worse, monomer scans are typically used to empirically determine whether a monomer satisfies the reactivity criteria, whereas the chemical properties are calculated computationally. As a result, one cannot easily choose moomers that satisfy all three criteria without using a combination of computational and synthetic techniques. A potential solution to this problem is to predict the reactivity of monomers using σ and ρ, thereby eliminating the monomer scans.

Even in the absence of absolute rate information, σ and ρ can be used to predict the range of reaction times as shown in Table 7. In this context, the exponential relationship between reaction rate and the σρ product is most apparent. From this simple analysis, one can readily select a "reactivity cutoff" for monomers using σ values that are readily available from the literature. Alternatively, using the predicted half-lives, one could extend the reaction times for less reactive monomers.

There are significant limitations to using σ values to perform monomer selection, and they all relate to the lack of a "universal" σ value that effectively describes both steric and electronic substituent effects. A number of methods for combining steric and electronic substituent effects have been reported (54–57), albeit in model systems, and the evaluation of these methods in the context of real-world chemistries is an obvious next step. The need for a robust method for predicting substituent effects will only increase as more chemists get involved in combinatorial chemistry and encounter the Hammett relationships intrinsic to library synthesis. As described in the next section, the use of competition experiments to determine new σ values appears to be a valuable method for addressing this issue.

Table 7 The Impact of σ and ρ on Reaction Time, Expressed as a Multiple of the Reaction Half-Life

		Half-life multiple			
Substituent	σ	ρ = −4.0	ρ = −2.0	ρ = 2.0	ρ = 4.0
p-MeO	−0.28	0.1	0.3	3.6	13.1
p-Me	−0.14	0.3	0.5	1.9	3.6
H	0.00	1.0	1.0	1.0	1.0
p-Cl	0.22	7.6	2.8	0.4	0.1
p-CO$_2$Me	0.44	57.5	7.6	0.1	0.02
p-NO$_2$	0.81	1737.8	41.7	0.02	0.0005

VIII. FUTURE APPLICATION: THE DETERMINATION OF NEW σ CONSTANTS VIA SOLID-PHASE HAMMETT RELATIONSHIPS

Although the symbol σ has been used throughout this chapter without subscripts or superscripts, a common notation has evolved to address the fact that multiple σ values have been reported for many substituents (32,33). Even though σ was originally defined as a "constant," the same substituent can provide different values for σ depending upon the system being studied. The notation describing different σ values is as follows:

1. Subscripts indicate the position of the substituent on a phenyl ring; thus σ_p describes a para substituent and σ_m describes a meta substituent.
2. Superscripts denote a special category of σ values that are to be used when the reaction center is capable of accepting (σ_p^+) or donating (σ_p^-) a pair of electrons to the substituent via conjugation. These superscripts are applicable only to para substituents.

In order to experimentally determine new σ constants that are directly comparable to the existing database of σ constants, one must either use the experimental system that provided the original data or demonstrate conclusively that an alternative system duplicates the established values. For example, Hammett's original σ values were defined by the dissociation constants of substituted benzoic acids in water at 25°C (27). Although this system is sufficient for many substituents, many substituted benzoic acids are insoluble under these conditions. As a result, alternative systems have been devised that circumvent the solubility problems yet provide reliable σ values for the previously reported substituents and substituents that were incompatible with the original system.

The competition experiments and combinatorial Hammett plots we have described in this chapter are ideally suited for the determination of σ values for new substituents. In particular, competition experiments using a reaction with a well-defined ρ value would allow us to directly compare the reactivity of a new substituent with substituents for which σ is already known. The ratio of products would then define σ because ρ is known. This approach imparts a high degree of validity to each new σ value, because it is based on a "head-to-head" comparison with a known substituent. Obviously, there are pitfalls to this approach—one wouldn't want to compare two substituents that might interact with one another, for instance—but the throughput advantages are difficult to resist.

The key hurdle to determining new σ values using solid-phase competition experiments is finding a reaction that shows a strong correlation with existing σ values. In other words, one would hope to construct a combinatorial Hammett plot with 20 substituents affording 20 linear relationships with an $R^2 > 0.95$ and an average ρ value with a standard error of less than 0.02. This Hammett plot would require only 190 competition experiments and yet would define an extremely robust system that could then be used to determine an unlimited number of σ values. Our current research efforts are directed at identifying a system that meets these criteria, and we are also collaborating with computational chemists to identify computational models for predicting new σ values.

IX. CONCLUSION

Even though the Hammett equation is taught in many undergraduate and graduate chemistry courses, chemists often fail to appreciate how often ρ and σ influence the chemical reactions they are conducting. Never has this been more apparent than with the synthesis of combinatorial libraries, in which virtually every monomer set contains a Hammett relationship waiting to be discovered.

Unfortunately, these Hammett relationships are usually discovered in a negative fashion: Monomers that look like they should work do not; some reactions go to completion, whereas others do not. However, if one is willing to take the time to characterize each solid-phase reaction in terms of its Hammett reaction constant ρ, many of these common scenarios can be avoided. Hopefully, the experimental techniques and applications described in this chapter will enable both combinatorial chemists and physical organic chemists to begin to utilize the Hammett equation to its fullest potential.

REFERENCES

1. RB Merrifield. J Am Chem Soc 85:2149, 1963.
2. GB Fields. Solid-phase peptide synthesis. In: R Rapley, JM Walker, eds. Molecular Biomethods. Totowa, NJ: Humana, 1998, pp 527–545.
3. B Seligmann, M Lebl, KS Lam. Solid-phase peptide synthesis, lead generation, and optimization. In: EM Gordon, JF Kerwin, Jr, eds. Combinatorial Chemistry and Molecular Diversity in Drug Discovery. New York: Wiley-Liss, 1998, pp 39–109.
4. G Barany, N Kneib-Cordonier, DG Mullen. Int J Peptide Protein Res 30:705, 1987.
5. SL Beaucage, RP Iyer. Tetrahedron 48:2223, 1992.
6. D Hudson. J Comb Chem 1:333, 1999.
7. D Hudson, J Comb Chem 1:403, 1999.
8. TW Greene, PGM Wuts. Protective Groups in Organic Synthesis. 3rd ed. New York: Wiley, 1999.
9. PM Hardy. Chem Ind (Lond) 1979: 617.
10. JH Jones. Amino Acid and Peptide Synthesis. Oxford, UK: Oxford Univ Press, 1992.
11. F Albericio, LA Carpino. Methods Enzymol 289:104, 1997.
12. LA Thompson, JA Ellman. Chem Rev 96:555, 1996.

13. JS Fruchtel, G Jung. Angew Chem Int Ed Engl 35:17, 1996.
14. PHH Hermken, HCJ Ottenheijm, D Rees. Tetrahedron 52:4527, 1996.
15. PHH Hermken, HCJ Ottenheijm, D Rees. Tetrahedron 53:5643, 1997.
16. S Booth, PHH Hermken, HCJ Ottenheijm, D Rees. Tetrahedron 54:15385, 1998.
17. JW Corbett. Org Prep Proced Int 30:489, 1998.
18. AW Czarnik. Biotechnol Bioeng (Comb Chem) 61:77, 1998.
19. B Merrifield. Br Polym J 16:173, 1984.
20. R Santini, MC Griffith, M Qi. Tetrahedron Lett 39:8951, 1998.
21. W Li, B Yan. J Org Chem 63:4092, 1998.
22. W Li, X Xiao, AW Czarnik. J Comb Chem 1:127, 1999.
23. B Yan, JB Fell, G Kumaravel. J Org Chem 61:7467, 1967.
24. B Yan. Acc Chem Res 31:621, 1998.
25. C Reichardt. Solvents and Solvent Effects in Organic Chemistry. 2nd ed. New York: Wiley, 1988.
26. G Illuminati. Tech Chem (NY) 8:159, 1976.
27. LP Hammett. J Am Chem Soc 59:96, 1937.
28. PR Wells. Linear Free Energy Relationships. New York: Academic Press, 1968.
29. C Hansch, A Leo. Exploring QSAR. Washington, DC: Am Chem Soc, 1995.
30. O Exner. Correlation Analysis of Chemical Data. New York: Plenum Press, 1988.
31. M Page, A Williams. Organic and Bio-Organic Mechanisms. Harlow, UK: Addison Wesley Longman, 1997.
32. O Exner. In: NB Chapman, J Shorter, eds. Correlation Analysis in Chemistry—Recent Advances. New York: Plenum Press, 1978, p 8.
33. C Hansch, A Leo. Substituent Constants for Correlation Analysis in Chemistry and Biology. New York: Wiley, 1979.
34. HH Jaffe. Chem Rev 53:191, 1953.
35. H Bartnicka, I Bojanowska, MK Kalinowski. Aust J Chem 46:31, 1992.
36. LT Kanerva, AM Klibanov. J Am Chem Soc 111:6864, 1989.
37. T Matsui, N Tokura. Bull Chem Soc Jpn 44:756, 1971.
38. Y Kondo, T Matsui, N Tokura. Bull Chem Soc Jpn 42:1037, 1969.
39. SW Gerritz, RP Trump, WJ Zuercher. J Am Chem Soc 122:6357, 2000.
40. PAS Smith, EM Bruckmann. J Org Chem 39:1047, 1974.
41. H Yamataka, T Takatsuka, T Hanafusa. J Org Chem 61:722, 1996.
42. H Yamataka, N Fujimura, Y Kawafuji, T Hanafusa. J Am Chem Soc 109:4305, 1987.
43. MM Diaz-Requejo, TR Belderrain, MC Nicasio, PJ Perez. Organometallics 19:285, 2000.
44. KJ Jensen, J Alsina, M Songster, J Vagner, F Albericio, G Barany. J Am Chem Soc 120:5441, 1998.
45. J Alsina, KJ Jensen, F Albericio, G Barany. Chem Eur J 5:2787, 1999.
46. NJ Maeji, AM Bray, RM Valerio, W Wang. Peptide Res 8:33, 1995.
47. NJ Maeji, RM Valerio, AM Bray, RA Campbell, HM Geysen. React Polym 22:203, 1994.
48. SW Gerritz, AM Sefler. J Comb Chem 2:39, 2000.
49. JA Linn, SW Gerritz, AL Handlon, CE Hyman, D Heyer. Tetrahedron Lett 40:2227, 1999.

50. M Green, J Berman. Tetrahedron Lett 31:5851, 1990.
51. L Kisfaludy, JE Roberts, RH Johnson, GL Mayers, J Kovacs. J Org Chem 35:3563, 1970.
52. B Penke, L Balaspiri, P Pallai, K Kovacs. Acta Phys Chem 20:471, 1974.
53. H Rink. Tetrahedron Lett 28:3787, 1987.
54. RW Taft. J Am Chem Soc 74:3120, 1952.
55. RW Taft. In: MS Newmann, ed. Steric Effects in Organic Chemistry. New York: Wiley, 1956, Chap 13.
56. T Fujita, T Nishioka. Prog Phys Org Chem 12:49, 1976.
57. DF DeTar, C Delahunty. J Am Chem Soc 105:2734, 1983.

13

Multispectral Imaging for the High Throughput Evaluation of Solid-Phase-Supported Chemical Libraries

Hicham Fenniri and Ryan M. McFadden
Purdue University, West Lafayette, Indiana

I. INTRODUCTION

Combinatorial synthetic methods allow for the preparation of large arrays of compounds as mixtures or individual entities (1–14). In the latter case, split synthesis has proven to be ideal in maximizing the number of compounds generated per synthetic step (15). Although the composition of chemical libraries produced using this method is predictable with a high level of confidence, the structural elucidation of potent members subsequent to an activity assay remains a challenge.*

To unravel the chemical nature of the active compounds several direct methods such as MAS-NMR (16), FTIR (17), MS (18), XPS (19), and X-ray diffraction (20) or indirect methods such as deconvolutive strategies (21–23), readout of chemical tags (24–27), bar codes (28), or radiofrequency signals (29,30) were developed. These screening and structural characterization phases have consistently been the bottleneck in the overall discovery process and still constitute in most cases a challenging task. This chapter summarizes our efforts toward the noninvasive and high throughput evaluation of spectroscopically self-encoded libraries using Fourier transform infrared (FTIR) and Raman multispectral imaging.

---

* Generally a colorimetric, fluorescence, or radiographic assay. See Refs. 1–14

II. FTIR IMAGING

Instrumentation developed to simultaneously acquire spatial and spectral information has in the past coupled traditional single-point spectroscopy with mapping techniques. A single-point spectrum is acquired at a known spatial location, the sample is moved to a contiguous location, and another spectrum is acquired. By incrementing the position of the sample over a region of interest and acquiring a single-point spectrum at each point, a chemical "map" of the sample is obtained (31). There are, however, significant drawbacks to this implementation; for example, spatial resolution for mapping techniques is affected by the mechanical movement of the sample. Also, the use of physical apertures to limit the area being investigated cuts back on signal levels, making it difficult to get good data from small areas. Furthermore, because the mapping measurements are obtained in a sequential manner, the mapping process becomes very time-consuming for large sample areas.

Based on newly developed array detector technology, spectroscopic imaging techniques bypass the problems associated with single-point mapping and simultaneously reveal the type, distribution, and relative abundance of chemical components within a particular system.* A vibrational spectrum is collected for each pixel on the array detector, creating a three-dimensional cube consisting of both spatially resolved spectra and wavelength-dependent images.

The image cube can be seen as a series of wavelength-resolved images or, alternatively, as a series of spatially resolved spectra, one for each point on the image. The complete integration of spatial and spectral information quite literally adds a new dimension to data analysis: The ability to explore the interdependence of spectral and spatial information is the basis for its unique capabilities, which differ qualitatively from simple point-by-point spectroscopy. In the infrared (mid- and near-infrared) and Raman domains, it is possible to obtain information about molecular structure and interactions among individual sample components.

A. FTIR Imaging Instrumentation

The experimental setup initially used in our study is a home-made instrument consisting of a step-scan FTIR spectrometer (Nicolet Magna 860), CaF_2 condensing lenses (25 mm diameter, infinite focal length), CaF_2 objective lenses (25 mm diameter, magnification 1.6), and a 64×64 mercury cadmium telluride (MCT) focal plane array (FPA) detector (Fig. 1).† This instrument is thus capable of

---

* FTIR imaging has recently emerged as a technique for the analysis of chemical and biological systems. See Refs. 32–34.
† This instrument is currently housed in the Purdue University FTIR Imaging Facility within the Department of Chemical Engineering. For a detailed description of this instrument see Ref. 35.

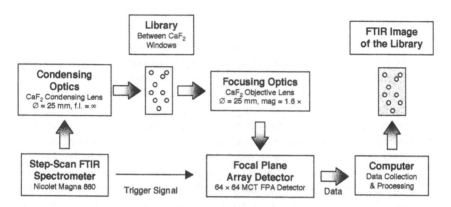

Figure 1 Block diagram of the FTIR imaging instrument. Infrared radiation modulated by the Nicolet Magna 860 step-scan spectrometer (external beam path width ~1 in. diameter) is focused through a set of condensing optics onto the sample plane that holds the chemical library. The radiation that passes through the sample is selectively absorbed by the library members. The light is then refocused onto the focal plane array (FPA) detector using focusing optics, and the data points of the interferogram for each pixel in the array are collected at each mirror position. The interferograms are then stored on a computer workstation, where they are processed to produce a set of spatially resolved infrared spectra.

collecting an image consisting of 4096 spatially resolved spectra in the 1000–4000 cm^{-1} spectral range with 8 cm^{-1} spectral resolution in a few minutes. With the specific optical setup used in the present study, the spatial resolution is under 50 μm and the field of view is 3 × 3 mm. However, the optics can be tailored for applications to smaller or larger samples. The resolution and data acquisition speed can be remarkably improved if larger focal plane array detectors are used.

Several advanced FTIR imaging instruments are currently on the market. For instance, Oracle IR (Spectral Dimensions Inc.), an FTIR spectral imaging system composed of a Nicolet Nexux 870 step-scan interferometer and an imaging microscope (256 × 256 MCT focal plane array) uses up to 65,536 detectors simultaneously. The Falcon 4M composed of Bruker's step-scan IRscope II and a 64 × 64 MCT focal plane array has a spectral range of 1–11 μm. Other detectors for the near-IR and mid-IR are also available. Finally, the Bio-Rad FTS 6000 Stingray imaging system with a free scanning range of 0–16 μm and a variety of detectors (narrowband MCT, broadband MCT, InSb) has an optimal field of view of 10 × 10 mm and a resolution of 3.1 μm. The resolution depends on the dimensions of the field of view (operator's choice). With the 256 × 256 MCT array, when the field of view is 0.8 × 0.8 mm the resolution is 3.1 μm, and when it is 10 × 10 mm the resolution is 40 μm.

B. Specific FTIR Imaging of Resin Beads

To test this methodology, we performed the identification of 130 µm TentaGel resin-bound Sharpless's $(DHQD)_2PHAL$ ligand in a mixture of sixty 90 µm Merrifield bead-supported amino acids comprising 11 natural amino acids.*

A mixture of the beads was placed on a polished CaF_2 window and swollen with methylene chloride. The solvent was allowed to evaporate, and another CaF_2 window was placed on top. Slight pressure was maintained in order to flatten the beads and prevent absorbance saturation effects. When these beads were placed in the field of view of the instrument, spectral information from each individual bead was collected in a single experiment, as illustrated in Figure 2. After imaging, the beads were recovered for further analysis and without damage by simply adding solvent, which caused them to resume their spherical shape. The raw intensity image (collected over all wavelengths) shows the spatial arrangement of all beads positioned between the CaF_2 windows (Fig. 2A). In order to identify the spatial position of beads with the $(DHQD)_2PHAL$ ligand, a spectral image was generated using the absorbance intensity from the 1620 cm^{-1} C$=$N stretching mode, unique to the ligand (Fig. 2B). The assignment was visually confirmed in this case because the beads carrying the ligand are larger and thus readily identifiable. Following the same procedure, other amino acids could also be specifically imaged using spectral features characteristic of each one of them. It is noteworthy that in spite of the very small amount of material attached to the beads (<1 nmol/bead), high quality IR spectra could be recorded in a short period of time.

Combinatorial libraries are not necessarily built on spherical resin beads. Flat matrices such as glass (36), cellulose paper (37), and others (5–14,38) are used as support material. FTIR imaging could provide a unique opportunity for in situ and real-time visualization of any chemical activity taking place on the surface of these supports. To illustrate the potential of multispectral imaging in this area, a silicon wafer photolithographically functionalized into a grid with oxidized regions (Si—O, lines) and alkyl-terminated regions (Si-alkyl, squares) was imaged using the Bio-Rad FTS 6000 Stingray. The FTIR image revealed, in a few seconds, the oxidized versus the freshly functionalized areas (Fig. 3). The $(SiO)_n$ layer in the nonporous oxidized area is undetectable.

---

* The resin-supported amino acids studied were Boc-Ala-O-Merrifield (0.9 mmol/g); Boc-Asn-O-Merrifield (0.6 mmol/g); Boc-Glu(OBzl)-O-Merrifield (0.8 mmol/g; Boc-Lys (2-Cl-Z)-O-Merrifield (0.5 mmol/g); Boc-Met-O-Merrifield (0.9 mmol/g); Boc-Phe-O-Merrifield (0.8 mmol/g); Boc-Pro-O-Merrifield (0.9 mmol/g); Boc-Thr(OBzl)-O-Merrifield (0.6 mmol/g); Boc-Trp-O-Merrifield (0.6 mmol/g); Boc-Tyr(2-Br-Z)-O-Merrifield (0.6 mmol/g); and Boc-Val-O-Merrifield (0.8 mmol/g). Their size was in the range of 100–200 mesh (~90 µm). The TentaGel resin used for the preparation of the resin-supported $(DHQD)_2PHAL$ ligand has an average size of 130 µm and a loading of 0.2 mmol/g.

(A)

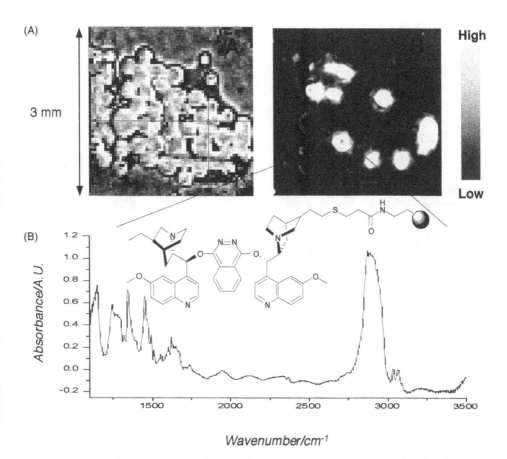

3 mm

High

Low

(B)

Figure 2 (A) Bright field image showing a collection of beads composed of 11 Merrifield-supported amino acids and TentaGel-(DHQD)₂PHAL. (B) FTIR image generated from the 1620 cm⁻¹ C=N stretching mode of the (DHQD)₂PHAL ligand.

III. DUAL RECURSIVE DECONVOLUTION*

This section introduces preliminary studies on the implementation of a screening method termed dual recursive deconvolution (DRED) that draws its strengths from chemical self-encoding techniques and deconvolution strategies. DRED operates through the iterative identification of the first and last building blocks of

---

* For more detailed information, see Ref. 39.

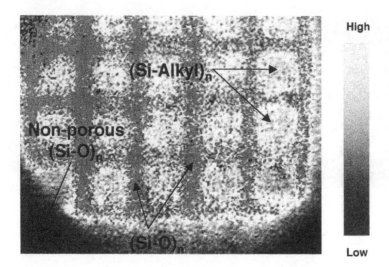

Figure 3 FTIR imaging of the oxidized versus alkyl terminated regions of a photolithographically functionalized porous silicon wafer. (Silicon supplied by Prof. JM Buriak.)

active members of combinatorial libraries generated through split synthesis. The
last building block can be readily obtained from pool screening after the last
coupling of the split synthesis, and the first position can be *encoded* in the vibrational fingerprint of the resin beads used.

Figure 4 outlines the key features of the DRED of a hypothetical 27-mem-
ber combinatorial library generated through split synthesis. The last step of this
process would generate three pools, each containing nine compounds. Screening
each of the pools separately would identify the best third position (21–23). The
key feature of the DRED method lies in the chemical nature of the beads used.
In this case each bead encodes and thus identifies the first randomized building
block. In the example of Figure 4, three spectroscopically distinguishable resin
beads encode building blocks A, F, and L at position 1. As a result the DRED
of a library would operate through the identification of the last (pool assay) and
first (encoded beads) randomized positions. This process could then be repeated
iteratively for the remaining unidentified positions until the entire sequence of the
active library members is unveiled. Remarkably, this exercise would dramatically
simplify the synthetic and screening efforts.

Table 1 summarizes the synthetic effort required for the DRED of combinatorial libraries varying in number of steps from three to six and using 10 or 20
building blocks. The number of steps for the preparation of a library using the
split synthesis (S_1) varies linearly whereas the size of the library (M) increases

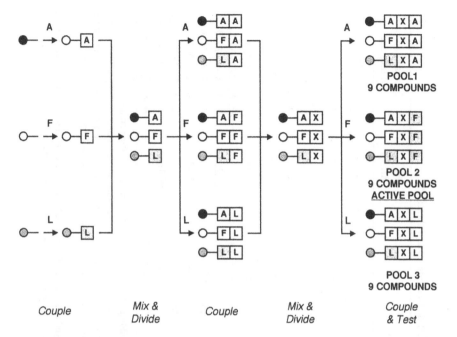

Figure 4 Illustration of the dual recursive deconvolution (DRED) strategy. Three spectroscopically distinguishable resin beads (black, white, and gray spheres) are used to encode the first building blocks (A, F, L) of the synthesis. X denotes any of the three building blocks A, F, or L. The last building block of an active member of the library could be revealed by pool assay after the last step of the split synthesis, and the first building block could be unveiled through multispectral imaging of the active beads. The gray boxes highlight the building blocks required for activity.

exponentially. The number of compounds synthesized per chemical step (R_1) increases rapidly as the library size increases, thereby highlighting the strength of the split synthesis method. Likewise, S_2 and R_2 show a similar trend except that in this case the ratio of compounds synthesized to the number of chemical steps includes the DRED synthetic steps and hence the full identification of the active member of the library. For instance, the synthesis and screening of a 64×10^6-member library would barely double the number of chemical steps required for the synthesis of the library using the split synthesis method (246 versus 120) and would involve only 20 spectroscopically distinguishable beads (DRED beads). The last column of Table 1 shows the general formula of the libraries and sublibraries to be synthesized with three to six building blocks or chemical transformations per member using both split synthesis and DRED.

Table 1 Synthetic Efforts for Dual Recursive Deconvolution (DRED) of Combinatorial Libraries[a]

L	N	$M\ (=N^L)$	S_1 $(=N\times L)$	R_1 $(=M/S_1)$	S_2	R_2 $(=M/S_2)$	Libraries to be synthesized[b]
3	10	1×10^3	30	33	42 (S_1+12)	24	1. A_1-X-A_3
	20	8×10^3	60	133	82 (S_1+22)	98	2. A_1-A_2-A_3
							3. A_1-A_2-A_3
4	10	1×10^4	40	250	62 (S_1+22)	161	1. A_1-X-X-A_4
	20	1.6×10^5	80	2,000	122 (S_1+42)	1,312	2. A_1-A_2-A_3-A_4
							3. A_1-A_2-A_3-A_4
5	10	1×10^5	50	2,000	96 ($S_1+32+14$)	1,042	1. A_1-X-X-X-A_5
	20	3.2×10^6	100	32,000	186 ($S_1+62+24$)	17,204	2. A_1-A_2-X-A_4-A_5
							3. A_1-A_2-A_3-A_4-A_5
							4. A_1-A_2-A_3-A_4-A_5
6	10	1×10^6	60	16,667	126 ($S_1+42+24$)	7,937	1. A_1-X-X-X-X-A_6
	20	64×10^6	120	533,333	246 ($S_1+82+44$)	260,163	2. A_1-A_2-X-X-A_5-A_6
							3. A_1-A_2-A_3-A_4-A_5-A_6
							4. A_1-A_2-A_3-A_4-A_5-A_6

[a] L is the number of synthetic steps per library member. N is the number of building blocks. $M=N^L$ is the library size using split synthesis. $S_1 = N\times L$ is the number of synthetic steps for the preparation of the library using the split synthesis method. R_1 corresponds to the number of compounds generated per synthetic step. S_2 is the total number of synthetic steps for library and sublibrary synthesis and for full identification of an active member of the library using DRED. R_2 is the number of compounds synthesized and screened per synthetic step. As indicated by this ratio, the larger the library the fewer the number of steps for its preparation and for the identification of the active members.

[b] General formula of the libraries and sublibraries to be synthesized using split synthesis and DRED for complete chemical identification of an active member of the library. X denoted randomized positions to be identified in subsequent sublibraries. A_n denotes randomized positions to be identified at that step. Bold A_n denotes positions identified in preceding sublibrary steps.

IV. NEAR-IR RAMAN IMAGING

To test the feasibility of this method a near-infrared Raman imaging (NIRIM) instrument was investigated (40) as a tool for the simultaneous identification of beads of various chemical compositions.* The NIRIM uses fiber bundle image compression (FIC) technology (41) to simultaneously collect a 3-D Raman spectral imaging data cube (λ–x–y) containing an optical spectrum (λ) at each spatial location (x, y) of a globally illuminated area. Thus, the light emitted from an entire imaged area may be dispersed and collected in a single charge-coupled device (CCD) optical array detector frame. Transformation of the 2-D CCD output into a 3-D spectral image requires simply mapping the CCD data into a format compatible with subsequent spectral image processing. The FIC method and closely related techniques have recently been applied to IR imaging (42), atomic emission imaging (43), and most recently to optical absorbance and fluorescence imaging (44). It should be noted that this is a real-time imaging technique as opposed to the previously reported step-scan methods (45–47), which require much longer times to generate an image of the sample and are thus impractical in the evaluation of large combinatorial libraries. For this reason, this technique is particularly suitable for the DRED methodology, because it combines expeditious multiple sample analysis with high spatial resolution (down to 1 µm) (40) and high chemical information content in the form of molecular vibrational fingerprints.

A. Raman Imaging Instrumentation

The NIRIM instrument (40) uses a near-infrared (NIR) external cavity narrowband, 400 mW, 785 nm diode laser (SDL-8630), which maximizes resolution and reduces sample fluorescence interference. The charge-coupled device (CCD) detector (Princeton Instruments LN/CCD-1024 EHRB) has a deep depletion, back-illuminated chip that is NIR-antireflection coated and roughened to virtually eliminate etaloning artifacts (quantum efficiency of 85% at 785 nm and 20% at 1050 nm). The NIRIM also uses a Kaiser Holoscop Imaging spectrograph with a 75 mm f/1.4 input lens and an 85 mm f/1.4 output lens. The image quality of this spectrograph is sufficient to image each 50 µm diameter fiber on a 2 × 2 pixel (about 54 × 54 µm) region of the CCD. Note that because the input and output focal lengths are not the same, the spectrograph has a magnification of 1.13, which restricts the number of FIC fibers that can be simultaneously detected to about 80 (representing a rectangular 8 × 10 fiber region at the collection end

---

* The NIRIM instrument is currently housed in the Purdue University Laser Facility within the Chemistry Department. *http://www.chem.purdue.edu/lfac/laser.html*

of the FIC fiber bundle). Larger spectral images are obtained simply by raster-scanning the sample over an array of adjacent rectangular regions and concatenating the resulting single-frame images to form a spectral image of an arbitrarily large area. An $N \times N$ image is assembled from $N \times N \times 80$ pixels; each pixel is in fact a 900 channel wide Raman spectrum, and the Raman shifts window is from 100 cm^{-1} to 1900 cm^{-1}. A review of all the remaining components of the NIRIM instrument, including mirrors, lenses, holographic filter, excitation fiber setup, and other design considerations is given in Ref. 40.

There is currently a commercial Raman imaging instrument composed of a Renishaw System 2000 and a Spectral Dimensions Raman imaging microscope. This instrument has a spectral range (Raman shift) of 100–6300 cm^{-1}, a spectral resolution of 8 cm^{-1} (with tunability of 1 cm^{-1}), a field of view of 200×200 µm, and a spatial resolution of 1 µm lateral ($\times 100$ objective).

B. Data Processing

The software used to either acquire or process the experimental data on the NIRIM instrument are pls_image.vi (data acquisition) (40), nirim.vi (3-D data cube acquisition) (40), and MultiSpec (spectral imaging analysis and classification) (48).* The MultiSpec program requires the user to first select known regions of the image and identify their composition (training fields). The program then uses built-in algorithms (operator's choice) to statistically determine the most likely chemical identity for each fiber's Raman output in the image. The image is then redisplayed with the fiber's Raman output color-coded as to their most likely chemical identity. In this study the images were analyzed using the spectral angle mapping (SAM) algorithm.

C. Near-IR Raman Imaging of Resin Beads

Vital to the success of the DRED method is the ability to identify resin beads based on their polymeric constituents and regardless of the chemical nature of their cargo. To address this dilemma, bead samples of 18 Merrifield resin beads carrying various protected amino acids† were placed on a sapphire single-crystal

---

* Inquiries about the MultiSpec software may be directed to Professor David Langrebe, School of Electrical and Computer Engineering, Purdue University, West Lafayette, IN 47907-1285. *http:// dynamo.ecn.purdue.edu/~biehl/multispec/*

† Ninety micrometer diameter polystyrene (1% DVB/PS) and 130 µm TentaGel-S-OH resin beads (Advanced ChemTech). The resin-supported amino acids studied (Advanced ChemTech) were Boc-Ala-O-Merriefield (0.9 mmol/g); Boc-Asn-O-Merrifield (0.6 mmol/g); Boc-Asp (OBzl)-O-Merrifield (1 mmol/g); Boc-Cys (Acm)-O-Merrifield (0.8 mmol/g); Boc-Gln-O-Merrifield (0.6 mmol/g); Boc-Glu (OBzl)-O-Merrifield (0.8 mmol/g; Boc-His (DNP)-O-Merrifield (0.6 mmol/g); Boc-Ile-O-Merrifield (0.9 mmol/g); Boc-Leu-O-Merrifield (1 mmol/g); Boc-Lys (2-Cl-Z)-O-Merrifield

[HEMEX (white), Crystal Systems, c-axis cut to eliminate fluorescence emission], positioned in the field of view of the NIRIM, and single-bead Raman spectra were recorded. Visual inspection of these spectra indicated that the spectral features were dominated by the matrix (polystyrene), and even background subtraction (unsubstituted polystyrene beads) did not reveal the spectral features of the loaded material. Hence, Raman imaging of PS-supported compounds is insensitive to the material loaded on the beads, at least up to 1.0 mmol of the amino acids studied per gram of resin. Interestingly, although the spectral features of the loaded material become detectable at much higher loadings, the resins' vibrations and their intensity remain essentially unaffected. An extreme illustration of this result is that of TentaGel beads, which are 30/70 (w/w) PS/poly(ethylene glycol) graft copolymer. The Raman spectrum of TentaGel shows vibrations specific to the poly(ethylene glycol) (PEG) component as well as the fingerprint transitions of PS, which have not been affected by the PEG component of the resin (compare Figs. 5A and 5C). In addition, the Raman spectra of polystyrene (PS) and TentaGel-S-OH beads displayed unique vibrations that were used to image and identify them selectively. Figure 5B shows a 5×5 FIC frames Raman image (25 frames, 50×40 individual spectra) of a mixture of PS and TentaGel resin beads, taken with the $10 \times$ objective, a global illumination laser power of ~ 400 mW, and an exposure time of 45 s per frame (total scan time 20 min). (This acquisition time could be decreased by at least an order of magnitude with a more powerful laser source and a more efficient optical setup.) The classified images (Figs. 5D–5F) were readily derived from the Raman data using the spectral angle mapper (SAM) algorithm of the spectral imaging software package MultiSpec (48). Transitions at 1277 cm^{-1} for TentaGel-S-OH (Fig. 5D) and at 1000 cm^{-1} or 1031 cm^{-1} for PS (Fig. 5F) were used to identify the corresponding beads. Figure 5E is an image in which both PS and TentaGel beads were specifically and concomitantly identified.

To further extend our ability to reliably identify resin beads based on unique differences in their chemical nature, we have also studied the following PS- and non-PS-based resins: (a) 4-bromo-PS, (b) 4-carboxy-PS, (c) PEG cross-linked Merrifield resin, (d) amino-PEGA, (e) HMBA-SPAR 50, and (f) SPAR 50.* Res-

---

(0.5 mmol/g); Boc-Met-O-Merrifield (0.9 mmol/g); Boc-Phe-O-Merrifield (0.8 mmol/g); Boc-Pro-O-Merrifield (0.9 mmol/g); Boc-Ser (OBzl)-O-Merrifield (0.6 mmol/g); Boc-Thr (OBzl)-O-Merrifield (0.6 mmol/g); Boc-Trp-O-Merrifield (0.6 mmol/g); Boc-Tyr (2-Br-Z)-O-Merrifield (0.6 mmol/g); and Boc-Val-O-Merrifield (0.8 mmol/g). The unsubstituted resins studied were TentaGel-S-OH (130 µm), 0.3 mmol/g, and hydroxymethyl-polystyrene (~90 µm), 1.1 mmol/g, 1% cross-linked. A detailed description of the design operation of the NIRIM, and the processing software can be found in Ref. 40.

* (a) 4-Bromo-PS 200–400 mesh (2.5 mmol/g, Chem-Impex International); (b) 4-carboxy-PS 100–200 mesh (3.5 mmol/g, Novabiochem); (c) PEG cross-linked PS 100–200 mesh (2 mmol/g, Advanced ChemTech); (d) amino-PEGA (0.4 mmol/g, Novabiochem); (e) HMBA-SPAR 50, 100–

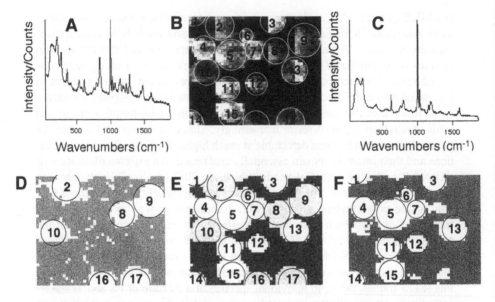

Figure 5 Experiments showing that the vibrations of the PS portion of TentaGel remains essentially unaffected by the major component (PEG) of the bead. Raman spectra of Tent-aGel-S-OH (A) and polystyrene (C) resin beads. Raman image of 5×5 FIC frames (50×40 FIC fibers and 15×12 μm per single image-pixel) of a mixture of polystyrene and Tent-aGel-S-OH recorded with a 10× objective, a global illumination laser power of ~400 mW per single-frame region, and a single-frame detector integration time of 45 (B). (See Ref. 40.) Classification images of TentaGel-S-OH (D), polystyrene (F), and a mixture of both types of resin beads (E) produced from the NIRIM image B.

ins a–c were chosen to establish that at least three additional PS-based resins can be readily distinguished (Fig. 6A). Resins d–f were chosen to establish the same conclusion for polyamide-based supports (Fig. 6F) and to demonstrate that they can also be readily differentiated from PS-based beads (Figs. 6C, 6D, and 6F–6H). One-milligram aliquots of the resins were combined in methanol, and a drop of this suspension was deposited on a sapphire crystal. After evaporation of the solvent the sapphire was placed in the field of view of the NIRIM. A library of single-bead near-IR Raman spectra of each of the resins was first recorded (Fig. 7), then several regions were arbitrarily selected for multispectral imaging. Figure 6 shows 5 × 5 frames (50 × 40 pixels) and 6 × 6 frames (60 × 48 pixels) in

200 mesh (polyacrylamide resin, 0.3 mmol/g, Advanced ChemTech); (f) SPAR 50, 200–400 mesh (0.8 mmol/g, Advanced ChemTech).

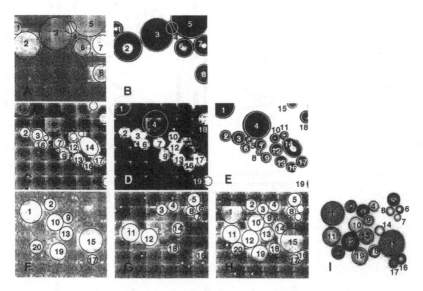

Figure 6 Series of experiments demonstrating the high reliability of multispectral imaging in identifying beads on the basis of their polymeric constituents. All the images were obtained from a statistical mixture of six different types of beads (see text). On the computer screen each group of beads is encoded with a different color. (A) Specific near-IR Raman imaging of 4-bromo-PS (beads 3 and 5, 1073 cm^{-1}), 4-carboxy-PS (beads 7 and 8, 637 cm^{-1}), and PEG cross-linked PS (beads 1, 2, 4, and 6, 703 cm^{-1}). (B) White light image of the beads imaged in A. (C) Specific near-IR Raman imaging of 4-bromo-PS (beads 1 and 4), 4-carboxy-PS (beads 2, 3, 6, 7, 9, 10, 12, 13, 16–19), and HMBA-SPAR 50 (beads 5, 8, and 15, 854 cm^{-1}). (D) Same image as in C, but only the beads with a PS backbone are visualized. In addition, the two PS-based resins were color-coded using specific vibrations to each of them (4-bromo-PS, beads 1 and 4: 4-carboxy-PS, beads 2, 3, 6, 7, 9, 10, 12, 13, 16–19). (E) White light image of the beads imaged in C and D. (F) Specific near-IR Raman imaging of PS-based resins (1, 2, 9, 10, 13, 15, 17, 19, and 20: 4-bromo-PS, 4-carboxy-PS, PEG cross-linked-PS). (G) Specific near-IR Raman imaging of polyamide-based resins (3–8, 11, 12, 14, 16, 18, amino-PEGA, HMBA-SPAR-50, SPAR-50). (H) Near-IR Raman image where PS- and polyamide-based resins were selectively and concomitantly identified (PS-based: same as F; polyamide-based: same as G). (I) White light image of the beads imaged in F–H. All the beads were identified by multispectral imaging as well as by single-bead microspectroscopy (4-bromo-PS: beads 1, 4, 15; 4-carboxy-PS: beads 9, 10, 13, 19; PEG cross-linked-PS: beads 2, 17, 20; amino-PEGA: beads 3, 5, 12, 16; HMBA-SPAR 50: beads 11, 14, 18; SPAR 50: beads 6–8).

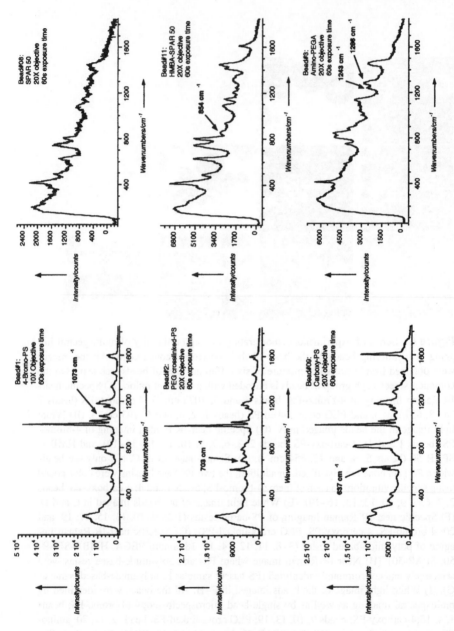

Figure 7 Single-bead Raman spectra of 4-bromo-PS, 4-carboxy-PS, PEG cross-linked-PS, amino-PEGA, HMBA-SPAR 50, and SPAR 50.

which all the beads were identified following the same procedure as in Figure 5. Because each bead is a collection of pixels and each pixel is a near-IR Raman spectrum of that area of the bead, comparison of these pixel spectra with the library of single-bead spectra recorded on the authentic samples confirmed the automated assignments. These results were reproducible regardless of the size and shape of the beads.

V. CONCLUSION AND PROSPECTS

Fourier transform infrared imaging provides a unique solution for the noninvasive, high throughput, and selective spectroscopic visualization of a library member and was shown here to single out one group of beads out of 60 composed of 11 different compounds, solely on the basis of their vibrational fingerprints. Because of its high sensitivity, rapidity in generating chemically specific images, and noninvasiveness, this technique is evidently amenable to study in situ reaction kinetics of an entire combinatorial library, in a multiplex format or as a mixture (49–52).* It could also be extended to the visualization of a hydridization (36) or receptor–ligand binding event, provided that the DNA or receptor investigated displays a unique IR signature that differentiates it from the library members.

The principles of dual recursive deconvolution of resin-supported combinatorial libraries have been proposed, and the feasibility of the key feature of this method, the identification of the first randomized position, has been demonstrated using near-IR Raman imaging of self-encoded resin beads. Any multispectral imaging technique could be applicable to the DRED method as long as the beads used display unique spectral features and provided the loaded material does not significantly alter their spectral signatures. For instance, secondary ion mass spectrometric (SIMS) imaging, (53) and FTIR imaging (H Fenniri, J Achkar, unpublished) are alternative approaches that we are currently investigating. As shown in this chapter, several commercially available and chemically distinct solid supports are eligible to explore the scope of DRED in combinatorial chemistry. However, because of inhomogeneities in their physical and chemical properties (e.g., swelling, porosity, size, reactivity), the design of a new class of resin beads to overcome this limitation has been deemed necessary. The added value of the DRED method is that it is a noninvasive screening technique, it does not require sophisticated equipment, it is inexpensive, and it does not involve the development of any encoding chemistry, because the DRED beads will soon become commercially available (54). Finally, the reliability of the imaging methods de-

---

* Infrared thermography is capable of simultaneously analyzing an entire library of catalysts but is unable to provide spectroscopic or kinetic information about the active beads.

scribed in this chapter is very high; it has been consistently accurate over the three years that we have been practicing resin bead multispectral imaging.

ACKNOWLEDGMENTS

This work was supported by Purdue University and the TRASK fund. HF is a Cottrell Scholar of Research Corporation. We thank Prof. Jillian M. Buriak and Dr. M. P. Stewart for a sample of porous silicon and Dr. Curtis Marcott (Procter & Gamble) for access to his Bio-Rad FTS 6000 Stingray.

REFERENCES

1. EM Gordon, JF Kerwin Jr, eds. Combinatorial Chemistry and Molecular Diversity in Drug Discovery. New York: Wiley, 1998.
2. AW Czarnik, SH DeWitt, eds. A Practical Guide to Combinatorial Chemistry. Washington, DC: Am Chem Soc, 1997.
3. Special Issue on Combinatorial Chemistry, Chem Rev 97, 1997.
4. H Fenniri, ed. Combinatorial Chemistry: A Practical Approach. Oxford, UK: Oxford Univ Press, 2000.
5. DR Liu, PG Schultz. Angew Chem Int Ed 38:36–54, 1999.
6. E Reddington, A Sapienza, B Gurau, R Viswanathan, S Sarangapani, ES Smotkin, T Mallouk. Science 280:1735–1737, 1998.
7. A Berkessel, DA Herault. Angew Chem Int Ed 38:102–105, 1999.
8. KL Ding, A Ishii, K Mikami. Angew Chem Int Ed 38:497–501, 1999.
9. MB Francis, EN Jacobsen. Angew Chem Int Ed 38:937–941, 1999.
10. BM Cole, KD Shimizu, CA Krueger, JPA Harrity, ML Snapper, AH Hoveyda. Angew Chem Int Ed 35:1668–1671, 1996.
11. AC Cooper, LH McAlexander, D-H Lee, RH Crabtree. J Am Chem Soc 120:9971–9972, 1998.
12. SM Senkan, S Ozturk. Angew Chem Int Ed 38:791–795, 1999.
13. PJ Cong, RD Doolen, Q Fan, DM Giaquinta, S Guan, EW McFarland, DM Poojary, L Self, HW Turner, WH Weinberg. Angew Chem Int Ed 38:484–488, 1999.
14. K Burgess, H-J Lim, AM Porte, GA Sulikowski. Angew Chem Int Ed 35:220–222, 1996.
15. Á Furka, LK Hamaker, ML Peterson. In: H Fenniri, ed. Combinatorial Chemistry: A Practical Approach. Oxford, UK: Oxford Univ Press, 2000, pp 1–32.
16. G Lippens, M Bourdonneau, C Dhalluin, R Warrass, T Richert, C Seetharaman, C Boutillon, M Piotto. Curr Org Chem 3:147–169, 1999 and references therein.
17. B Yan, H-U Gremlich, S Moss, GM Coppola, Q Sun, L Liu. J Comb Chem 1:46–54, 1999 and references therein.
18. V Swali, GJ Langley, M Bradley. Curr Opin Chem Biol 3:337–341, 1999 and references therein.

19. S-E Yoon, Y-D Gong, J-S Seo, MM Sung, SS Lee, Y Kim. J Comb Chem 1:177–180, 1999.
20. J Kein, CW Lehmann, H-W Schmidt, WF Maier. Angew Chem Int Ed 37:3369–3372, 1998.
21. A Nefzi, JM Ostresh, RA Houghten. Chem Rev 97:449–472, 1997 and references therein.
22. DL Boger, W Chai, Q Jin. J Am Chem Soc 120:7220–7225, 1998.
23. MC Pirrung, J Chen. J Am Chem Soc 117:1240–1245, 1995.
24. WC Still. Acc Chem Res 29:155–163, 1996.
25. MC Needels, DG Jones, EH Tate, GL Heinkel, LM Kochersperger, WJ Dower, RW Barrett, MA Gallop. Proc Natl Acad Sci USA 90:10700–10704, 1993.
26. KS Lam, M Lebl, V Krchnak. Chem Rev 97:411–448, 1997.
27. WL Fitch, TA Baer, WW Chen, F Holden, CP Holmes, D MacLean, N Shah, E Sullivan, M Tang, P Waybourn, SM Fisher, CA Miller, LR Snyder. J Comb Chem 1:188–194, 1999 and references therein.
28. C-Y Xiao, CF Zhao, H Potash, MP Nova. Angew Chem Int Ed 36:780–782, 1997.
29. KC Nicolaou, X-Y Xiao, Z Parandoosh, A Senyei, MP Nova. Angew Chem Int Ed 34:2289–2291, 1995.
30. X-Y Xiao, KC Nicolaou. In: H Fenniri ed. Combinatorial Chemistry: A Practical Approach. Oxford, UK: Oxford Univ Press, 2000, pp 75–94.
31. WJ Haap, TB Walk, G Jung. Angew Chem Int Ed 37:3311–3314, 1998.
32. LH Kidder, VF Kalasinsky, JL Luke, IW Levin, EN Lewis. Nature Med 3:235–237, 1997.
33. S-J Oh, JL Koenig. Anal Chem 70:1768–1772, 1998.
34. M Fisher, CD Tran. Anal Chem 71:2255–2261, 1999.
35. CM Snively, S Katzenberger, G Oskarsdottir, J Lauterbach. Opt Lett 24:1841–1843, 1999.
36. GH McGall, AD Barone, M Diggelmann, SPA Fodor, E Gentalen, N Ngo. J Am Chem Soc 119:5081–5090, 1997.
37. H Wenschuh, H Gausepohl, L Germeroth, M Ulbricht, H Matuschewski, A Kramer, R Volkmer-Engert, N Heine, T Ast, D Scharn, J Schneider-Mergener. In: H Fenniri, ed. Combinatorial Chemistry: A Practical Approach. Oxford, UK: Oxford Univ Press, 2000, pp 95–116.
38. BE Baker, NJ Kline, PJ Treado, MJ Natan. J Am Chem Soc 118:8721–8722, 1996.
39. H Fenniri, HG Hedderich, KS Haber, J Achkar, B Taylor, D BenAmotz. Angew Chem Int Ed 39:4483–4485, 2000.
40. AD Gift, J Ma, KS Haber, BL McClain, D Ben-Amotz. J Raman Spectrosc 30:757–765, 1999.
41. J Ma, D Ben-Amotz. Appl Spectrosc 51:1845–1848, 1997.
42. H Suto. Infrared Phys Technol 38:261, 1997.
43. MP Nelson, WC Bell, ML McLester, ML Myrick. Appl Spectrosc 52:179–186, 1998.
44. BL McClain, J Ma, D Ben-Amotz. Appl Spectrosc 53:1118–1122, 2000.
45. PJ Treado, MD Morris. Appl Spectrosc Rev 29:1–38, 1994.
46. GJ Puppels, M Grond, J Greve. Appl Spectrosc 47:1256–1267, 1993.
47. L Markwort, B Kip, ED Silva, B Roussel. Appl Spectrosc 49:1411–1430, 1995.

48. L Biehl, D Langrebe. MultiSpec: A Tool for Multispectral Image Data Analysis. Pecora 13, Sioux Falls, August 1996.
49. FC Moates, M Somani, J Annamalai, JT Richardson, D Luss, RC Willson. Ind Eng Chem Res 35:4801–4803, 1996.
50. SJ Taylor, JP Morken. Science 280:267–270, 1998.
51. A Holzwarth, SW Schmidt, WF Maier. Angew Chem Int Ed 37:2644–2647, 1998.
52. MT Reetz, MH Becker, M Liebl, A Furstner. Angew Chem Int Ed 39:1236–1239, 2000.
53. CL Brummel, JC Vickerman, SA Carr, ME Hemling, GD Roberts, W Johnson, D Gaitanopoulos, SJ Benkovic, N Winograd. Anal Chem 68:237–242, 1996.
54. H Fenniri, PCT 60/180, 939.
55. MP Stewart, JM Buriak. Adv Mater 12:859–869, 2000.

14

Microwave Applications to the Optimization of Combinatorial Synthesis

Joy Jing-Yi Yang and Yen-Ho Chu
National Chung Cheng University, Chia-Yi, Taiwan, Republic of China

Shui-Tein Chen
Institute of Biological Chemistry, Academia Sinica, Taipei, Taiwan, Republic of China

I. INTRODUCTION

The purpose of this chapter is to provide useful details concerning the application of microwaves to the optimization of combinatorial preparation of small-molecule organic compounds in solutions as well as on solid supports. Our results presented here demonstrate that the presence of microwaves greatly improves the rate of chemical reactions, which is critical to high throughput library synthesis, not only with a large reduction of reaction time but also without altering the final yield (1–11). This chapter also discusses the use of capillary electrophoresis (CE) in assessing stereoisomers formed during the combinatorial synthesis. In combinatorial chemistry, issues such as reaction stereoselectivity and possible racemization during reactions are rarely mentioned in the literature, and these are vital to the quality of the libraries prepared (12). Using tetrahydro-β-carbolines as examples, we demonstrate the usefulness of CE in enantiomeric separation of stereoisomers and rapid quantitative measurement of their corresponding isomer ratios as well as in assessing reaction racemization for its overall potential in library optimization.

    Combinatorial chemistry has significantly changed the discovery processes for new drugs, efficient organic/inorganic catalysts, and novel materials by screening and testing vast numbers of possibilities (13,14). To fully realize the promise of combinatorial chemistry, it typically requires high throughput synthesis, characterization, and screening of libraries. In library synthesis, the strategy of either parallel or split synthesis readily generates large numbers of small-molecule organic compounds (15). As to the characterization of libraries, a number of analytical techniques including HPLC and MS have conveniently addressed the purity and chemical integrity of the compound libraries (16). For high throughput screening of libraries, activity-based assays or direct binding assays are often the methods of choice to increase the speed of the lead discovery process. Although many reactions originally developed for solution chemistry have been exploited for library generation on solid support, to obtain an optimized reaction condition for every library compound with the same or similar reaction rates may not be straightforward, due to the presence of heterogeneous microenvironments on resins and a wide range of chemical reactivities from the starting materials. Various reaction issues such as the selection of resin, linker, temperature, reaction time, solvent, and catalyst all need to be evaluated, and this optimization process for synthesizing a compound library can take months to complete, whereas the final synthesis may take only weeks. The development of rapid and convenient methods would therefore be of value to speed up the reaction optimization process. Of these methods, "chemistry in a microwave," first reported in 1986 and primarily driven by the need for faster and cleaner reactions, offers the greatest potential in expediting synthetic organic chemistry (1–11).

Microwave dielectric heating has been routinely employed in industrial and food processing, but its application to chemical transformations emerged only recently. The organic reactions that benefit most from microwaves are obviously those that have marginal rates under conventional conditions. For example, the use of microwaves has significantly reduced the time required for the complete acid hydrolysis of peptides and proteins from 24 h to 5 min (17,18). Also, the Diels–Alder reaction of anthracene with maleic anhydride typically requires a reflux of 90 min as carried out conventionally, but with microwaves it is complete in 90 s (19). We report here the use of microwaves for the optimization of combinatorial preparation of tetrahydro-β-carbolines both in solutions and on solid supports. Our results show that microwaves in open reaction vessels greatly accelerate the reactions with a rate enhancement of, in one example, more than 4000-fold.

II. COMBINATORIAL SYNTHESIS IN A MICROWAVE

Tetrahydro-β-carboline is an important building block for synthesizing various indole and isoquinoline alkaloids, and natural products that have the carboline structure often exhibit important medicinal properties. For instance, demethoxyfumitremorgin C and cyclotryprostatins have been reported as mammalian cell cycle inhibitors, and tetrahydro-β-carboline-3-carboxylic acid derivatives were demonstrated to inhibit monoamine oxidase and to interfere with benzodiazepine receptors (20). This broad spectrum of biological activity and the rigid heterocyclic skeleton of the unique pharmacophore obviously are excellent targets for combinatorial applications. Using the Pictet–Spengler reaction (Table 1), substituted tetrahydro-β-carbolines can be directly prepared from tryptophan reaction with various aldehydes and ketones (21). Reactions involving aliphatic aldehydes

Table 1 1,2,3,4-Tetrahydro-β-Carbolines from the Pictet–Spengler Reaction in Toluene Under the Conditions of Room Temperature, Reflux, and Microwaves

	Room temp.		Reflux		Microwave[b]	
R$_1$COR$_2$	Reaction time	Isolated yield[c]	Reaction time	Isolated yield[c]	Reaction time	Isolated yield[c]
Benzaldehyde	10 h	71%	<10 min	99%	15 s	83%
p-Nitrobenzaldehyde	24 h	82%	<10 min	78%	20 s	74%
Salicylaldehyde	>9 days	0%	20 min	52%	90 s	65%
1-Naphthaldehyde	15 h	70%	<10 min	70%	30 s	69%
2-Butanone	24 days	81%	50 min	74%	15 min	67%[d,e]
3-Methyl-2-butanone	>60 days	88%[f]	18 h	N.A.[g]	3.5 h	N.A.[g]
			>15 days[h]	92%[f]	>36 h[i]	62%[f]
3-Pentanone	52 days	72%	5 h	56%[f]	2.5 h	57%[f]

[a] The reaction condition: L-tryptophan, 51 mg (0.25 mmol); aldehyde or ketone, 1.0 mmol; TFA, 5% (v/v); toluene, 2.5 mL.

[b] The microwave-assisted Pictet–Spengler reaction was carried out in an open vessel and temperature-controlled at 100°C (300 W, Synthewave 402, Prolabo).

[c] Isolated yield; approximately 1:1 diastereomeric trans/cis ratios were obtained throughout the study of this work.

[d] Longer reaction times increased, but insignificantly, the yield of the reaction (70% for 30 min; 73% for 60 min).

[e] Only 38% conversion was observed if the Pictet–Spengler reaction was carried out in DMF, a very efficient coupler of microwaves, for 15 min.

[f] The tetrahydro-β-carbolines were the only products, and the progress of the reaction was analyzed by [18]C HPLC.

[g] Not available; the starting tryptophan was completely consumed, but the overall reaction was intractably complicated.

[h] The reaction was carried out by refluxing in CHCl$_3$.

[i] The reaction was performed using the setting of 20% power.

are generally complete in hours, and aromatic aldehydes and some ketones give slow reactions at ambient temperatures. Arylketones react sluggishly or do not react at all with tryptophan. The reaction usually requires acid catalysts [e.g., 1–10% trifluoroacetic acid, (TFA)] and typically produces mixtures of stereoisomers in moderate to excellent yields. The stereochemistry of diastereomeric products, cis- and trans-1,3-disubstituted tetrahydro-β-carbolines, can be readily determined by ^{13}C NMR following the method developed by Cook and coworkers (22). The chemical signals corresponding to carbon atoms 1 and 3 in the trans isomers are shifted upfield relative to the cis isomers, presumably due to the unfavorable 1,3-diaxial interactions that are present in the trans isomers but are absent from the cis isomers (22). Integration of the peaks in both ^{1}H and ^{13}C spectra afford the ratio of cis and trans diastereomers for each compound in the library.

Results presented in Tables 1 and 2 clearly show that when microwaves are used the Pictet–Spengler reactions both in solution and on solid support can be readily accelerated and therefore conveniently optimized. For example, in Table 1, microwave reactions of tryptophan with benzaldehyde and p-nitrobenzaldehyde were complete in 15 and 20 s respectively, whereas at ambient temperature the reactions took 10 and 24 h, respectively. These correspond to a 2400-fold

Table 2 1,2,3,4-Tetrahydro-β-Carbolines from the Pictet–Spengler Reaction on Wang Resin Under the Conditions of Room Temperature, Reflux, and Microwaves

	Room temp.		Reflux		Microwave[b]	
R_1COR_2	Reaction time	Isolated yield[c]	Reaction time	Isolated yield[c,d]	Reaction time	Isolated yield[c]
Benzaldehyde	72 h	88%	10 min	63%	4 min	68%
Acetaldehyde	1.5 h	83%	10 min	50%	1 min	64%
Phenylacetaldehyde	3 h	81%	10 min	61%	3 min	57%

[a] The reaction condition: L-tryptophan-loaded Wang resin (capacity 0.5 mmol/g), 100 mg (0.05 mmol); aldehyde, 0.2 mmol; TFA, 1% (v/v); toluene, 2.5 mL.

[b] The microwave-assisted Pictet–Spengler reaction was temperature-controlled at 100°C (300 W, Synthewave 402, Prolabo).

[c] Isolated yield; approximately 1:1 diastereomeric trans/cis ratios were obtained throughout the study of this work.

[d] Under refluxing condition, small portions of tryptophan and the tetrahydro-β-carboline product were found cleaved and accumulated in filtrate.

and 4320-fold rate enhancement, respectively. Most significantly, salicylaldehyde did not react with tryptophan after 9 days at ambient temperature,* but under microwaves the reaction was complete in 90 s (Table 1). For aliphatic ketones, the exceedingly slow Pictet–Spengler reactions at room temperature (in days) were impractical for library preparation; using microwaves in open reaction vessels, however, the time required to synthesize tetrahydro-β-carbolines became acceptable (hours). Also, it appeared that microwave heating is far better than conventional reflux in terms of the reaction acceleration (see below for possible explanation). Instead of toluene solvent for this work, the use of dimethylformamide, a very efficient coupler of microwaves, took a rather longer time to carry out the Pictet–Spengler reaction (e.g., only 38% conversion, instead of 100% for toluene, was observed if the microwave-mediated tryptophan reaction with 2-butanone was carried out at 100°C for 15 min). Reactions carried out from room temperature, reflux, and microwaves gave an approximately 1:1 ratio of trans to cis diastereomers and were all in moderate to excellent isolated yields.

The Pictet–Spengler reaction was recently exploited for library generation of tetrahydro-β-carbolines on solid support (23–27). Table 2 demonstrates that microwaves surely accelerate the Pictet–Spengler reaction on Wang resin and significantly shorten its reaction time from hours at ambient temperature to minutes. By integrating with commercial automated synthesis instruments, this microwave-mediated solid-phase combinatorial synthesis would be attractive to scientists whose area of research is combinatorial chemistry. We used a catalytic amount (1%, v/v) of TFA in this work because the Wang resin linker used in this procedure is stable to only low (<5%) TFA concentrations. Higher TFA concentrations may be employed to speed up the Pictet–Spengler reaction if other TFA-stable solid supports such as Kaiser oxime resin are chosen. In all of the solid-phase reactions that were carried out that involved heat (i.e., reflux and microwaves), the chemical yields were somewhat lower than those of the corresponding reactions at ambient temperature. This is not totally unreasonable, because the presence of heat may facilitate, at least in principle, the cleavage of tryptophan or the products from the resin. In addition to using the Wang resin, we are currently in the process of conducting combinatorial synthesis using other solid supports.

III. LIBRARY CHARACTERIZATION AND OPTIMIZATION BY CAPILLARY ELECTROPHORESIS

To synthesize library compounds in a combinatorial manner, it is always desirable to establish a generalized experimental protocol that requires only a short reaction

---

* The starting tryptophan was completely consumed and converted to the intermediate Schiff base. No Pictet–Spengler reaction was observed after 9 days.

time. For this reason, slow reactions including some Pictet–Spengler reactions studied here are often treated with an increase of temperature to accelerate the reaction. However, under these heated conditions, unwanted recemization and different stereoselectivity during reactions can occur, and thus an efficient analytical method is needed to monitor the reaction stereoselectivity and measure the relative amounts of the stereoisomers formed under various reaction conditions. Methods frequently employed for stereoisomer separations include HPLC, thin-layer chromatography (TLC), gas chromatography (GC), and, more recently, capillary electrophoresis (CE). CE is a high resolution and fast separation technique that measures the electrophoretic mobility of a charged species in the presence of an electric field gradient (28,29). Compared with HPLC in the separation of stereoisomers, this CE method offers low cost (e.g., the selector molecule is used

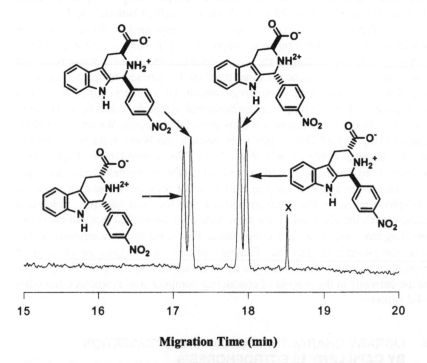

Figure 1 Stereoisomer separation of substituted tetrahydro-β-carbolines using capillary electrophoresis in the presence of β-cyclodextrin. Integration of the peaks can readily yield the ratio of cis and trans isomers for each of the compounds. Buffer used: 192 mM glycine, 25 mM Tris base, 100 mM sodium dodecyl sulfate, 56 mM β-cyclodextrin (pH 8.5). Capillary: uncoated fused silica, 127 cm total length, 120 cm effective length, 50 μm inner diameter. CE conditions: 30 kV, 70 μA, 280 nm, 25°C. (Y.-H Chu, unpublished results.)

as an additive without chemical immobilization), high efficiency in separation, rapid method development and optimization, and easy application to stereoisomer studies. In a previous study, we used cyclodextrins as additives in the electrophoresis buffer and employed micellar electrokinetic chromatography (MEKC), a special form of CE, to resolve the tetrahydro-β-carboline compounds (12). We and others demonstrated that, under reflux conditions, the reaction of tryptophan with aldehydes yielded the corresponding tetrahydro-β-carbolines with notable racemization (12,30–32). MEKC is a useful application of CE in which the mobile phase consists of a charged micelle and a CE selector such as cyclodextrin in the electrophoresis buffer. Separation of stereoisomers is achieved by differential partitioning of enantiomers between the micelle and the selector (i.e., different binding affinities of both enantiomers to the CE selector), resulting in differential migration (33–37). Figure 1 shows that, using MEKC, β-cyclodextrin in a Tris-glycine buffer containing SDS readily furnished the separation of all four possible tetrahyro-β-carboline compounds derived from tryptophan and p-nitrobenzaldehyde. This CE method is well suited for stereoisomer analysis and is likely to facilitate the rapid optimization of experimental conditions for the synthesis of library compounds.

Using MEKC, our preliminary result on fast microwaves of the Pictet–Spengler reaction indicates that the racemization during reactions was found to be minimal, if any. Similar microwave effects have been reported in the literature; no apparent racemization was obtained from the microwave-mediated chemical synthesis and acid hydrolysis of peptides and proteins (17,18). Extensive work is under way to further evaluate the utility of microwaves in combinatorial synthesis, especially on solid supports.

IV. SUMMARY

Our results demonstrate that microwave heating is totally different from conventional heating and, as can be seen from Tables 1 and 2, that microwaves greatly improve the rate of the Pictet–Spengler reaction with the benefits of short reaction time (more energy efficiency), clean reaction, and no significant alteration of the final chemical yield. It has been suggested that the effect of rate acceleration is attributable to a dipolar polarization, that the rotation of molecules in an attempt to align themselves with applied microwave field produces specific effects that cannot be achieved by conventional reflux. This nonthermal effect most likely originates from lowering the activation energy of the reactions either through the storage of microwave energy as vibrational energy of some functional groups of a molecule (an enthalpy effect) or by alignment of molecules (an entropy effect).

It is our opinion that the use of microwaves for combinatorial synthesis is still in its early development. Including our results presented in this chapter, many useful organic reactions proceed more rapidly and in comparable yield when microwaves are used rather than conventional heating procedures. Not surprisingly, we anticipate that this area of research will continue to be the focus of extensive study.

In this chapter, we have also demonstrated the usefulness of capillary electrophoresis in characterizing the stereoisomers of substituted tetrahydro-β-carbolines, determining isomer ratios, and assessing possible reaction racemization during library synthesis. This CE method should be applicable to the optimization of libraries of small organic molecules. With both microwaves and CE as useful tools for optimizing library synthesis, the development of a stereospecific route to molecular diversity will surely increase the integrity of materials available for study and offer more potent compounds for drug discovery.

REFERENCES

1. N Elander, JR Jones, S-Y Lu, S Stone-Elander. Chem Soc Rev 29:239–249, 2000.
2. A Fini, A Breccia. Pure Appl Chem 71:573–579, 1999.
3. JJV Eynde, D Rutot. Tetrahedron 55:2687–2694, 1999.
4. RN Gedye, JB Wei. Can J Chem 76:525–532, 1998.
5. S-T Chen, P-H Tseng, H-M Yu, C-Y Wu, K-F Hsiao, S-H Wu, K-T Wang. J Chin Chem Soc 44:169–182, 1997.
6. SA Galema. Chem Soc Rev 26:233–238, 1997.
7. AK Bose, BK Banik, N Jayaraman, MS Manhas. CHEMTECH 1997:18–24;
8. S Caddick. Tetrahedron 51:10403–10432, 1995.
9. CR Strauss, RW Trainor. Aust J Chem 48:1665–1692, 1995.
10. G Majetich, R Hicks. J Microwave Power Electromagn Energy 30:27–45, 1995.
11. RA Abramovitch. Org Prep Proced Int 23:683–711, 1991.
12. CC Cheng, Y-H Chu. J Comb Chem 1:461–466, 1999.
13. B Jandeleit, DJ Schaefer, TS Powers, HW Turner, WH Weinberg. Angew Chem Int Ed 38:2494–2532, 1999.
14. AW Czarnik, S Dewitt, eds. A Practical Guide to Combinatorial Chemistry, Washington, DC: Am Chem Soc, 1997.
15. BA Bunin. The Combinatorial Index. San Diego: Academic Press, 1998.
16. B Yan. Analytical Methods in Combinatorial Chemistry. Lancaster, PA: Technomic, 2000.
17. S-T Chen, P-H Tseng, H-M Yu, C-Y Wu, K-F Hsiao, S-H Wu, K-T Wang. J Chin Chem Soc 44:169–182, 1997.
18. S-T Chen, S-H Chiou, Y-H Chu, K-T Wang. Int J Peptide Protein Res 30:572–576, 1987.
19. SS Bari, AK Bose, AG Chaudhary, MS Manhas, VS Raju, EW Robb. J Chem Ed 69:938–939, 1992.
20. M Toyota, M Ihara. Nat Prod Rep 15:327–340, 1998.
21. ED Cox, JM Cook. Chem Rev 95:1797–1842, 1995.

22. J Sandrin, D Soerens, JM Cook. Heterocycles 4:1249–1255, 1976.
23. PP Fantauzzi, KM Yager. Tetrahedron Lett 39:1291–1294, 1998.
24. JP Mayer, D Bankaitis-Davis, J Zhang, G Beaton, K Bjergarde, CM Anderson, BA Goodman, CJ Herrera. Tetrahedron Lett 37:5633–5636, 1996.
25. L Yang, L Guo. Tetrahedron Lett 37:5041–5044, 1996.
26. R Mohan, Y-L Chou, M Morrissey. Tetrahedron Lett 37:3963–3966, 1996.
27. K Kaljuste, A Unden. Tetrahedron Lett 36:9211–9214, 1995.
28. BL Karger, Y-H Chu, F Foret. Annu Rev Biophys Biomol Struct 24:579–610, 1995.
29. JP Landers. Handbook of Capillary Electrophoresis. Boca Raton, FL: CRC Press, 1994.
30. PD Bailey, MH Moore, KM Morgan, DI Smith, JM Vernon. Tetrahedron Lett 35:3587–3588, 1994.
31. PD Bailey, SP Hollinshead, NR McLay, K Morgan, SJ Palmer, SN Prince, CD Reynolds, SD Wood. J Chem Soc Perkin Trans I 1993: 431–439.
32. PD Bailey. Tetrahedron Lett 28:5181–5184, 1987.
33. Y-H Chu, CC Cheng. Cell Mol Life Sci 54:663–683, 1998.
34. L Zhang, S Jin, X Zhou, J-L Gu, R-N Fu. Chromatographia 42:385–388, 1996.
35. B Chankvetadze, G Endresz, G Blaschke. Chem Soc Rev 1996:141–153.
36. S Fanali, M Cristalli, R Vespalec, P Bocek. In: A Crambach, MJ Dunn, BJ Radola, eds. Advances in Electrophoresis. New York: VCH, 1994, pp 1–86.
37. K Otsuka, S Terabe. In: NA Guzman, ed. Capillary Electrophoresis Technology; New York: Marcel Dekker, 1993, pp 617–629.

15
Optimizing Solution-Phase Synthesis Using Solid-Phase Techniques

Michael G. Organ and Debasis Mallik
York University, Toronto, Ontario, Canada

I. INTRODUCTION

It is truly interesting to follow the undulating development of the field of small-molecule solid-phase organic synthesis (SPOS) since the seminal work of Leznoff in the early 1970s (1). It has traversed the gamut of intellectual scrutiny as being "innovative," "esoteric," and, more recently, "powerful" with its application in high throughput synthesis (HTPS). In fact, from 1994 to 1998 the pharmaceutical and agrochemical industries were polarized toward the use of SPOS for the rapid preparation of diverse molecular libraries. This led to the development of start-up companies whose synthetic strategies (e.g., Affymax) or equipment line (e.g., Irori) were based primarily or exclusively on SPOS. More recently, the pendulum has again swung with respect to SPOS, only this time the movement is not necessarily away from it. Rather, solution-phase synthesis is being used increasingly to prepare molecular libraries because of adaptations of SPOS to facilitate reactions in the solution phase.

In truth, the once clear distinction between what constituted solid-phase and solution-phase synthesis is becoming obscured. For example, "catch-and-release" strategies have been developed whereby one or more solution-phase reactions can be carried out on a substrate and the product is then selectively "fished out" of the mixture onto a solid (2). During this process, atoms are added to the substrate through the formation of covalent bonds, so it is not simply a purification. Is such a process SPOS, or is it solution-phase synthesis?

For the purpose of this review, it is important to establish some definitions to address this issue. SPOS applies to a synthetic sequence whereby the starting material is loaded onto a solid support in the first step and is carried through to the end of the sequence on the solid, after which it is cleaved and purified as necessary. All other sequences will be considered solution-phase, including those in which intermediates are temporarily bound to a solid support. This would include the catch-and-release situation described above. Processes that are mediated by a catalyst on solid support will also be considered solution-phase. In such processes, the catalyst is typically used substoichiometrically and is returned intact after each cycle. Solid-supported reagents will be considered solution-phase in their application as well, and this will include species that are consumed in the transformation, e.g., nucleophiles or electrophiles, as well as those that are consumed but do not contribute atoms to the product (e.g., the conversion of a phosphine to a phosphine oxide in a Mitsonobu reaction). There are also entities that are necessary in at least a 1:1 ratio with the starting materials to promote the reaction but are returned intact, e.g., chiral auxiliaries. In such cases, the starting material is in solution prior to the transformation and the product remains in solution after the transformation.

The focus of this review is on the use of solid-supported reagents and catalysts to assist solution-phase reactions to facilitate high throughput synthesis. The work is grouped according to mechanistic class. Not all of the work cited herein necessarily pertains to, or culminates in, molecular library preparation. Some of the adaptations of SPOS to solution-phase chemistry are in their early stages and have yet to be used for this application. It is the potential for such use that merits their inclusion in this review. Although catalysts are discussed with reference to their application, there is no attention given to catalyst discovery by HTPS, which is a field unto itself. Further, "supports" are not limited to polymeric material but extend to other materials, including, for example, silica, charcoal, and glass.

II. BACKGROUND

Solution-phase synthesis has been around for literally centuries, and until the 1990s it was the normal way a medicinal chemist would work to provide potential small-molecule lead compounds. A good bench chemist using conventional solution-phase synthesis could produce anywhere between one and 100 compounds per year depending on their complexity. The development of reliable high throughput biological assays in the mid-1980s increased the demand for compounds. Now a biochemist could screen the entire year's collection of compounds produced by a synthetic chemist in one operation. The need to increase the output of the synthetic chemist pressed the development of HTPS, which encompasses parallel and mixture syntheses, to include combinatorial synthesis.

In truth, HTPS focuses as much on technology as it does on reactions per se and could be characterized loosely as expedited analytical chemistry and sample handling. The traditional bottleneck and expense in synthesis have been reaction workup and the purification and identification of products. HTPS shifted the focus in synthetic strategy from the actual transformations themselves to the purification. This was accelerated by developments in computer-controlled robotics and parallel sample-handling stations. In order for HTPS to be effective and worth the investment, the chemistry involved must be reliable and predictable. HTPS is intended to be automated and "hands-off." Even modern equipment does not have the ability to interpret each reaction on an individual basis in order to make an adjustment to push a particular reaction vessel to completion.

In order for solution-phase reactions to be run in an HTPS format, which is amenable to automation, a number of considerations have to be addressed. Traditional solution-phase reactions often require liquid–liquid extraction, which still remains a major issue for synthesizer engineers to contend with. Further, most reactions, even atomically economical ones, usually require an excess of one reaction partner to push the transformation to completion. Unless that partner is volatile, purifying it away from the desired product can pose a serious problem for HTPS. SPOS deals with these issues by immobilizing one reaction partner on a solid and allowing the other partner(s) that are in solution to be used in excess and removed by filtration. This also alleviates the concern over workup and extraction. By-products still remain a concern in SPOS, as do incomplete reactions. This material will contaminate the product and make some form of secondary purification necessary following product cleavage from the polymer. These problems are amplified manyfold in standard solution-phase synthesis where a crude prepurification by filtration is not possible. However, adaptations from SPOS to solution phase can address these issues and introduce purification by filtration to the solution-phase arsenal.

The principal adaptations from SPOS to solution include solid-supported reagents, solid-supported catalysts, and solid-supported scavenging agents that remove excess starting materials and, in some cases, by-products. The latter has been reviewed thoroughly quite recently and therefore is not discussed here (3). In this review, the development and use of supported reagents and catalysts are discussed with examples from the literature since 1998.

III. SOLID-SUPPORTED REAGENTS

Despite the significant attention given to solid-supported reagents for HTPS in the last few years, it is far from a new concept (4). In truth, Patchornik and his coworkers were the first to disclose an alternative strategy to solid-phase peptide synthesis in 1966 that used solid-supported active esters of N-blocked amino acids (used in excess relative to the growing chain) to prepare peptides in solution (5). They cited a number of perceived advantages over Merrifield's solid-phase strategy, not the least of which was the potential to readily analyze and purify intermediates if necessary during the synthesis. This points to a major advantage of solution-phase chemistry, that being the ability to assay the reaction directly without having to detach the product from the solid during the course of the transformation. A number of new methods have been developed to analyze reaction progress on the solid, including NMR, Raman, and IR spectroscopy, but they are not general at this stage for HTPS. Thus, as Patchornik and coworkers pointed out almost 35 years ago, solid-supported reagents can be used in large excess to drive reactions to completion and to obtain products of suitable purity by simple filtration. This reverse strategy meant that the same benefits could be afforded to solution-phase as to solid-phase synthesis without the additional step of cleaving products from the support, in addition to the analysis benefits.

For solid-supported reagents to be useful for HTPS, a number of general criteria must be met. The functional groups must maintain sufficient reactivity after they are immobilized to work for a wide variety of starting materials to ensure that parallel reactions yield the desired product of the transformation reliably and in a timely fashion. The reagent must be stable, and the pertinent groups must not become detached from the solid or otherwise break down prior to the desired transformation. Finally, any by-products of the reagent should remain solid-associated following the transformation to avoid the need for subsequent purification following filtration, or much of the practical benefit (and elegance) will be lost.

For this review, an entity is considered to be a solid-supported reagent if it is used stoichiometrically and contributes atoms to the product or is derivatized to a by-product during the transformation. This includes nucleophiles, electrophiles, oxidants, and reductants.

A. Solid-Supported Electrophiles

The development of stable electrophilic agents with active leaving groups is clearly one of the growth areas for solution-phase HTPS. The reagents developed thus far are primarily based on stabilized anions, e.g., sulfonamides, or neutral leaving groups, e.g., an amine from a quaternary ammonium salt. Further, most developments have been in the area of C—X bond-forming reactions, although some new C—C bond-forming reagents are now surfacing.

1. C—X Bond Formation

a. Amide Formation. Following in the footsteps of Patchornik's work, most amide bond formations involve activated carboxylic acid derivatives on solid support. However, a number of different leaving group platforms have been developed to effect this transformation. Three different platforms are outlined in

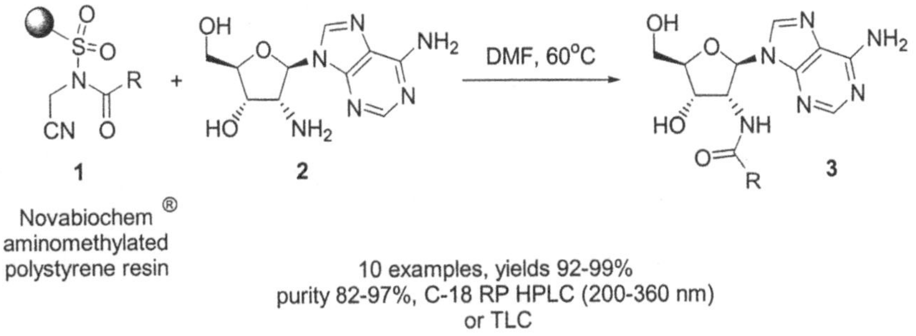

1 Novabiochem ® aminomethylated polystyrene resin

2

10 examples, yields 92-99% purity 82-97%, C-18 RP HPLC (200-360 nm) or TLC

3

Scheme 1

Schemes 1–3. In Scheme 1, a sulfonamide-based leaving group on aminomethyl-ated polystyrene **1** is used in the preparation of a small library of nucleosides **3**, as demonstrated by Link and coworkers (6).

The other two examples demonstrate the use of N-oxide-based leaving groups. The example in Scheme 2 uses Merrifield-based or ArgoPore-based N-hydroxysuccinimide resins **4**, both of which provided excellent yields of amides **6** (7). The work in Scheme 3 is an example of a resin-to-resin transfer where both the reagent and the starting material are on resin. In this case, N-hydroxysuc-cinimide **9** was used as a chaperone molecule to transfer the desired moiety from the reagent **7** to the starting material **11** (8).

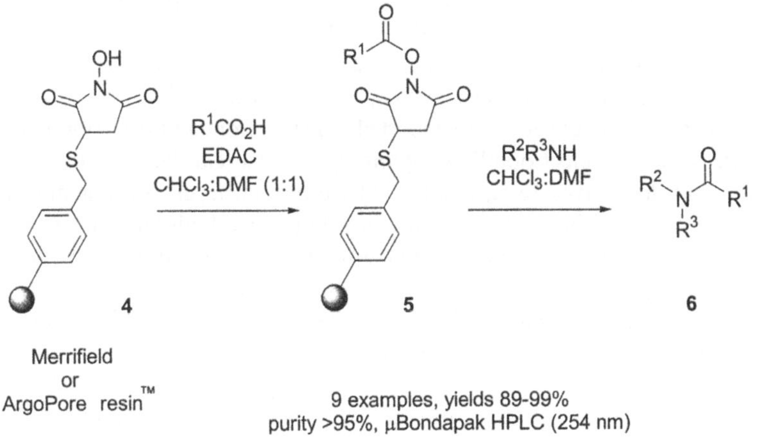

4

Merrifield or ArgoPore resin™

5

6

9 examples, yields 89-99% purity >95%, μBondapak HPLC (254 nm)

Scheme 2

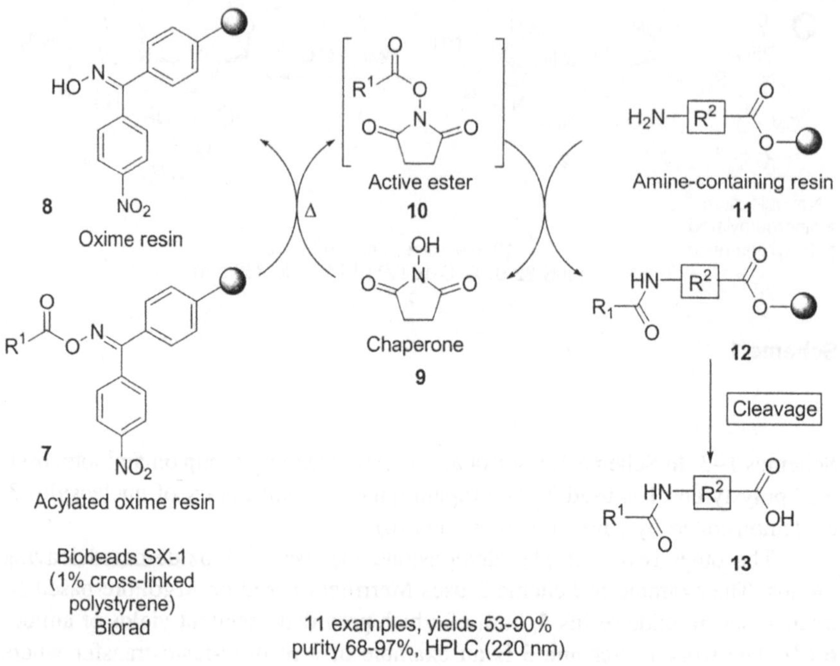

8 NO_2
Oxime resin

Active ester
10

Chaperone
9

H₂N—R² (Amine-containing resin)
11

Acylated oxime resin
7 NO_2

HN—R²
R₁
12

| Cleavage |

HN—R²
R¹
13

Biobeads SX-1
(1% cross-linked
polystyrene)
Biorad

11 examples, yields 53-90%
purity 68-97%, HPLC (220 nm)

Scheme 3

b. Sulfonamide Formation. Closely related to amide bond formation is the creation of sulfonamide bonds. In the example shown in Scheme 4, Ley et al. (9) opted for a polymer-supported DMAP sulfinating platform **15**. Supported DMAP was developed much earlier and in fact has been available for a variety of tasks as a supported reagent for approximately 20 years (10).

An especially useful series of supported reagents from a purification point of view are supported electrophilic coupling agents that can be used to make carboxylic acid derivatives [e.g., amides (11), esters (12), and sulfonamides]. Here both the acid and the nucleophile are in solution and both the coupler and its urea by-product are solid-supported, rendering the transformation easily purifiable by filtration. The urea by-product is similar in many respects to the acid derivative product, making purification sometimes a challenge for the homogeneous version of this reaction. In the example shown in Scheme 5, Sturino and Labelle demonstrate the use of polymer-bound (through an ammonium linkage) 1-(3-dimethylaminopropyl)-3-ethylcarbodiimide (P-EDC) **19** (13) to effect the synthesis of a variety of acylsulfonamides **20** (14).

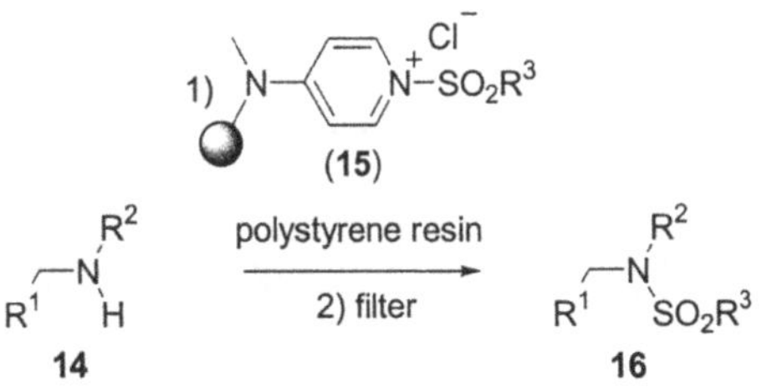

(15)

14

polystyrene resin

2) filter

16

7 sulfonyl pyridinium chlorides, yields >90%
purity >90%, GLC and ^{1}H NMR spectroscopy

Scheme 4

    c.   Ether Formation.   Janda and coworkers (15) developed a PEG-sup-ported triflating agent **23** that has been proven successful in capping both lithium enolates to produce enol ethers **22** and phenols to provide phenyl ethers **26** (Scheme 6). They demonstrated that the solid-supported reagent had greater sta-bility than the corresponding homogeneous reagent **24**. On standing in air for 48 h, **23** was less than 10% hydrolyzed (shown by ^{19}F NMR spectroscopy).

    A polymer-supported Mitsonobu reagent **28** (16) was employed by Georg and coworkers (17) to produce carbon-based aryl ethers **29** (Scheme 7). One of the truly challenging practical issues of reactions employing triphenylphosphine, which frequently produces phosphine oxide as a by-product, is the purification. In this case, the phosphine oxide by-product remains on the resin and is removed by filtration. However, the method still requires short-path silica gel chromatogra-

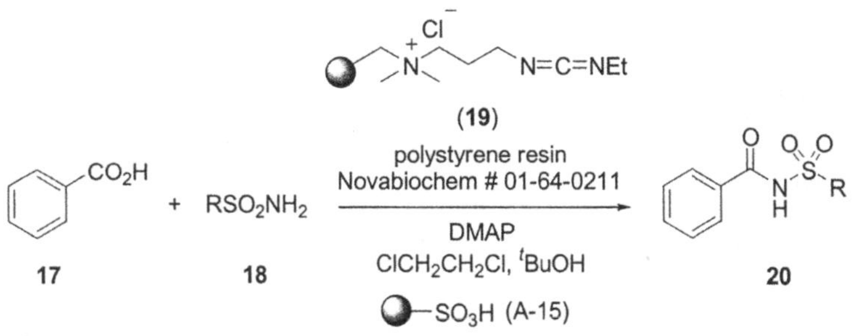

(19)

17 + RSO_2NH_2

18

polystyrene resin
Novabiochem # 01-64-0211

DMAP
$ClCH_2CH_2Cl$, ^{t}BuOH

—SO_3H (A-15)

20

25 examples, yields 56-81%
purity 85-92%, HPLC (255 nm)

Scheme 5

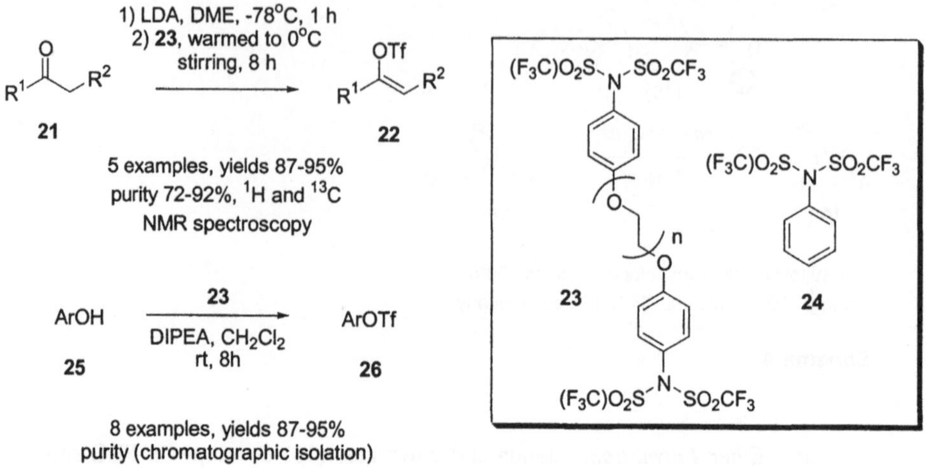

1) LDA, DME, -78°C, 1 h
2) **23**, warmed to 0°C
stirring, 8 h

21 → **22**

5 examples, yields 87-95%
purity 72-92%, ^{1}H and ^{13}C
NMR spectroscopy

ArOH $\xrightarrow[\text{DIPEA, CH}_2\text{Cl}_2]{\textbf{23}}$ ArOTf

25   rt, 8h   **26**

8 examples, yields 87-95%
purity (chromatographic isolation)

Scheme 6

phy and liquid–liquid extraction in some cases following filtration, indicating that this procedure, as is, would have limited use in the production of larger libraries.

  d. Tertiary Amine Formation. Janda's group developed a two-polymer system for the preparation of tertiary amines via a catch-and-release strategy (18). In this process (Scheme 8), the insoluble electrophilic regenerated Michael acceptor (REM) resin **31** is attacked by a secondary amine, producing the resultant Mannich base **32**. Treatment with an electrophile quaternarizes the amine **33**, rendering it susceptible to β-elimination. Then the salt is treated with Hünigs

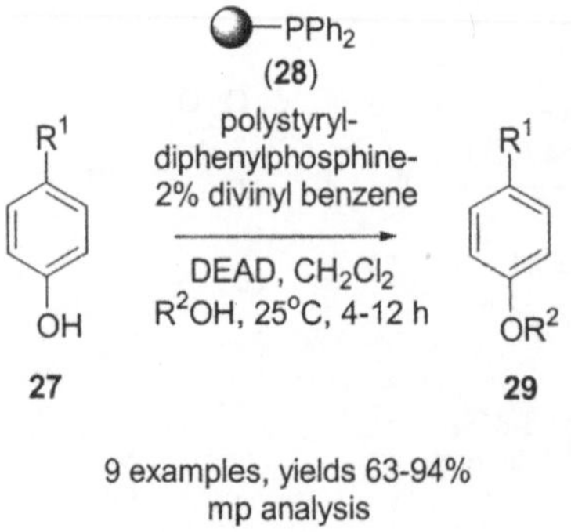

⬤—PPh$_2$
(**28**)
polystyryl-
diphenylphosphine-
2% divinyl benzene
$\xrightarrow[\text{R}^2\text{OH, 25°C, 4-12 h}]{\text{DEAD, CH}_2\text{Cl}_2}$

27 → **29**

9 examples, yields 63-94%
mp analysis

Scheme 7

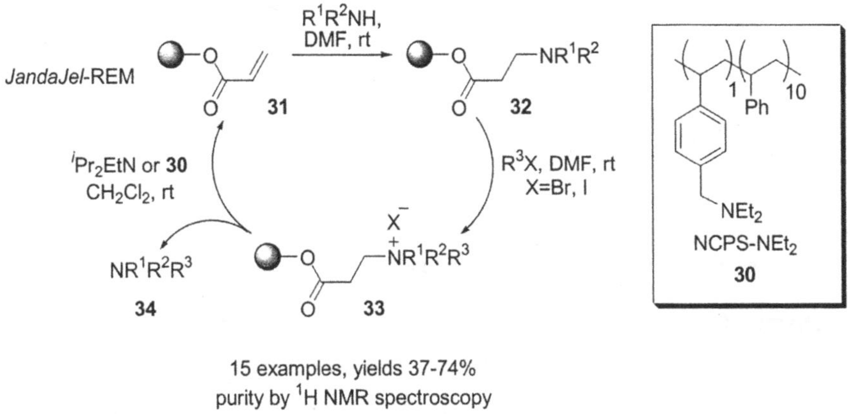

15 examples, yields 37-74%
purity by ^{1}H NMR spectroscopy

Scheme 8

base to release the tertiary amine product **34**. This process was used earlier by others, and in this case the product usually requires purification from the salt by-products following filtration (19). Janda's adaptation uses a soluble non-cross-linked polystyrene (NCPS) base **30** to act as a proton sponge during the elimination step. Thus, once precipitated, both polymers are easily filtered away from the desired amine product.

  e. *Bromination and Chlorination.* Electrophilic halogen sources, such as the succinimide series (i.e., NCS, NBS, and NIS), and perhalogens, such as hydrogen bromide perbromide, have had successful adaptations to the solid support. The example in Scheme 9 illustrates the use of chloro-p-toluenesulfonyl resin **36** to chlorinate ketone **35** to provide α-chloroketone product **37** (20).

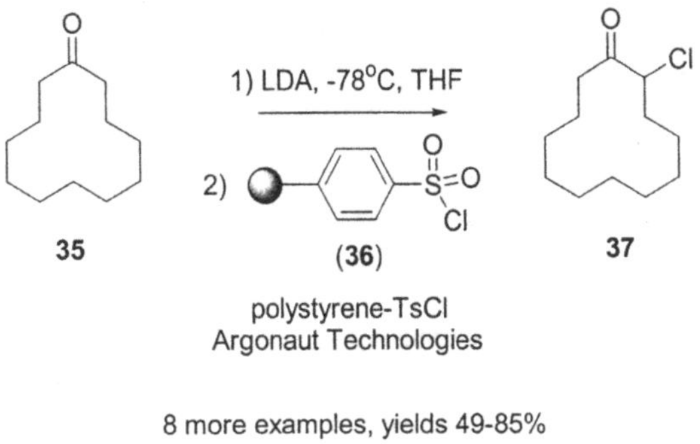

polystyrene-TsCl
Argonaut Technologies

8 more examples, yields 49-85%

Scheme 9

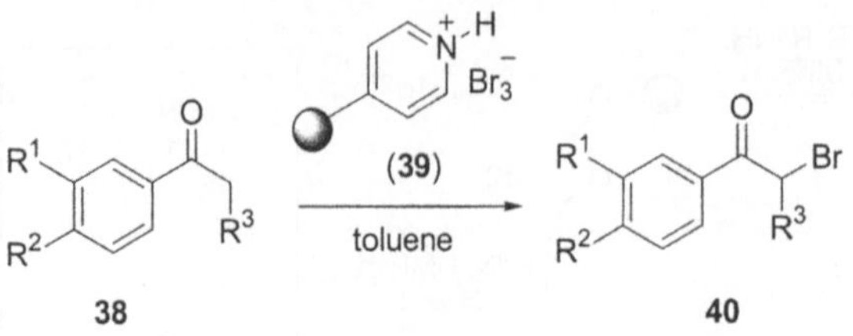

38   40

31 examples, yields 30-99%
purity 70 - >95%, ^{1}H-NMR spectroscopy

Scheme 10

The supported pyridinium bromide perbromide reagent **39** used by Ley and coworkers (21) (Scheme 10) was first developed by Fréchet et al. (22) in 1977. In the Scheme 10 reaction the reagent is used effectively to prepare a small sublibrary of α-bromoketones **40**, and the bromide later serves as a leaving group to yield a variety of α-phenoxyketones.

f. Halohydrin Formation. Epoxides have become a starting material of choice in synthesis, in large part because of the significant effort that has gone into their preparation in high optical purity (23). Further, significant work has been done with excellent results thus far for the asymmetric opening of *meso*-epoxides. Also, the regioselectivity of unsymmetrical epoxide opening is well understood, albeit substrate-specific. Afonso et al. (24) developed a polymer-supported equivalent to triphenylphosphonium bromide (i.e., **42**) to open epoxides (Scheme 11). They probed the mechanism of this transformation using ^{31}P

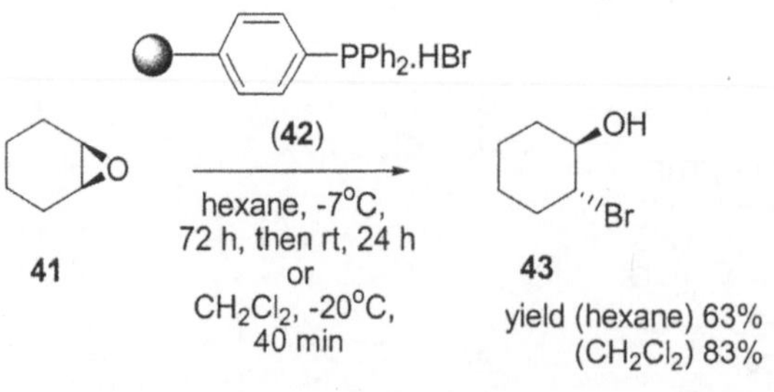

41   43

hexane, -7°C,
72 h, then rt, 24 h
or
CH$_2$Cl$_2$, -20°C,
40 min

yield (hexane) 63%
(CH$_2$Cl$_2$) 83%

7 more examples, yields 16-97% (varying reaction conditions)
purity by IR, ^{1}H-NMR and HRMS analysis

Scheme 11

NMR spectroscopy and chiral phosphine sources that consistently yielded racemic products. This led them to conclude that HBr is the sole participant that is released slowly by the phosphine and that protonation of the epoxide precedes opening. In any case, they reported significantly easier purification using the polymer-supported phosphine.

g.   Acetoxyhalogenation and Azidoiodination.   A useful transformation to obtain contiguous placement of functionality with rigorous stereospecificity is electrophilic 1,2-cohalogenation. The first example of a polystyrene-supported di(acyloxy)halogenate(I) **45**, shown in Scheme 12, was used to prepare iodoacetate and iodoazide derivatives of glycals **44** (i.e., **46** and **47**) (25). These reagents were prepared by treating the trimethylammonium polystyrene iodide (or bromide) with PhI(OAc)$_2$ in methylene chloride for 2 h. Kirschning et al. report that the resins are stable even after extensive washing.

h.   Selenenylation.   Supported electrophilic selenium reagents have been developed as safety catches that are at the interface of solution- and solid-phase chemistry (26). In the example presented in Scheme 13, supported selenenyl bromide **48** initiates an oxyselenenylation to produce selenium-substituted pyrans **50** (27). The product is not released in this step to constitute a solution-phase reaction by the definition of this review; however, it is released subsequently, making it a catch-and-release reaction.

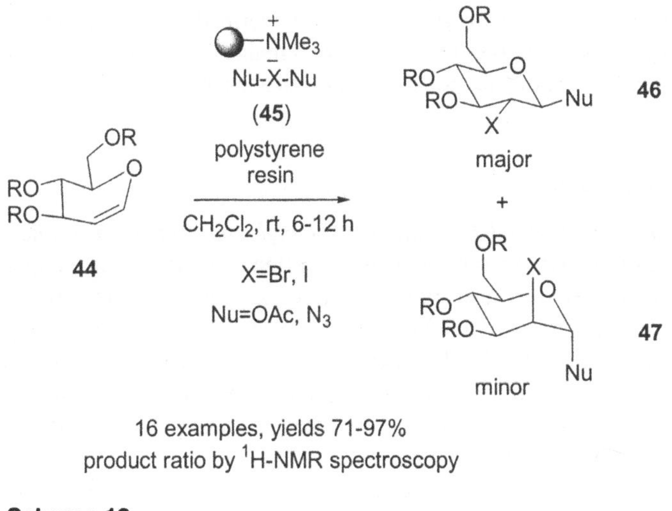

16 examples, yields 71-97%
product ratio by ^{1}H-NMR spectroscopy

Scheme 12

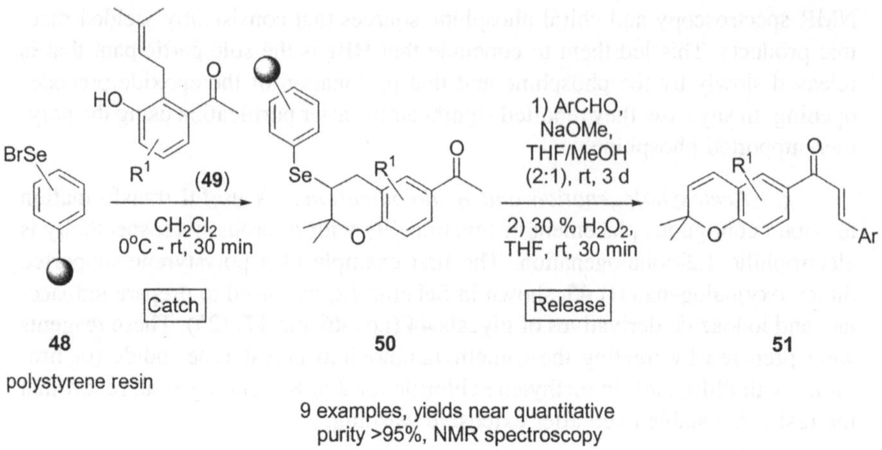

Scheme 13

B. Solid-Supported Nucleophiles

There are far fewer solution-phase applications of supported nucleophiles in HTPS than their electrophilic counterparts because, in general, it is easier and perhaps more obvious to prepare reactive intermediates in solution than on a solid (28). There are numerous SPOS examples that employ neutral nucleophiles (e.g., an amine) and fewer examples that involve a reactive intermediate (i.e., an anion) on the solid. In these cases, the product frequently remains on the solid, making these reactions excellent candidates for catch-and-release strategies in solution-phase applications.

1.   C—X Bond Formation

a.   C—O and C—N Bond Formation of Phenols and Activated Amines. Although many amine-based resins have been used effectively as proton sponges for various reactions, they may also play the role of activator for nucleophilic substitutions. The example shown in Scheme 14 demonstrates the power of using the filterable activator/base 1,3,4,6,7,8-hexahydro-2*H*-pyrimido[1,2-*c*]pyrimid-ine on polystyrene (PTBD resin, **53**) for the substitution reaction to produce **54** (29). This transformation was the first in a three-step sequence that required no in-line purification other than simple filtration.

The example in Scheme 15 illustrates the use of PTBD resin for the forma-tion of a C—N bond from activated imides **56** and **58** and α-bromoketones **55** (30). Of course, to be a true activator the pK_a of the phenol (~10, H_2O reference)

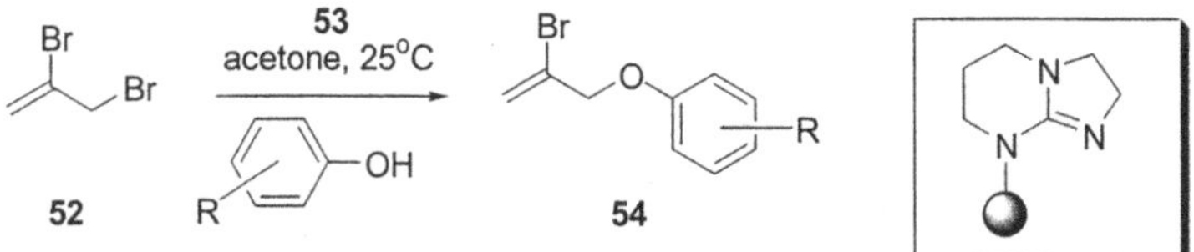

4 examples, quantitative conversion by ^{1}H-NMR spectroscopy

Scheme 14

or activated amine (~<10, DMSO reference) must be lower than that of the corresponding protonated PTBD urethane (~13.6, H_2O reference).

2. C—C Bond Formation

There are examples of anions being generated on solid and subsequently quenched with electrophiles. Although many of these are SPOS examples, they are also amenable to solution-phase catch-and-release strategies and will therefore be outlined here. The first example in Scheme 16 is a catch-and-release aldol condensation of a zinc enolate that was transmetallated from the corresponding lithium derivative that had been generated from ester **61** using LDA (31).

In a particularly striking example, Teitze and Steimmetz (32) generated a Weiler dianion on resin **65**, which was reacted once with an electrophile to form **66**; the dianion was regenerated to form **67** and then reacted a second time with soluble electrophiles (Scheme 17).

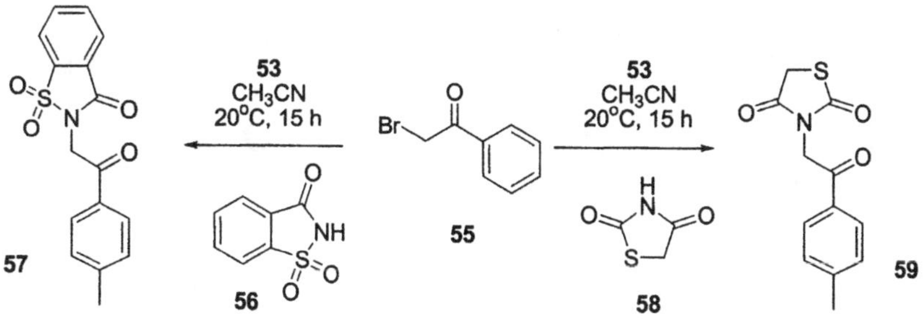

yield 39%
purity 87%

yield 65%
purity 99%

Scheme 15

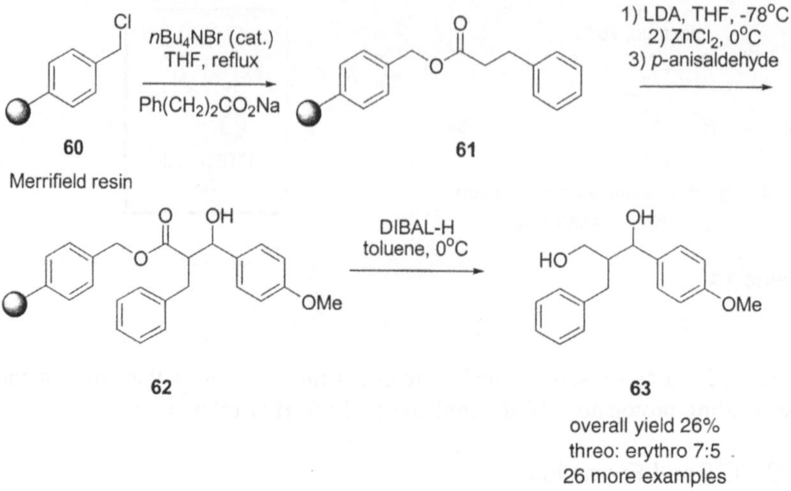

Scheme 16

There are also examples in the literature that deal with the generation of highly reactive unstabilized anions on resin that were captured with soluble electrophiles. In the first case, a metal–halogen exchange generated the Grignard reagent on solid support **72**, which was then quenched with a variety of aldehydes and subsequently released with concomitant cyclization to produce the cyclic ether products **74** (Scheme 18) (33).

Janda and coworkers (34) reported the first example of a polymer-supported directed ortho metallation (DOM) of compound **76** that was quenched with an aldehyde to produce alcohol product **77** (Scheme 19). Subsequent heating of the

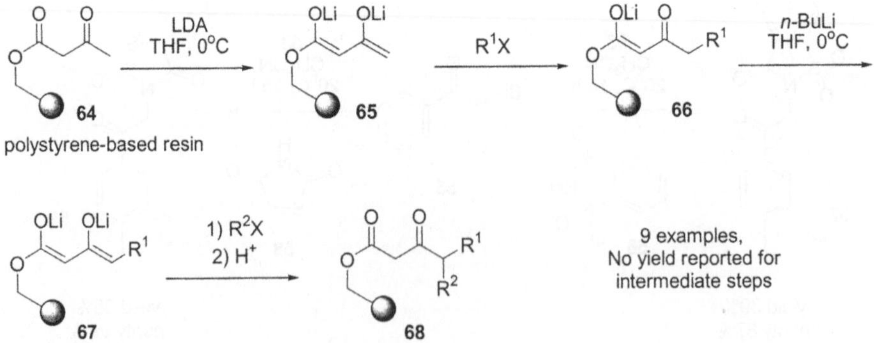

Scheme 17

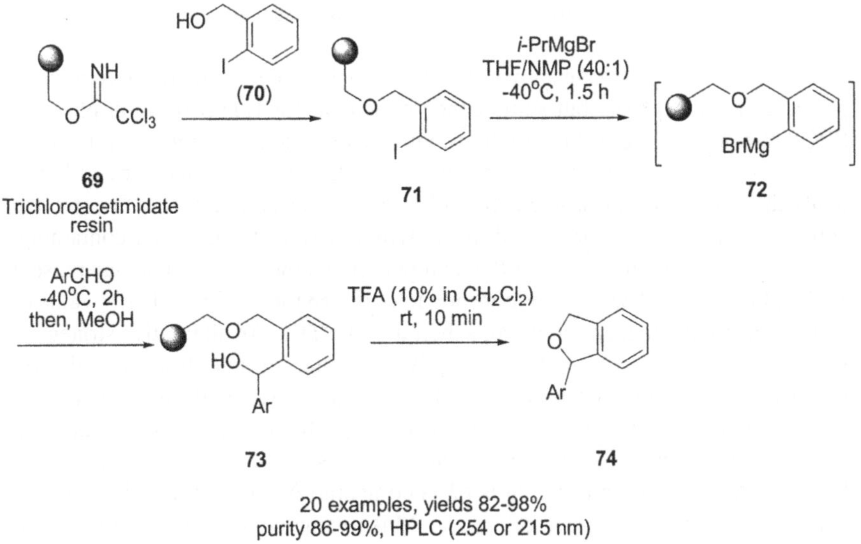

Scheme 18

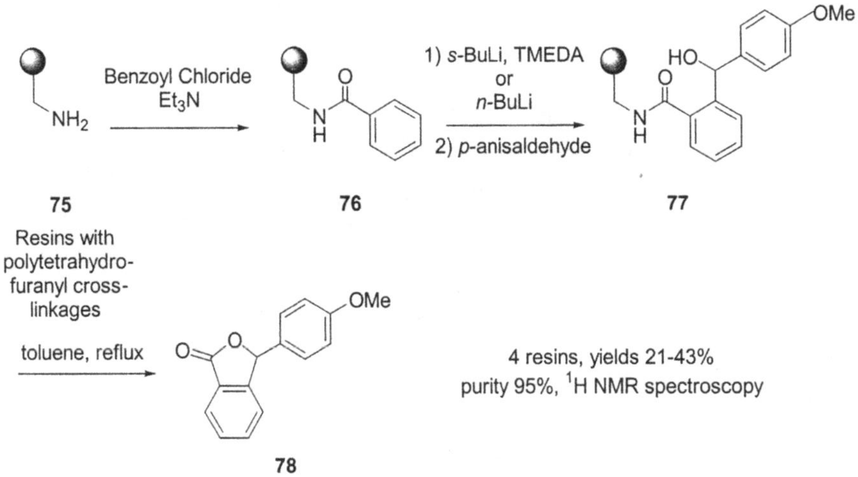

20 examples, yields 82-98%
purity 86-99%, HPLC (254 or 215 nm)

Scheme 19

4 resins, yields 21-43%
purity 95%, ^{1}H NMR spectroscopy

solid-supported product in toluene led to the formation of lactone **78** with simultaneous release from the resin.

These examples have shown that it is possible to generate reactive nucleophiles on a solid phase, but some concerns remain for HTPS solution-phase or SPOS chemistry. One of the problems with generating active anions on resins is that many resins are inherently difficult to keep dry; thus the quenching of kinetically derived anions prior to electrophilic capture is a serious issue. Second, the active reagent has to be generated in situ with a base (often nitrogen-containing, e.g., LDA), which introduces another compound into the mixture that may present purification issues. These issues conspire to diminish the ''off-the-shelf'' value of supported anionic reagents and limits their use in HTPS. An alternative strategy is to create shelf-stable masked anion equivalents that are not strongly moisture-sensitive and do not have base-derived by-products. Although this is harder to imagine with nonstabilized anions, there are a number of enolate surrogates that can and have been employed. The example from Kobayashi et al. (35) shown in Scheme 20 to make aminoalcohols **82** is comparatively recent, but this concept was first employed over 25 years ago by Mukaiyama et al. (36), who performed scandium-catalyzed aldol condensations on a solid support.

Another enolate equivalent is an enamine, and a group of stable and effective reagents was developed by Aznar et al. (37). Scheme 21 illustrates both the formation of a sublibrary of solid-supported enamines **83** and their use in subsequent additions to isocyanates **84** and Michael acceptors **87**.

The above two examples are much more practical alternatives to creating group I metal salts *in situ* and are legitimate candidates as shelf-stable anion equivalents in a bottle.

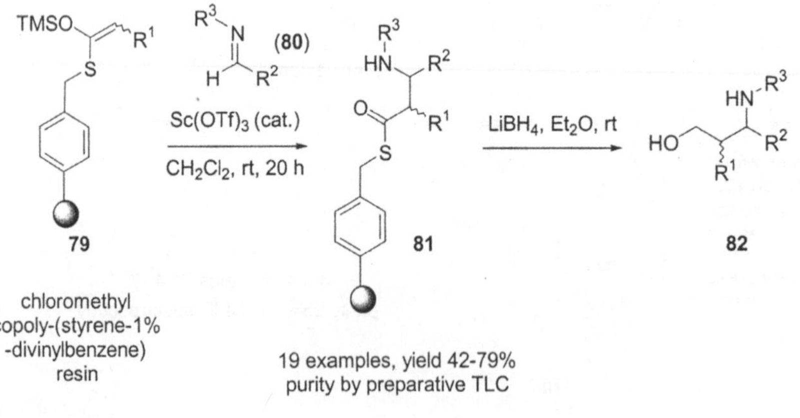
TMSO
R1
S
R3
N
(80)
H
R2
Sc(OTf)3 (cat.)
CH2Cl2, rt, 20 h
R3
HN
R2
O
R1
S
LiBH4, Et2O, rt
HN
R3
HO
R2
R1
79
81
82
chloromethyl
copoly-(styrene-1%
-divinylbenzene)
resin
19 examples, yield 42-79%
purity by preparative TLC

Scheme 20

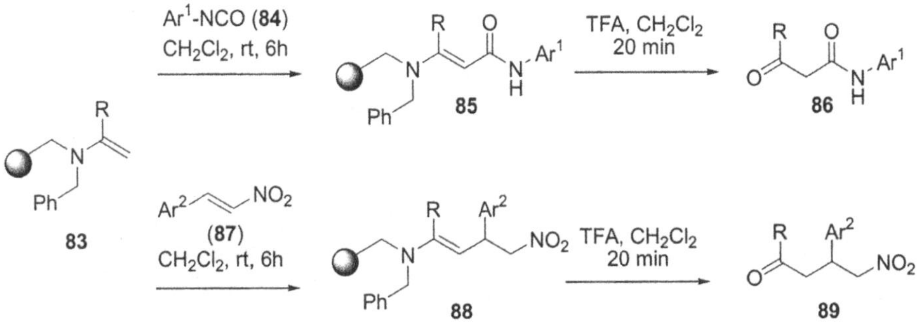

Wang resin

19 examples, yields 45-77%
purity 65-91%, NMR and MS

Scheme 21

It has been mentioned a number of times in this review that phosphine-based and phosphine oxide–based by-products are inherently challenging to purify away from the desired product. This is often the case for the variants of the Wittig olefination reaction, especially in an HTPS situation. A couple of related approaches have been developed to render the phosphine oxide by-product of this reaction polymer-bound so that it can be separated from the desired product by simple filtration (38). This approach (Scheme 22) offers a double advantage over straight solution-phase chemistry in that the resin-bound phosphate by-product will chelate the quaternary salt of the base required to generate the active Wittig reagent, thus removing it concurrently (39). Whereas Barrett's approach required a fairly reactive base, i.e., *tert*-butyltetramethylguanidine (Barton's base), to achieve consistently good yields, the free base is low boiling (bp 82°C), and any excess can be removed readily under vacuum.

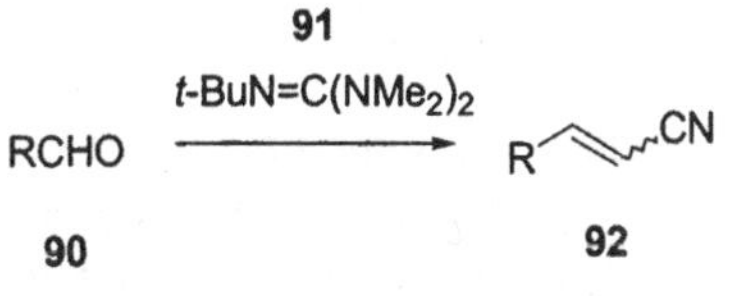

9 examples, yields 85-95%
E and *Z* mixtures, purity >95%,
GCMS

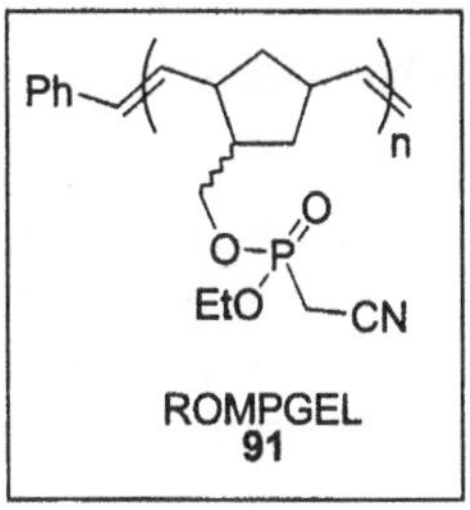

Scheme 22

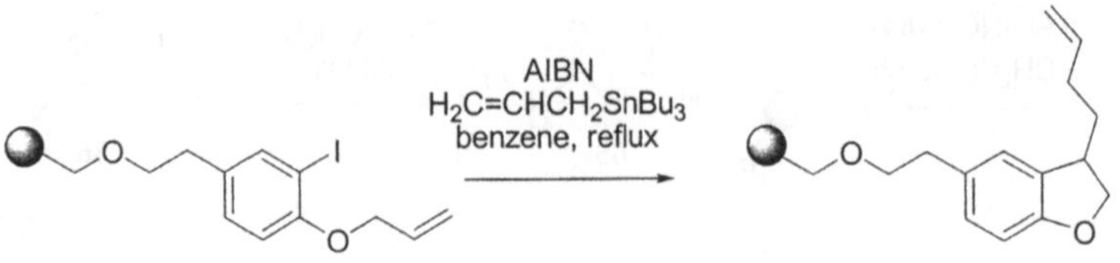

93

Polystyrene beads

94

2 more examples
^{1}H NMR conversions 10-49%

Scheme 23

C. Solid-Supported Radicals and Radical Acceptors

Although one could imagine that supporting a radical substrate on a solid might actually enhance intramolecular reactions by limiting access to potential soluble quenching sources,* it is not clear whether this would help or hinder intermolecular examples where the donor is on the solid and the acceptor is in solution. The example in Scheme 23 demonstrates an intra- and intermolecular radical sequence to generate the final product (41). In this case, the allylstannane was used in very high excess, presumably to prevent quenching at the stage of the cyclized radical product derived from **93**.

Zhu and Ganesan (42) reported an interesting intermolecular sequence where the radical donor (i.e., intermediate from **97**) is in solution and the acceptor is on solid (**96**) (Scheme 24). When the reaction was conducted on polystyrene resin in particular, the yields were comparable to those obtained under homogeneous reaction conditions. The results obtained thus far involving solid-supported radical reactions have been more demonstrative rather than applied to significant library synthesis in either solid or solution phase.

D. Solid-Supported Oxidations

1. C—H Bond Oxidations

Hypervalent iodine species have been used as oxidants for some time now (43). The reagents and by-products are considerably more environmentally friendly and less toxic then their metal-based counterparts (e.g., chromium, selenium), making their use attractive. The first polymer-supported hypervalent iodine species were generated more than 25 years ago and were shown to have synthetic utility (44).

---

* For the creation of a small library of indolines using an SPOS radical cyclization, see Ref. 40.

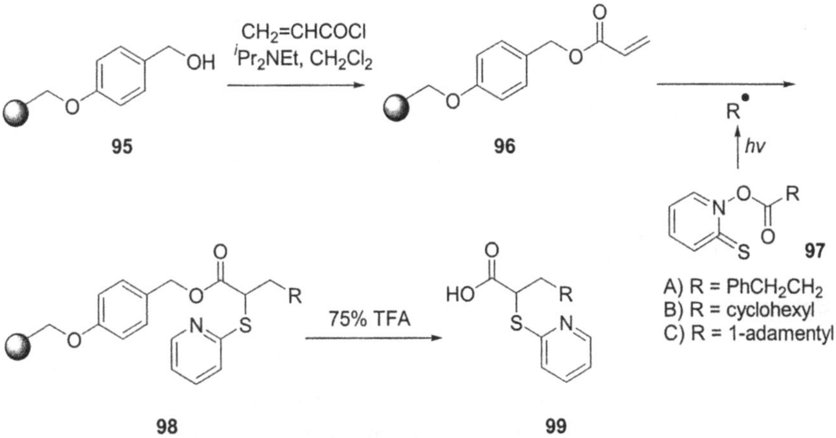

A) R = PhCH$_2$CH$_2$
B) R = cyclohexyl
C) R = 1-adamentyl

Scheme 24

Isolated yields (%)

Products	Polystyrene Wang Resin	Solution Phase Precendent	Tentagel Wang Resin
A	94	98	76
B	91	95	76
C	63	94	45

More recently, these reagents have received a great deal of attention for their potential use in HTPS. Togo et al. (45) investigated the use of polymer-supported diacetoxyiodobenzene **101** to iodinate electron-rich aromatics **100** (Scheme 25). In the same report (45), the authors also disclose an oxidative 1,2-aryl migration of alkyl aryl ketones **103** using reagent **101** (Scheme 26), which, overall, represents a

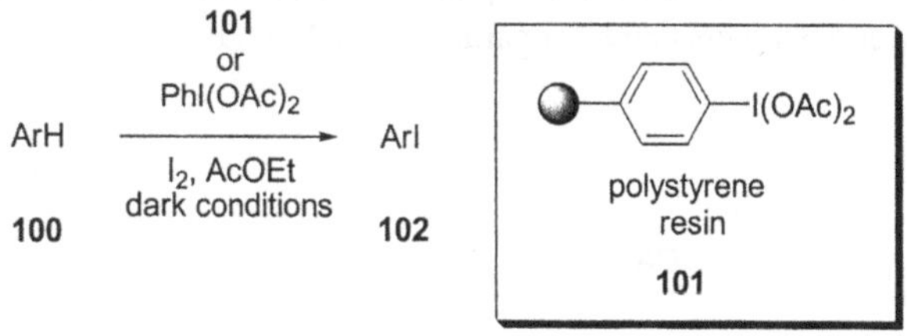

9 examples, yields* 56-99%
*over iodination can be controlled
by varying reaction conditions

Scheme 25

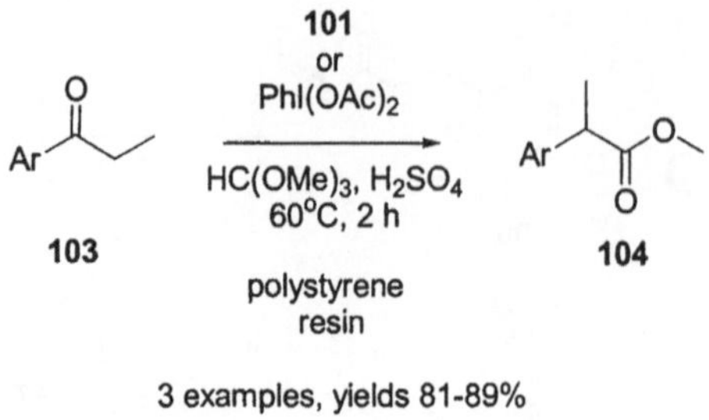

101
or
PhI(OAc)₂

$$\text{HC(OMe)}_3, \text{H}_2\text{SO}_4$$
$$60°C, 2 \text{ h}$$

polystyrene
resin

3 examples, yields 81-89%

Scheme 26

C—H bond oxidation. They also demonstrate that the reagent is regenerated readily and thus could be recycled many times with little negative impact on the yields of their transformations.

The groups of Chen (46) and Ley (47) simultaneously reported a wider variety of oxidations using **101**. One of the reactions reported by both groups was the oxidation of acetophenone derivatives **105** to form acyloins or the corresponding acetate **106** as shown in the example in Scheme 27 reported by Ley et al. (47).

2. C—O Bond Oxidations

Clearly, the major use of oxidants in organic synthesis is in the oxidation of aliphatic alcohols to form ketones and aldehydes. These reactions are frequently nasty to work up, because emulsions caused by metal particulates hinder liquid–liquid extraction. Any approach that eliminates extraction will not only greatly propel HTPS but would also be welcomed in traditional synthesis.

Chromium is a commonly used metal for oxidation chemistry. However, as mentioned above, chromium workups are perhaps the worst of all the transition metal workups for emulsions. Frequently, the biphasic mixture must be filtered before the organic and aqueous layers can be separated, which would make HTPS

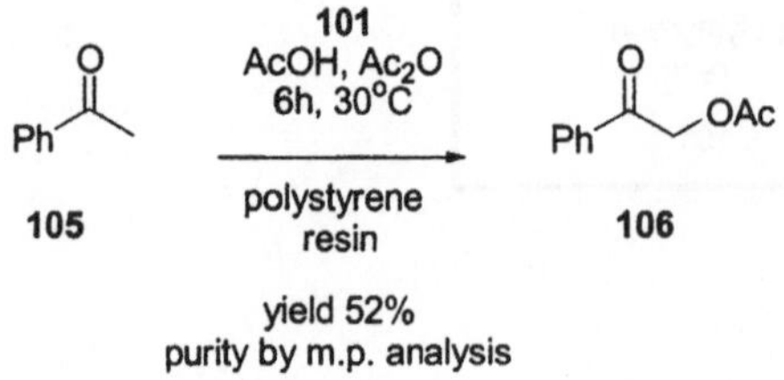

101
AcOH, Ac₂O
6h, 30°C

polystyrene
resin

yield 52%
purity by m.p. analysis

Scheme 27

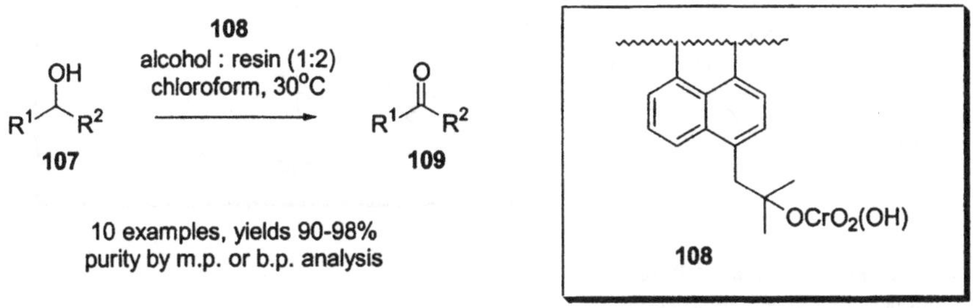

Scheme 28

impossible using such reagents. A number of supported chromium reagents have been developed. Geethakumari and Sreekumar (48) developed a chromic acid oxidant **108** that is linked covalently to the support via a C—O bond. This reagent has proven quite efficacious in the oxidation of secondary alcohols **107** to provide ketones **109** and primary alcohols **107** to selectively produce aldehydes **109** (Scheme 28). The same group also generated oxidants that are akin to PCC on resin. In this case, a pyrazole was attached to the resin and one amine was quaternarized to generate the counterion for the chromate **111**. This reagent has also been used effectively for the production of aldehydes and ketones **112** (Scheme 29) (49).

Certainly, nonmetal oxidants have also made a significant contribution to C—O bond oxidation. Despite the nasty odors associated with the Moffatt–Swern oxidation, it is still a favored method to oxidize primary alcohols to aldehydes. Vederas and coworkers at the University of Alberta (50) developed a polymeric sulfoxide based on poly(ethylene glycol) (**114**) (Scheme 30). The reduced polymer can be regenerated, and, perhaps most important, there are no volatile sulfides liberated in the process.

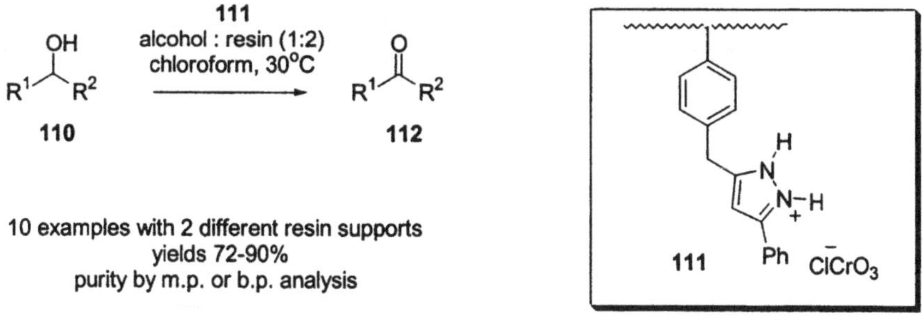

Scheme 29

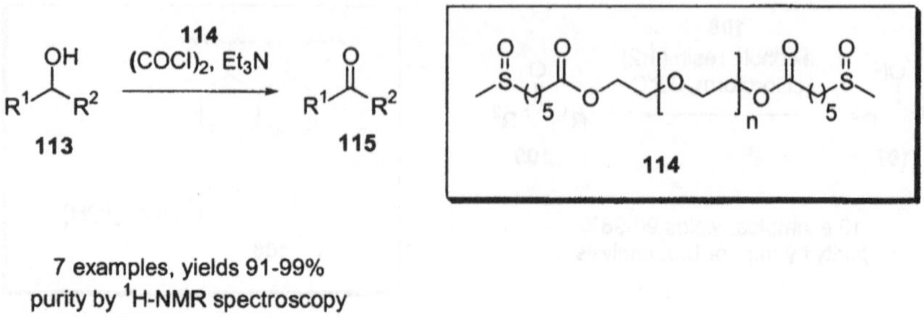

7 examples, yields 91-99%
purity by ¹H-NMR spectroscopy

Scheme 30

N-Oxides are also effective C—O bond oxidizers, and Boll and Fey (51) introduced a solid-supported TEMPO equivalent by mounting 3,3,5,5-tetra-methyl-4-azapiperidinone N-oxide onto aminomethyl-functionalized silica via reductive amination (**117**) (Scheme 31). In this case they oxidized a range of primary and secondary alcohols **116** using sodium hypochlorite as a co-oxidant.

Recently, a variety of halide-based oxidants were adapted to the solid and used effectively in the oxidation of alcohols. The reactions in Scheme 32 were reported to proceed cleanly and in excellent yields using the supported perbro-mide reagent **120** (52).

In a second set of examples (Scheme 33), hypervalent iodide anions were affixed to solid support via a quaternary ammonium salt (**125–127**), and these oxidizers were used quite effectively to produce a variety of carbonyl products **128** from the corresponding alcohols **124** (53).

It is well known that hypervalent iodine is an effective oxidizer of phenol derivatives to produce the corresponding quinone structure, and it has been shown that polymer-supported diacetoxyiodobenzene **101** is equally as efficacious as its homogeneous counterparts to effect this transformation. The examples in Scheme

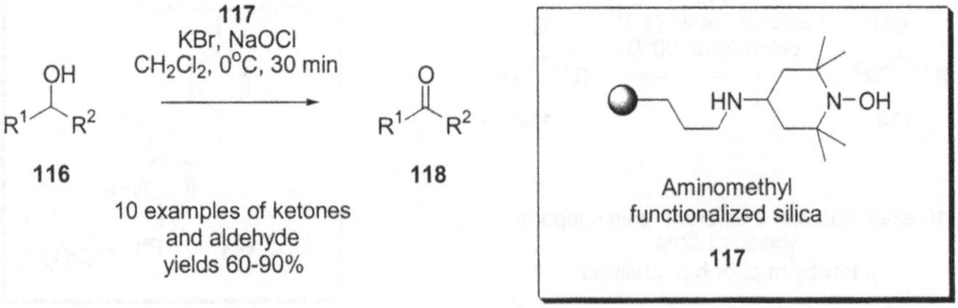

10 examples of ketones
and aldehyde
yields 60-90%

Aminomethyl
functionalized silica
117

Scheme 31

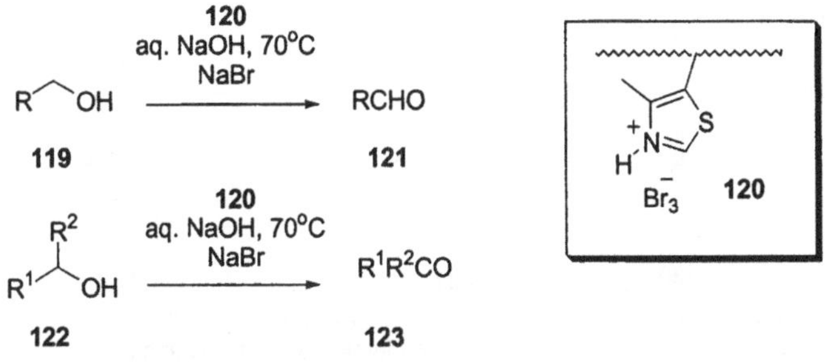

R~~~OH
119

120
aq. NaOH, 70°C
NaBr
———————→

RCHO
121

R¹~~~(R²)OH
122

120
aq. NaOH, 70°C
NaBr
———————→

R¹R²CO
123

22 examples, yields 81->99%

Scheme 32

34 illustrate such oxidations as well as an oxidative cyclization to produce spiro-cyclic quinones **132** (47). Ley et al. (47) commented that the polymeric reagent produced consistently good yields and excellent purity compared with homogeneous reagents used on analogous substrates in the literature (54) and reiterated Togo et al.'s findings (45) that **101** was readily recyclable.

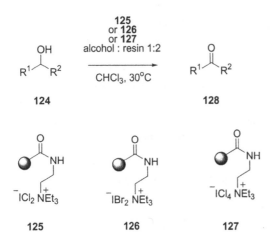

OH
R¹$\stackrel{}{\diagdown}$R²
124

125
or **126**
or **127**
alcohol : resin 1:2
———————→
CHCl₃, 30°C

O
R¹$\stackrel{\parallel}{C}$R²
128

125
$^-$ICl₂ $^+$NEt₃

126
$^-$IBr₂ $^+$NEt₃

127
$^-$ICl₄ $^+$NEt₃

9 examples for 3 different resins each
yields (**125**) 61-93%, (**126**) 72-92%, (**127**) 71-89%
purity by m.p. analysis

Scheme 33

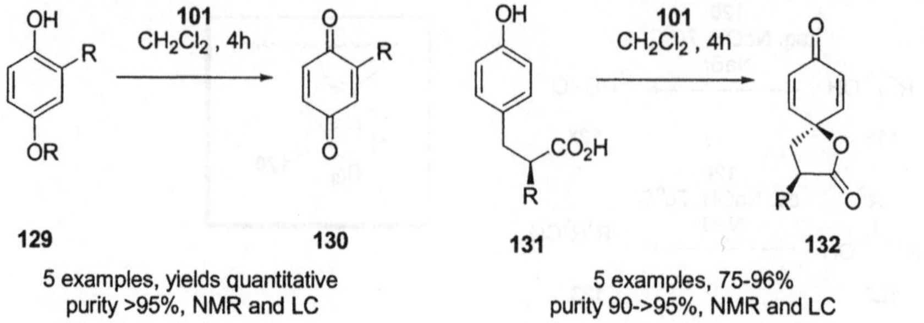
101
CH_2Cl_2, 4h
129
130
5 examples, yields quantitative
purity >95%, NMR and LC
101
CH_2Cl_2, 4h
131
132
5 examples, 75-96%
purity 90->95%, NMR and LC

Scheme 34

3. Sulfur, Phosphorus, and Nitrogen Oxidation

The polymer-supported diacetoxyiodobenzene reagent **101** discussed in the sections above has also been shown to be effective in a range of heteroatomic oxidations (Scheme 35) (46). Sulfides **133** have been oxidized effectively to disulfides and to sulfoxides **134** with no observed overoxidation to the corresponding sulfone using this oxidant. Phosphorus (e.g., **135**) has similarly been oxidized to

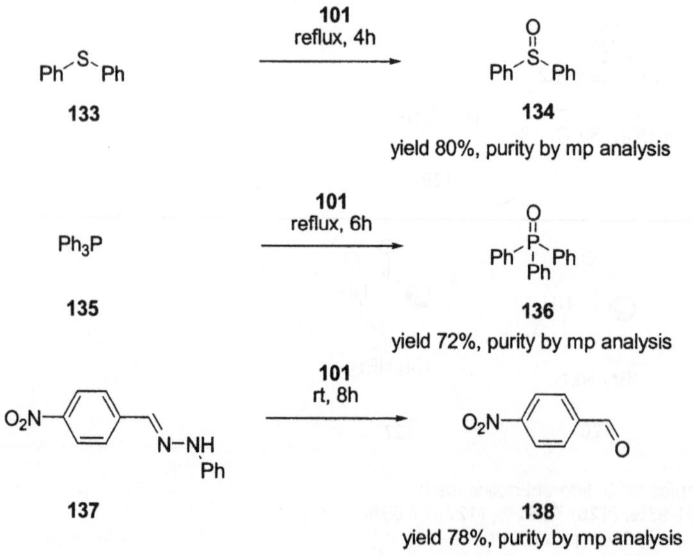
101
reflux, 4h
133
134
yield 80%, purity by mp analysis
101
reflux, 6h
Ph_3P
135
136
yield 72%, purity by mp analysis
101
rt, 8h
137
138
yield 78%, purity by mp analysis

Scheme 35

the phosphine oxide **136**, and hydrazones **137** have been reacted to form the corresponding carbonyls **138**.

4. C=C Bond Oxidations (Epoxidations)

Dimethyldioxirane (DMD) is a popular reagent for epoxidizing olefins. The reagent itself has a short shelf life and is best generated in situ from Oxone and used directly. Recently, Organ et al. (29) used this reagent to prepare a solution-phase library of epoxides that were subsequently aminated to make β-aminoalcohols. A highly polar medium is required to form the active oxidizer, and water is frequently the cosolvent. The library of epoxides created by Organ et al. required acetone/water cosolvent, which led to the requirement for in-line liquid–liquid extraction using an array of diatomaceous earth columns. A solid-supported DMD analog was prepared by Marples and coworkers (55), who developed an acetonitrile/EDTA system to allow the oxirane to form on the polymer (Scheme 36). They prepared two polymer-supported oxirane precursors, one built onto TentaGel S-Br resin **140** and the second prepared from hydroxymethyl resin **141**. The supported runs were performed opposite the equivalent nonsupported trifluoromethyl acetophenone derivatives. The yields reported using **140** and **141** are variable but display some generality over a variety of substrates **139**. Further,

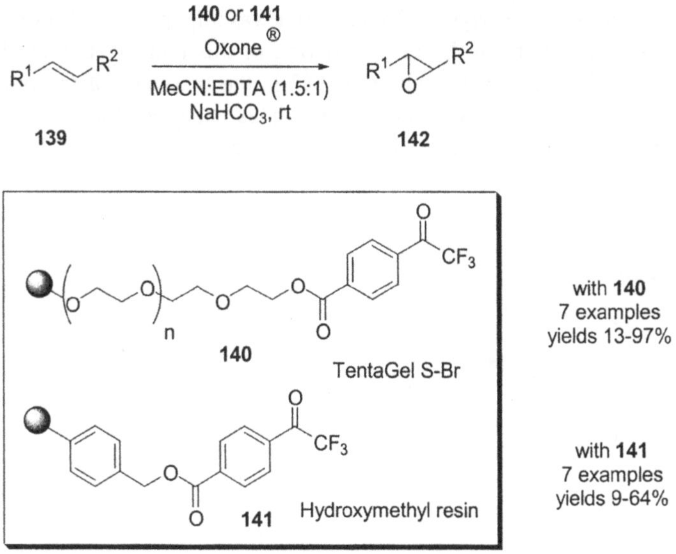

Scheme 36

the supported yields were typically lower than those of the equivalent nonsupported run.

E. Solid-Supported Reductions

1. C—X Bond Reductions

Despite health and environmental risks, tin-based reagents are still commonly employed in reduction reactions. In recognition of this, supported tin hydride derivatives have been under noticeable investigation for the last decade (56). Dumartin et al. (57) took this to the next level by developing an Amberlite XE 305-supported stannyl hydride **144** that is used in catalytic amounts and regenerated in situ with sodium borohydride (Scheme 37). In this case, the authors only reduced 1-bromoadamantane **145** as a model for their study, but they did report significantly lower amounts of tin (10,000-fold in some cases) in the reduced product **146** compared with reactions performed homogeneously.

2. C=O and C=NR Bond Reductions

The aluminum- and boron-based hydride reduction series are obvious candidates for adaptation to the solid phase because they are "-ate" complexes, requiring a cationic partner for hooking them to a support. Indeed, supported cyanoborohydride **148** was developed over 20 years ago (58), and more recently Ley et al. (59) reported a simple preparation of **148** by filtering a solution of sodium cyanoborohydride through Amberlyst A-26 that, after washing, provided 2 mmol of BH_3CN^- ion per gram of resin. Using a variation of this reagent, Ley et al. (60) diastereoselectively reduced the ketone in **147** in the final step of the production of dihydroepimaritidine **149** (Scheme 38). This examples nicely demonstrates the use of a polymer-supported reagent in a complex total synthesis (60).

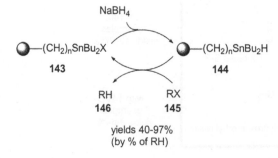

Scheme 37

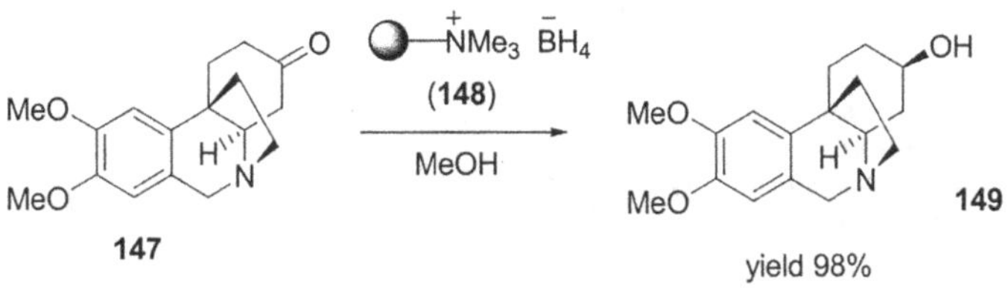

Scheme 38

Probably the greatest use of supported borohydrides is reductive amination by the pharmaceutical industry. The example by Hodges and his coworkers at Warner Lambert (now Pfizer Pharmaceuticals) (Scheme 39) is only one of a plethora of supported reductive aminations in the literature being used in HTPS (61). The supported aldehyde is added in the third step simply to scavenge the excess amine that is used to drive the reaction to completion.

3. Electron-Transfer-Mediated Reductions Followed by C—C Bond Formation

Reductions that involve the simultaneous oxidation of a fully reduced metal are useful in synthesis but are often a challenge to carry out practically. Such procedures might include dissolving metal reductions, organic molecule reducers (e.g., sodium naphthalide), or amalgams. The more recent use of samarium-based reductants (although not Sm^0) has received increasing attention, but they too are challenging to work with, and often unreproducible or inconsistent results are observed. In any case, none of the above methods would be suitable for an HTPS

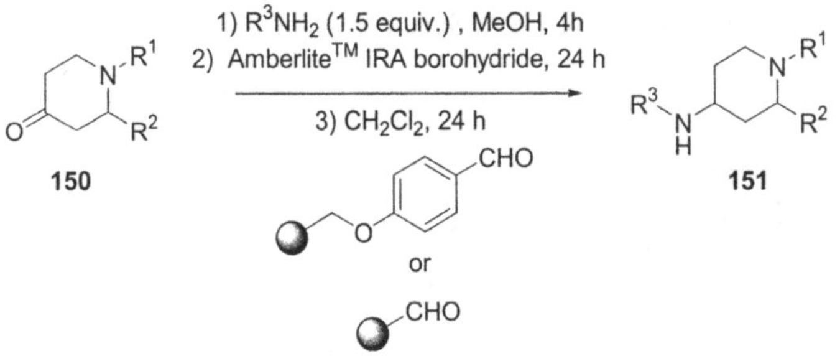

5 different amines were used
in a library preparation for this intermediate

Scheme 39

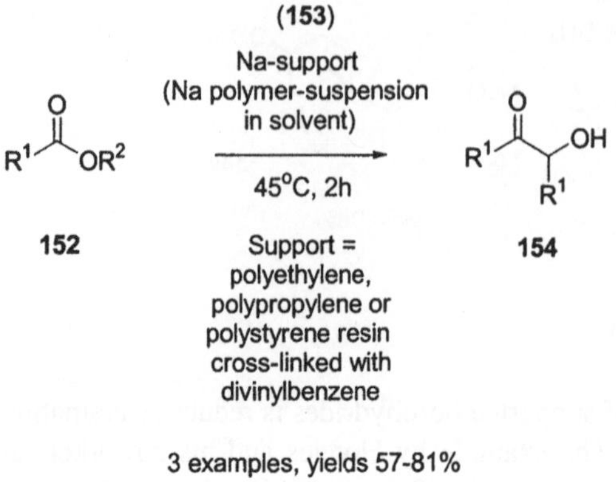

(153)

Na-support
(Na polymer-suspension
in solvent)

$R^1 \overset{O}{\underset{}{\Large\|}} OR^2$ $\xrightarrow{\hspace{2cm}}$ $R^1 \overset{O}{\underset{R^1}{\Large\|}} OH$

45°C, 2h

152

Support =
polyethylene,
polypropylene or
polystyrene resin
cross-linked with
divinylbenzene

154

3 examples, yields 57-81%

Scheme 40

application, because of either the chemistry itself or the inability to handle such sensitive reagents in a parallel synthesis format. The ability to produce a readily dispensable yet filterable source of a fully reduced metal would greatly enhance the usefulness of this reduction technique (62).

Makosza et al. (63) reported the development of "high-surface alkali metals" that have been deposited on various supports including sodium chloride, polyethylene, polypropylene, and cross-linked polystyrene from their solutions in ammonia. These alkali-metal-doped polymers can be stored as stable suspensions under argon in inert solvents. In stability studies, these reagents were shown to lose minimal activity after 3 months of storage at 10°C. The authors report that once the titer has been determined, the suspensions need only be shaken to homogeneity and the suspension removed by a wide-bore needle. The use of these reagents is discussed below in reference to Schemes 40–42.

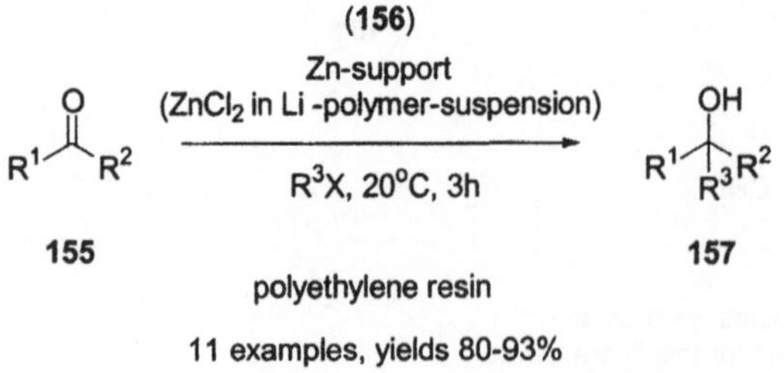

(156)

Zn-support
(ZnCl$_2$ in Li -polymer-suspension)

$R^1 \overset{O}{\underset{}{\Large\|}} R^2$ $\xrightarrow{\hspace{2cm}}$ $R^1 \overset{OH}{\underset{R^3 R^2}{\Large\|}}$

R^{3}X, 20°C, 3h

155

157

polyethylene resin

11 examples, yields 80-93%

Scheme 41

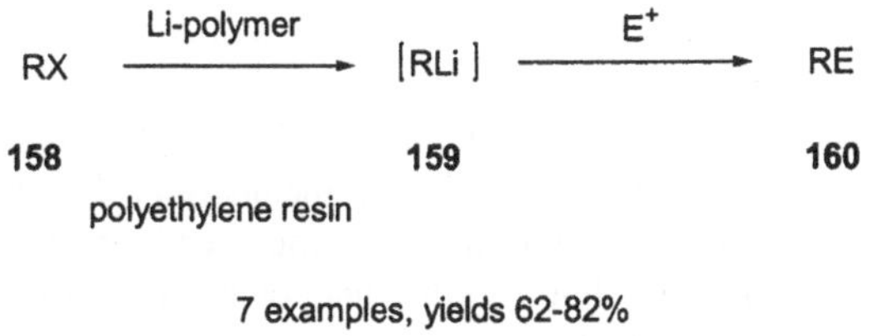

RX $\xrightarrow{\text{Li-polymer}}$ [RLi] $\xrightarrow{\text{E}^+}$ RE

158   **159**   **160**

    polyethylene resin

        7 examples, yields 62-82%

Scheme 42

    a. *C=O Bond Reduction.* Acyloin reactions, both intramolecular and intermolecular, have been reported using high surface sodium **153** (Scheme 40) employing a variety of types of supports and solvents.

    Barbier reactions were also shown to be promoted effectively by supported active zinc **156**, as were a variety of Reformatski reactions (Scheme 41) to form alcohol products **157**.

    b. *C—X Bond Reduction.* Makosza and coworkers also used high-surface lithium polymer **160** to reduce halides **158** to provide the corresponding organolithium intermediate **159**, which then underwent substitution with a number of electrophiles (E$^+$) to give rise to products **160** (Scheme 42).

IV. SOLID-SUPPORTED CATALYSTS

Two areas of considerable interest in HTPS are the use of supported catalysts in solution-phase synthesis and the development of such catalysts themselves. Homogeneous catalysts, not unlike excess reagents or by-products, are often left behind following reaction workup, causing purification problems. Adapting them to the solid support allows for their removal by simple filtration. In the next two subsections, the use of supported catalysts in HTPS, or in synthesis in general that could be applied to HTPS, is highlighted. This review includes both metal-based and nonmetal-based catalysts. There is no discussion of catalyst development per se, but rather the focus is on the use of supported catalysts in solution-phase synthesis and issues pertaining to their handling.

A. Metal-Catalyzed Reactions

Despite many published mechanisms for the wide variety of known metal-catalyzed reactions, catalysis remains among the least well understood areas of synthetic chemistry. Coupling that with the adaptation of traditionally homogeneous catalysts to the solid will ensure that supported catalysis will continue to draw the research attention of synthetic chemists for some time. Mechanistically, heterogeneous and homogeneous catalysis can be quite distinct from each other. For example, hydrogenation using supported late transition metals (e.g., Pd on charcoal) is quite different from the corresponding homogeneous versions. Not only are the proposed mechanisms different, but also the functional group compatibility between the two methods varies. Heterogeneous catalysis is much more susceptible to poisoning with sulfur, for example, whereas the ability of solubilized metals to undergo more facile ligand–ligand exchange reduces this concern. Thus, unlike reagents (see Sec. III), the nature of the solid (polymer vs. nonpolymer, insoluble polymer vs. soluble polymer, etc.) is much more important for predicting the generality of a new supported catalyst, which is so important for HTPS.

1. Metal-Based Catalysts in C—C Bond Formation

The fastest expanding area of catalysis with HTPS applications is the reaction of transition metal sigma-bonded complexes to form C—C bonds (i.e., coupling reactions). This area, which includes Suzuki, Stille, and Sonogashira couplings, has been developing steadily since the early 1970s, and the vast majority is homogeneous catalysis using phosphine-based ligands. Triphenylphosphine, a common ligand for Pt, Pd, Rh, and Ru, is particularly challenging to remove during purification, making it less than ideal for solution-phase synthesis. However, along with arsenes, phosphines are among the best ligands for catalysis. These sigma donor ligands also back-bond well with late transition metals to maintain their solubility and yet are readily exchangeable, thus allowing catalysis to take place. Therein lies the challenge for supported catalysts: to find a support that maintains the catalytic activity of the metal while not allowing the metal to leach off the support. This could cause blacking out of the metal, which, of course, affects turnover and yield, but also introduces metals into the molecular library, which can affect the reliability of biological screening results of these compounds.

The metal that has clearly attracted the most attention in coupling reactions using supported catalysts is palladium. There have been numerous reports of supported Pd-catalyzed Suzuki couplings, Heck reactions, Stille couplings, carbonylations, and allylic substitutions (examples of these are discussed below). In light of the number of examples to report in this area, the reports are summarized in Table 1 and the catalysts are summarized in Figure 1.

Entry 1 in Table 1 illustrates the use of supported triphenylphosphine Pd equivalent **161** in Suzuki reactions with primarily dialkylboron-substituted alkenes **174** (and some boronic acids) with bromide and triflate coupling partners

Table 1 Carbon–Carbon Bond Formation Using Supported Catalysts in Organic Solvents

Entry	Examples	Yield	Ref.
1	**161** 80°C, 2h NaOEt R^1BY_2 + R^2X ⟶ R^1-R^2 + **161** **174** **175** **176** Reuse 13 examples	78-96%	64
2	**161** PhB(OH)$_2$, Na$_2$CO$_3$ toluene:EtOH:H$_2$O (10:1:1) 80°C, 24 h **177** ⟶ **178** 5 more examples	56-98%	66
3	**162** Na$_2$CO$_3$ MeOH:H$_2$O (1:1) Δ, 5h **179** + (HO)$_2$B⟶CO$_2$H **(180)** ⟶ **181** 1 example	95%	67
4	**164 or 164** TEA, DMF 115°C, 2.5-6.5 h **182** + ⟶R **(183)** ⟶ **184** 2 examples	85-91%	69
5	**165** H$_2$, 90% aq. ethanol:heptane (1:1) 70°C **185** ⟶ **186** 1 example	Not reported	70
6	**166** R$_2$NH, TEA EtOH : heptane (1:1) 70°C, 11-14 h **187** OAc ⟶ **188** NR$_2$ 2 examples	Not reported	70

Table 1 Continued

7	R^1 —X **189** $\xrightarrow[\substack{(190)}]{\substack{167 \\ NaOAc \\ \diagdown R^2}}$ R^1 $\diagup\diagdown R^2$ **191** + R^1 R^2 **192** 43 examples with 2 different catalyst systems	38-99%	71
8	R^1 —X **193** $\xrightarrow[\substack{R^2 \diagdown \\ (194)}]{\substack{169 \\ TEA, toluene \\ 100°C}}$ R^1 $\diagup\diagdown R^2$ **195** 7 examples	55-87%	72
9	R^1 —X **196** $\xrightarrow[\substack{\equiv\!-R^2 \\ (197)}]{\substack{169 \\ Et_2NH, toluene \\ 60°C}}$ R^1 $\equiv R^2$ **198** 4 examples	58-87%	72
10	$\diagdown\diagup$OAc **199** $\xrightarrow[\substack{toluene, 100°C, 16 h \\ 1 example}]{\substack{169 \\ CH_2(CO_2Me)_2 \\ (C_4H_8N)_3P=N^tBu}}$ $\diagdown\diagup$ $\substack{CO_2Me \\ CO_2Me}$ **200**	56%	72
11	HO_2C —I **201** $\xrightarrow[\substack{\diagdown CO_2H \\ (202) \\ 1 example}]{\substack{171 \\ K_2CO_3, H_2O, reflux, 2 h}}$ HO_2C $\diagup\diagdown CO_2H$ **203**	90%	73
12	**204** $\xrightarrow[\substack{\diagdown CO_2Me \\ (205) \\ 1 example}]{\substack{172 \\ TEA, MeOH \\ reflux, 17 h}}$ **206** $\diagup\diagdown CO_2Me$	81%	73
13	$\substack{OAc \\ Ph \diagdown\diagup Ph}$ **207** $\xrightarrow[\substack{DMSO, 22°C, 24 h \\ 1 example}]{\substack{173 \\ NaCH(CO_2Me)_2}}$ $\substack{CH(CO_2Me)_2 \\ Ph \diagdown\diagup Ph}$ **208**	85%	73

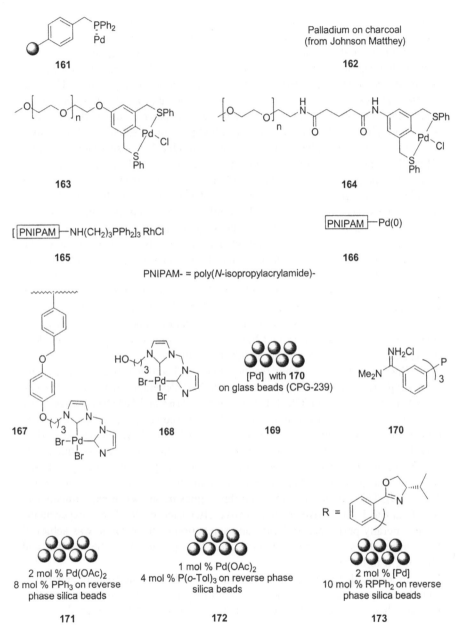

Figure 1 Supported metal catalysts for use in organic solvents.

175 (64). Although the catalyst is not new, its preparation is reported.* Each reaction in the report was repeated using (PPh₃)₄Pd to evaluate the efficacy of the supported catalyst, and in most cases **161** provided noticeably higher yields than its homogeneous counterpart. In Table 1, entry 2 illustrates more Suzuki reactions performed using **161** to couple electron-poor bromopyridines **177** (and bromobenzenes substituted with electron-withdrawing groups) that often provide poor yields (66). Fenger and Le Drain (66) also compared results with (PPh₃)₄Pd, and, although variable, the supported catalyst performed equally well overall.

In an interesting alternative to the quasi-soluble **161**, researchers at Smith-Kline Beecham (67) reported a truly heterogeneous Suzuki reaction (Table 1, entry 3) to prepare **181** using Pd/C **162**. They report carrying out this process on a 24 M scale. It is interesting to ponder what the active species actually is in this relatively basic and operationally simple system. Buchecker and coworkers at Hoffmann-La Roche performed some very telling experiments to elucidate these questions.† They set up similar Suzuki reactions using Pd/C and halted the reaction halfway. The heterogeneous mixture was then filtered and left to incubate. The reaction ceased; however, when the Pd/C was added back to the same mixture, it again proceeded to completion. This would indicate that the support is vital and that the metal does not dissociate from the surface during the catalytic cycle. There are reactions such as enyne cyclizations that do proceed without ligands with Pd black, and in fact ligands slow these reactions down. This does not appear to be the case for the Suzuki reaction.

Heck coupling is another reaction that has been investigated extensively with supported catalysts. Bergbreiter et al. (69) developed new poly(ethylene glycol)-bound tridentate SCS–Pd(II) complexes (**163** and **164**) to catalyze Heck reactions. The first-generation catalyst **163** was shown to be quite efficient in converting a small number of substrates (i.e., **182** and **183**) into Heck product **184**. However, this support was found to be labile, and Pd black was found contaminating the products. The second-generation catalyst **164**, which was based on amide linkages rather than ethers, was found to be much more stable (Table 1, entry 4). This catalyst was recycled three times in one set of experiments and was found to be as active or more active after each cycle. These experiments were generally performed in DMF, and separation of the catalyst was achieved by pouring the soluble mixture into ether, thus precipitating the catalyst and allowing it to be recovered.

In a related study, Bergbreiter et al. (70) addressed the issue of separating soluble catalysts from the product. As with the previous example, precipitation is the method most often employed to remove a soluble catalyst (or product); this

---

* For original reports on supported TPP, See Ref. 65a. For preparation also, see Ref. 65b.
† For an earlier report of Suzuki reaction using Pd/C, see Ref. 68.

can have its own problems. In addition to adding a manipulation step, crashing a polymer out of solution rapidly can entrap compounds that could contaminate subsequent runs if the catalyst is reused. Another strategy is to have the catalyst in one phase and the product (at least) in a different phase. Phase transfer chemistry is quite well established, but reactions are often sluggish because the reagent, substrate, and catalyst are not in the same phase. To address this problem, Bergbreiter et al. applied the concept of "thermomorphism," which refers to a change in the solubility of one phase in another as a function of temperature. In this case, they used a poly(N-isopropyl acrylamide) polymer equipped with diphenylphosphino moieties to serve as binding sites for Rh (**165**) and Pd (**166**). This polymeric system is soluble in aqueous ethanol but not in heptane, which formed a bilayer in hydrogenation and allylic substitution experiments (Table 1, entries 5 and 6, respectively). Hydrogenation did not proceed at all at 22°C when the mixture was biphasic, but when the mixture was heated to 70°C, a homogeneous solution resulted and a linear uptake of H_2 was observed over the course of an hour. The catalyst was recycled four times by cooling the hydrogenation mixture back to 22°C, assaying, adding more 1-dodecene substrate, and reheating it to 70°C. There was no appreciable loss of catalyst activity throughout this process. Although this idea deals nicely with removing the catalyst from the phase containing the product, to be practically applicable to HTPS the issue of phase separation (liquid–liquid extraction) would have to be dealt with first.

Herrmann and coworkers, who have been working for some time in the area of divalent group IV elements, developed a polystyrene-supported (Wang resin) biscarbene Pd(II) catalyst **167** (71). They performed Heck reactions with both **167** and the homogeneous catalyst **168** (Table 1, entry 7). The results of the two sets of experiments were very similar, which raised concern that the results seen in the supported runs were in fact from a solvated catalyst (i.e., from leached Pd) and not from **167**. Like the experiment with Pd/C discussed above, one experimental run was filtered at 20% conversion and the filtrate was monitored for activity. No additional conversion was observed; which suggests that the Pd stays associated with the support throughout. That being said, elemental analysis of **167** after one use demonstrated significant Pd loss (1.0% w/w Pd down to 0.4% w/w). Following that, Pd loss was minimal in subsequent recycled uses of the catalyst, and Herrmann et al. attributed the significant loss in the first cycle to the initial catalyst activation [i.e., reduction of Pd(II) to Pd(0)]. In any case, this issue would have to be addressed before such a catalyst would be useful for HTPS when biologically active species are being produced and screened.

Williams and coworkers published a couple of reports of the use of nonpolymer-supported Pd catalysts. In one instance (72) guanidinium phosphine complexes **170** were loaded onto glass beads to form **169**, and this catalyst was used to effect a range of Pd-catalyzed reactions including Heck and Sonogashira couplings and allylic substitutions (Table 1, entries 8, 9, and 10, respectively). The reactions reported had good, although not excellent, yields. A careful analysis was conducted to determine the amount of leaching in each of the different reaction types after single and repeated use. The amount of Pd leached into the product ranged between 0.01% and 0.30% Pd relative to the level loaded on the catalyst.

In the second study, Williams and coworkers (73) formed a catalytically active Pd material by suspending reverse-phase silica (Davisil 300) in cyclohexane, adding TPP and Pd(OAc), and then simply removing the solvent in vacuo. This surface-adsorbed catalyst 171 was found to be sluggish in some preliminary studies of Heck reactions, and changing the ligand from TPP to tri-o-tolylphosphine 172 improved the turnover significantly (Table 1, entries 11 and 12) (73). These authors also reported an asymmetric allylic alkylation experiment using catalyst 163 (Table 1, entry 13), illustrating some further utility of this supported catalyst approach. It is not clear from the report that the authors performed control experiments to verify that the active catalyst was in fact support-associated, although they do indicate that Pd leaching levels are low. They state that the catalyst was recycled, which further hints that the Pd stays associated with the surface of the silica, providing, of course, that the reactions are worked up by simple filtration, but this aspect was not made clear in the report. In any case, the major benefit associated with the reverse-phase Pd catalyst according to the report was that very polar substrates could be employed whereas other catalyst systems presumably proved ineffectual.

The adaptation of organic transformations to aqueous media is an area of synthesis that continues to attract a great deal of attention, primarily for health and environmental considerations. It could have a tremendous impact on HTPS if one considers the practicality of performing sensitive reactions (e.g., air- and moisture-free) on a large parallel scale. Aqueous adapted reactions could reduce the dependence on enclosed reaction blocks, which have a number of drawbacks including leaks and cross-contamination. Inaccessibility to the reaction once it has begun makes it very difficult to monitor progress or to add reagents to push sluggish reactions to completion.

A number of Pd supports (see Fig. 2 and Table 2) have been developed that are compatible with aqueous conditions, and, although the issue of liquid–liquid separation must be addressed, these preliminary results suggest that aqueous high throughput catalytic reactions could be on the horizon. Process chemists at Eli Lilly and Co. have developed an air- and moisture-stable resin-bound Pd catalyst 209 for Suzuki reactions (Table 2, entry 1) (74). Deloxan, a supported thiourea, is based on a highly cross-linked macroporous polysilane backbone that swells very little in any solvent including water. The active catalyst was easily prepared by treating Deloxan with $Pd(OAc)_2$ in methanol and then rinsing thoroughly with water and methanol to remove the unbound metal. It was reported

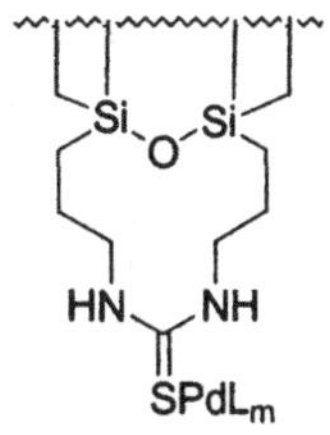

209

PVP-Pd nanoparticle catalyst

210

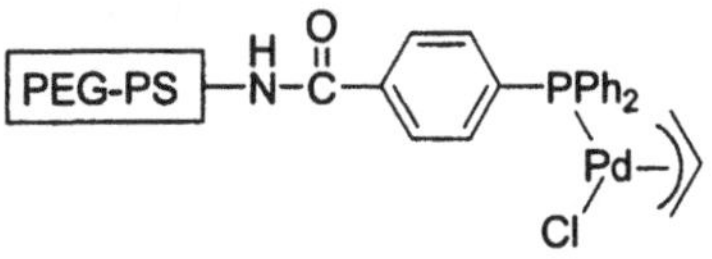

211

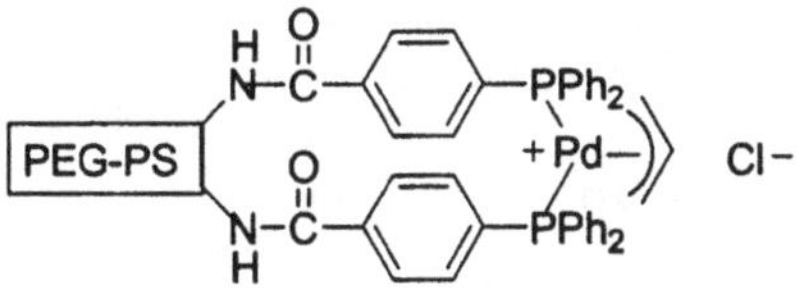

212

PEG-PS = poly-ethylene glycol based polymer support

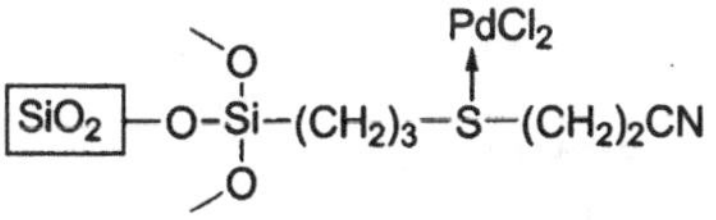

213

Figure 2 Supported metal catalysts for use in aqueous media.

(74) that **209** performed at a turnover rate similar to that of $(PPh_3)_4Pd$ under homogeneous conditions and that **209** could be reused two or three times, after which activity dropped significantly. Residual Pd in the crude **216** was reported to be below 3 ppm.

The El-Sayed group at Georgia Tech have also performed Suzuki reactions in aqueous solution (Table 2, entry 2) (75). In this case, they developed Pd nanoparticles stabilized by poly(*N*-vinyl-2-pyrrolidone) (PVP) **210**, which form a colloid in solvent. They monitored the reactions using fluorescence spectroscopy by measuring the appearance of biaryl **219** from boronic acid **217** and iodide **218**. From these results they concluded that the reaction takes place at the surface of the Pd nanoparticles, although they did indicate that significant Pd black was deposited on the bottom of the vessel following the coupling reaction. Thus, such a catalyst would need to be improved for applications in the synthesis of biologically relevant molecules for screening purposes.

Table 2 Carbon–Carbon Bond Formation Using Supported Catalysts in Aqueous Media

Entry	Examples	Yield	Ref
1	**209** ^{i}PrOH, H$_2$O, K$_2$CO$_3$ 80°C, 3 h **214** **215** **216** 11 examples	27-97%	74
2	**210** 40% EtOH, reflux **217** **218** **219** 6 examples	26-95%	75
3	**211** aq. KOH, 25°C or NaBPh$_4$ **220** **221** **222** 8 examples	66-91%	78
4	**211** PhB(OH)$_2$, K$_2$CO$_3$, or NaBPh$_4$ H$_2$O, 25°C **223** **224** 5 examples	29-99%	78
5	**211 or 212** NuH or NuNa K$_2$CO$_3$, H$_2$O, rt, 12 h **225** **226** 6 examples	2-100%	79
6	**211** aq. KOH **227** **228** **229** 1 example	92%	79
7	**211** CO (1 atm) aq. alkaline, 25°C **230** **231** 9 examples	45-100%	80
8	R^{1}SnBu$_3$ + R^{2}I **213** CH$_3$CN, H$_2$O (4:1) 80°C, 5h R^1–R^2 **232** **233** **234** 12 examples	71-89%	81

Phosphines are still the most commonly used ligands for Pd, including supported ligands, and there have been attempts to develop such catalysts for use in aqueous media.* Uozumi et al. (77) reported a grafted poly(ethylene glycol) (PEG)/polyethylene (PE) copolymer (PEG-PS resin) with diphenyl-phosphino moieties affixed through an amide linkage to the end of the PEG chains (PEP) to serve as metal-binding sites. The final stable catalyst **211** was prepared readily by mixing PEG-PS resin with [PdCl(π-C$_3$H$_5$)]$_2$ in dichloro-methane, filtering, and washing away the excess metal precursor with additional dichloromethane. The resulting complex was shown to have good swelling properties in water and organic solvents such as dichloromethane or THF. The same authors (78) went on to use this catalyst system in a variety of Pd-cata-lyzed reactions that were run in water alone with no cosolvent, thus constitu-ting true aqueous transformations. Suzuki couplings were conducted using **211** with aryl iodides **220** and arylboronic acids **221** (Table 2, entry 3) or allylic acetates **223** (Table 2, entry 4) (78). Allylic substitutions on **225** (Table 2, entry 5) were successfully carried out with quite a range of nucleophiles, includ-ing sodium azide, sodium sulfinate, malonate-type anions, and amino acids (*N*-alkylation) (79). One demonstrative Heck reaction was carried out with iodo-benzene **227** and acrylic acid **228** to prepare cinnamic acid **229** (Table 2, entry 6) (79).

Uozomi and Watanabe (80) reported successful carbonylation reactions us-ing **211**, which is conceptually interesting (Table 2, entry 7). In this case, they developed a four-phase system: gaseous (CO), organic (the starting material **230**), solid (the catalyst **211**), and aqueous (the solvent water, containing the salt of product **231**). They propose that the CO and substrate diffuse into the catalyst, where carbonylation takes place, and then the readily available water simply adds to the resultant acyl-Pd species. The basic reaction conditions deprotonate the benzoic acid product **231**, making it water-soluble, which is key for recycling. Filtration of the reaction mixture removes the product (which is subsequently acidified and recovered), leaving behind the catalyst, which can be reconstituted with a fresh basic water solution. Addition of new organic substrate and reintro-duction of CO reconstructs the system; this cycle has been repeated 30 times with no apparent loss of activity of the catalyst. In addition to the aryl substrates, the system also appears to be well capable of dealing with vinyl halide starting materials (e.g., bromostyrene to cinnamic acid).

Kang et al. (81) developed sulfide-based poly[3-(2-cyanoethylsulfanyl)pro-pylsiloxane palladium] **213**, which has been shown to be a reliable catalyst in Stille couplings (Table 2, entry 8). They report that the catalyst is recyclable, although they mention reusing it only twice. Nonetheless, the catalyst has suc-

---

* For recent examples of Pd-phosphine catalysts used in aqueous media, see Refs. 76a and 76b.

cessfully coupled iodides **233** with a variety of Stille partners including aryl-, alkenyl-, and alkynylstannanes **232**.

Hydroformylation is an attractive way of introducing a reactive carbonyl group, especially since successful asymmetric versions of the procedure emerged. Polymer-supported chiral catalysts were used for this application as early as the late 1970s (82). Nozaki's group has been active in this area for a number of years, and they developed an effective homogeneous asymmetric procedure using (*R*, *S*)-BINAPHOS (83). They adopted this ligand to the solid by affixing one of the BINAP rings with a vinyl group and subjecting it to radical copolymerization with styrene in the presence of divinylbenzenes (see **236**). They performed a number of mechanistic studies on the hydroformylation of styrene, including studies on the order of catalyst preparation, the effect of cross-linking, and catalyst recycling. In general, the ratio of *iso* (**237**) to *normal* (**238**) products (see Scheme 43) was about 8.5 : 1.5, and the entantiomeric excess values (*ee*'s) of the *iso* product were excellent (≳90%). A number of other substrates were tested, and the results are summarized in Scheme 43 (84).

A different metal-binding system on solid support was proposed for hydroformylation by Andersen et al. who suggested that phosphine-based ligands suffer

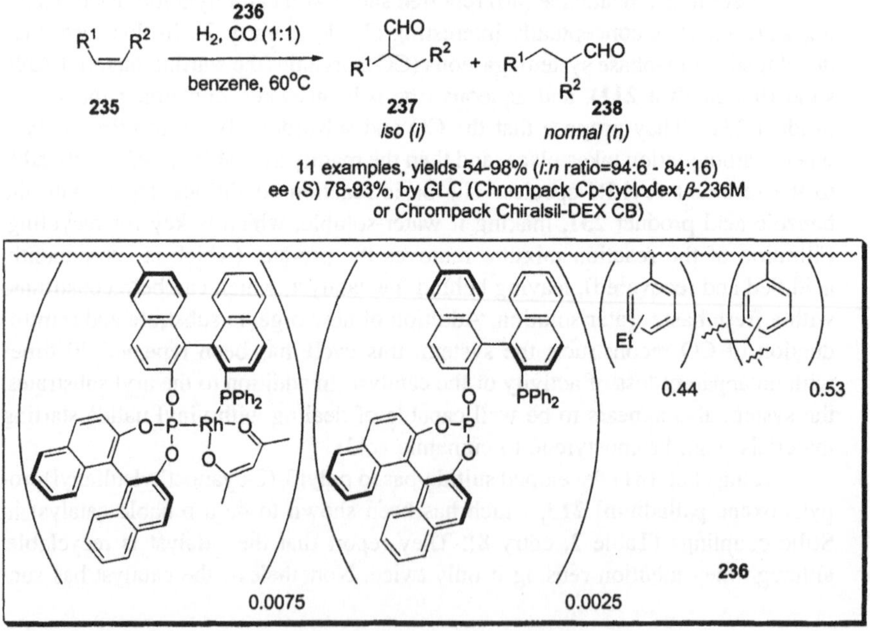
236
H_2, CO (1:1)
benzene, 60°C
235
237
iso (i)
238
normal (n)
11 examples, yields 54-98% (i:n ratio=94:6 - 84:16)
ee (S) 78-93%, by GLC (Chrompack Cp-cyclodex β-236M
or Chrompack Chiralsil-DEX CB)
0.0075
0.0025
0.44
0.53
236

Scheme 43

significant metal dissociation. They also expressed concern over the side reactions, such as oxidation, which would prevent the phosphine P atom from serving as an electron donor. Their supported design was based on the homogeneous dirhodium tetraacetate complex, which imposed binding geometries of 90° on Rh that should tightly hold the catalytically active metal. The final catalyst **241** (Scheme 44) was used to hydroformylate 1-hexene **239**, and it showed mixed results (85). The ratio of *normal* **242** to *iso* **243** hydroformylation products (1:1) was not nearly as pronounced as in the above example, and significant amounts of double-bond isomerization **240** were also observed. However, these are early results, and the authors suggest that the rigidity imposed by the supported system can generate unique results not obtainable using the corresponding homogeneous system(s).

Ring-closing metathesis (RCM) is fast becoming the method of choice for constructing cycloalkenes of various sizes (86,87a) (for a review, see Ref. 87b). It has been used effectively in total synthesis (88,89) and also to prepare libraries of epothilone E analogs (90). The efforts of Herrmann in the area of carbene ligands for catalytically active metals spurred others to try to adapt a carbene ligand system to Ru for the purpose of improving the stability of the Grubb's catalyst in RCM (91,92).

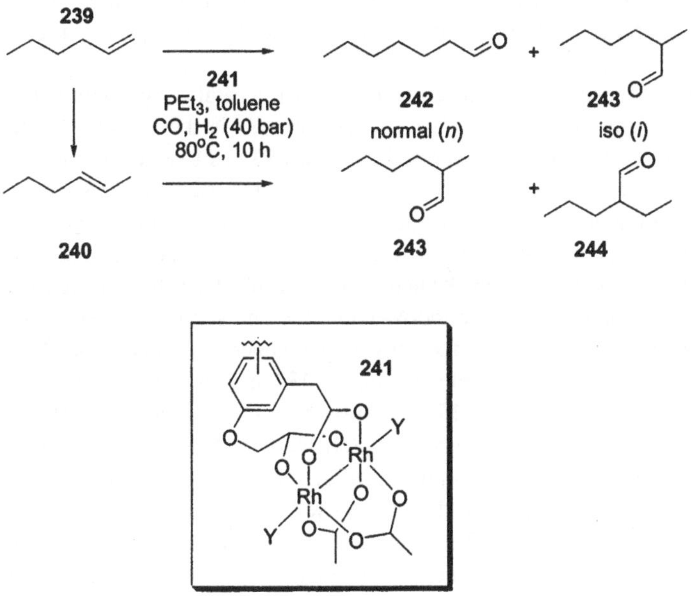

Scheme 44

The combined efforts of Barrett and Procopiou (from the Glaxo group, Stevenage) have produced a couple of reports in the area of supported metathesis catalysts, the latest of which is based on such carbene-based ligands. They have created what they call a "boomerang" catalyst whereby solid-supported Ru (through a carbene linkage) first dissociates from the solid where it does its metathesis work. Then, after the metathesis products are generated, the catalyst, ligands and all, is then effectively "scavenged" out of the solution back to the support. In the past, clean removal of Grubb's catalyst 250 from reaction products has proven to be problematic. Therefore, this is potentially a very important advancement for RCM applications in HTPS. The first-generation catalyst 248 (93) was modified by replacing one of the phosphine moieties by an Arduengo carbene 251 to provide 247. The corresponding homogeneous catalyst 249 was found to be very stable compared with Grubb's catalyst 250. There was no visible degradation of 249 when it was heated at 60°C in toluene for 14 days, whereas 250 showed significant decomposition after sitting in methylene chloride at room temperature for 1 h. The catalysts were tested on a variety of substrates (245), (Scheme 45), and 247 was shown to generally outperform 248 in the initial reaction; in subsequent reuses of 247 it was found to be much more resilient. It was also demonstrated that when 5 mol% of TPP was added to the reactions, the longevity of 247 was significantly enhanced, although the thought of putting TPP into HTPS reactions is not a welcome one from a purification point of view. In general, it does not seem that 247 can be used with efficacy beyond three recyclings under the reaction conditions that were deemed to be most optimal in their study (94).

Cobalt has been proven to be a useful metal for catalyzing a number of reactions. Perhaps the best known Co-catalyzed process is the Pauson–Khand reaction, and a phosphine-based supported catalyst 253 has been developed for this process. The example in Scheme 46 demonstrates the use of this catalyst to convert a small number of enyne substrates 252 into bicyclic enones 254 (95).

Kerr's group has also been very active in the area of supported ligand design for Co for use in the Pauson–Khand reaction (96). They have developed systems based on supported morpholine-N-oxide 258 and supported methyl sulfide 257 and tested their scope. The former catalyst, generated on ArgoGel-NH₂, gave good to excellent yields of cyclopentenones 259 and 262 from a variety of alkenes (256 and 261) and alkynes (255 and 260), respectively, and it was shown to have no loss of activity after five reuses (Scheme 47). This ligand system is under consideration to promote asymmetric transformations, and the modular design of its construction means that the ligand itself is a candidate for optimization by combinatorial synthesis. The supported methyl sulfide Co catalyst 257 was also shown to be efficacious with the same reactions and could also be reused many times (97).

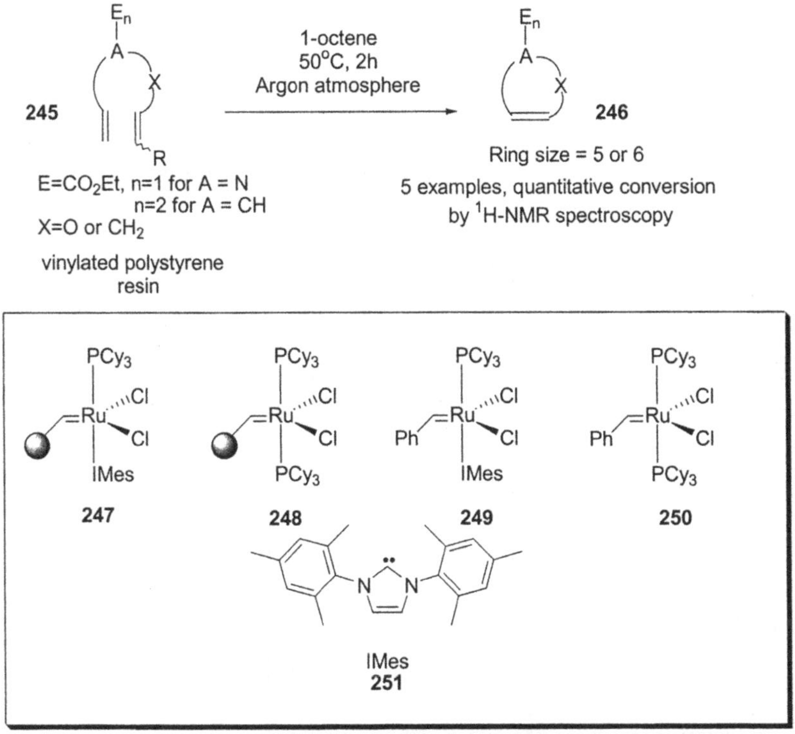

245
1-octene
50°C, 2h
Argon atmosphere
246
E=CO2Et, n=1 for A = N
n=2 for A = CH
X=O or CH2
vinylated polystyrene resin
Ring size = 5 or 6
5 examples, quantitative conversion by 1H-NMR spectroscopy
PCy3
Cl
Ru
IMes
Ph
247
248
249
250
IMes
251

Scheme 45

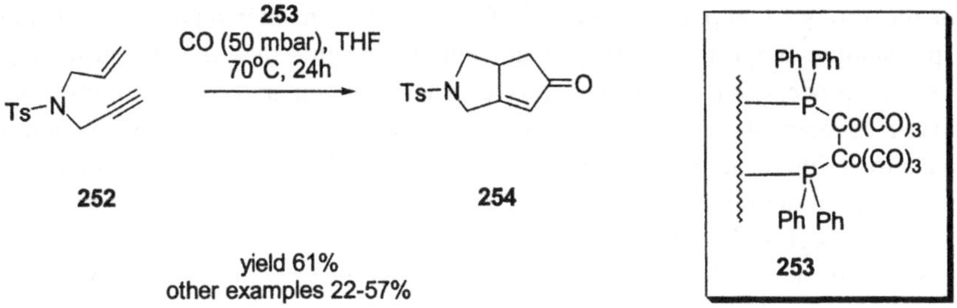

253
CO (50 mbar), THF
70°C, 24h
Ts–N
O
252
254
yield 61%
other examples 22-57%
Ph Ph
P
Co(CO)3
Co(CO)3
253

Scheme 46

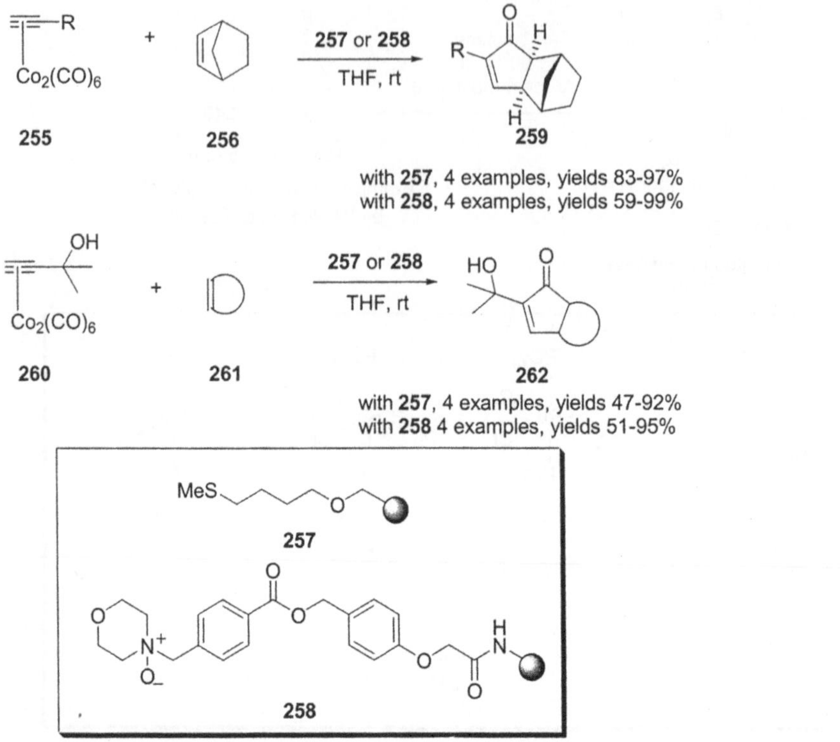

with **257**, 4 examples, yields 83-97%
with **258**, 4 examples, yields 59-99%

with **257**, 4 examples, yields 47-92%
with **258** 4 examples, yields 51-95%

Scheme 47

In the last few years, Iqbal and his group have been working on the use
of Co to catalyze multicomponent condensation reactions (98). They developed
an interesting coupling between a ketone (or ketoester) **263**, an aldehyde **264**,
and acetonitrile in the presence of acetyl chloride to produce beta-amino carbonyl
(or dicarbonyl) compounds **266** and **267**. One of their later reports (99) details
a polyaniline-supported Co(II) acetate catalyst **265** that they developed to catalyze this process that has proven very efficacious compared to the homogenous
system (Scheme 48). In addition to reducing the aqueous workup and extraction
to a simple filtration, they demonstrated superior yields and diastereoselectivities
in each example reported.

Although radical reactions are among the oldest studied and have proven
very useful for certain synthetic applications such as bromination, they have not
been widely used in C—C bond-forming reactions until somewhat recently (100).
Likely, the main reason for this has been the difficulty in controlling the radical
clock and disproportionation. For molecular library preparations, where purifica-

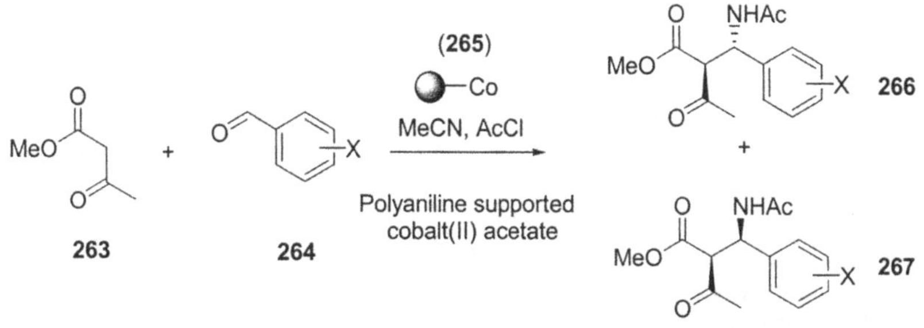

7 examples
yields (homogeneous : 31-52%
polymer supported : 51-68%)
Better diastereoselectivity with polymer support

Scheme 48

tion is of prime importance, it is understandable that radical reactions have not been used, or at least not used significantly. Interestingly, there are a number of recent SPOS reports of C—C bond-forming radical-based reactions highlighting intramolecular (101), intermolecular (102), and intra/intermolecular variants (103). Of course, supported substrates offer both advantages and disadvantages for radical chemistry. Being heterogeneous and on the solid phase, substrates undergoing intramolecular reactions may be favored over those undergoing intermolecular side reactions such as simple quenching. Conversely, reactions that are designed to be intermolecular could be compromised, especially in cases where the initial radical species is formed in solution and has to react with a partner in a different phase, i.e., on the solid.

Reductive, radical-induced cyclizations have certainly been used effectively in synthesis (100). Nonreductive, radical-induced cycloisomerization reactions, also called atomic transfer radical cyclization (ATRC), have gained in popularity because the halide is not lost in the process, which opens the door for further chemistry with that functional group. A number of transition metals have been used to mediate this process, including Ru (104), Fe (105), and Cu (106). Clark and his group (107) developed a family of N-alkyl-2-pyridyl-methanimine ligands to coordinate Cu where the complex is tethered to silica via the imine nitrogen. The most reactive catalyst **269** was shown to be effective in catalyzing the ATRC reaction for a range of substrates (**268**) (Scheme 49). Although it was shown to be significantly less reactive than the corresponding homogeneous runs with **270**, **269** afforded the usual benefit of filtration workup, and it could be recycled, although with diminishing activity with each reuse. During these recyclings, the catalyst slowly turned from dark brown to green, indicating that the Cu had been oxidized, presumably to $CuCl_2$.

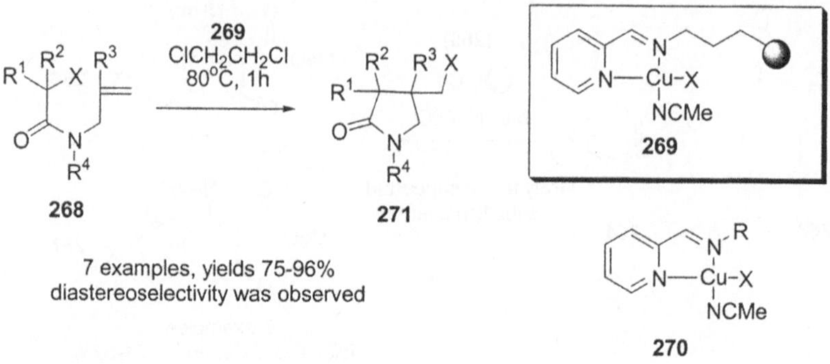

Scheme 49

The area of solid-supported Lewis acids is not new, but certainly it is attracting renewed interest with the potential to apply such catalysts to HTPS (108). Metals that are attracting much attention in this area include those in the lanthanide family, and there are some very practical reasons for this interest. These metals are often complexed to water, meaning that they are not only water-stable but are also potentially useful as catalysts for reactions performed in aqueous media. In any case, they are stable and quite robust in applications where the catalyst cannot be handled easily under strictly inert conditions, such as parallel synthesis. Kobayashi has been quite active in this area and has developed a couple of scandium-based supported Lewis acids. The polyethylene-based catalyst **275** was used effectively to catalyze multicomponent reactions, including the imine addition and cycloaddition examples, such as the one shown in Scheme 50 (109).

Kobayashi's group also developed a microencapsulated scandium-based catalyst **279** that was produced by coating $Sc(OTf)_2$ with polystyrene. This cata-

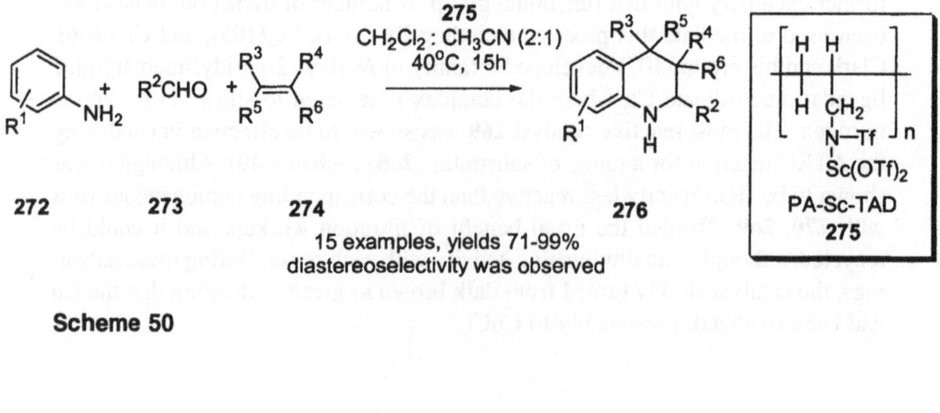

Scheme 50

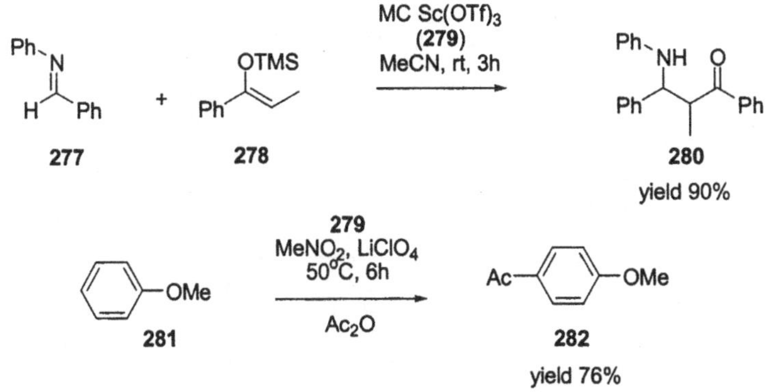

Scheme 51

lyst was used to promote Mukaiyama aldol–type chemistry to produce products **276** and even Friedel–Crafts acylation chemistry (products **282**) quite effectively (Scheme 51) (110). The catalyst was reused several times with no appreciable loss of activity.

More recently, Nagayama and Kobayashi (111) reported a supported Sc(OTf)₂ catalyst where the metal is affixed to the ligand via a benzenesulfonate–scandium bond. From the polystyrene core, the ligand was constructed with an "alkyl aromatic spacer" to create "hydrophobic reaction fields" to allow the catalyst to be used in water. This catalyst **285** was used to promote allylstannane **284** additions to a variety of carbonyl compounds **283** in straight water solvent (Scheme 52). In a brief survey, **285** also was also used to promote a Mukaiyama aldol condensation, a Diels–Alder reaction, and a Strecker-type reaction (111). All reactions were carried out in water, and the yields were excellent. The mole percent of catalyst used was generally between 1.6 and 3.2, it could be recovered readily by simple filtration following the reaction, and **285** could be reused many times with no apparent loss of activity.

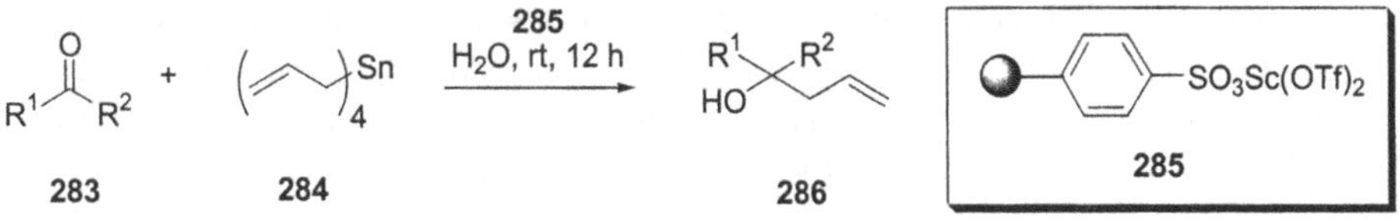

9 examples, yields 72% to quantitative

Scheme 52

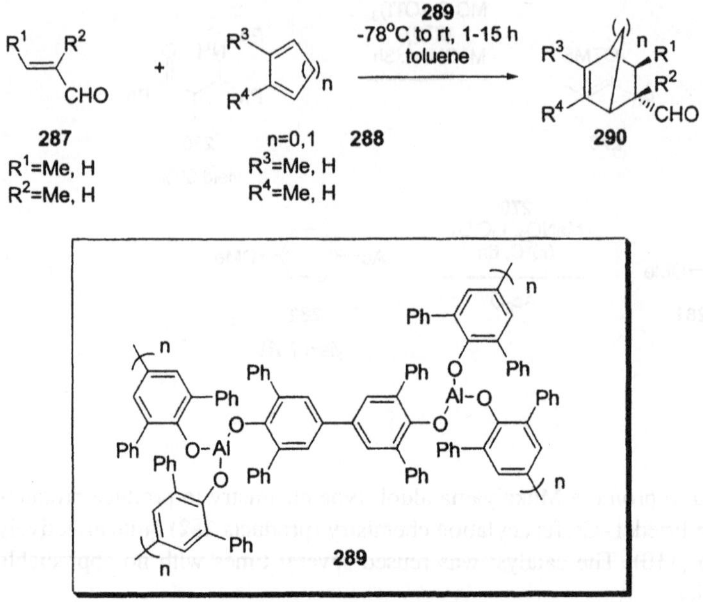

287
R^1=Me, H
R^2=Me, H

n=0,1   288
R^3=Me, H
R^4=Me, H

289
-78°C to rt, 1-15 h
toluene

290

289

9 examples, yields 40->99%
moderate *endo:exo* ratio
(yields are isolated or from ^{1}H-NMR analysis)

Scheme 53

Aluminum is also a commonly used metal for Lewis acid applications. Yamamoto and coworkers (112) created an aluminum phenoxide based polymer system **289** to promote Diels–Alder cycloaddition reactions. The yields (Scheme 53) were acceptable to excellent, but the endo/exo selectivity was not consistently good.

Although asymmetric reactions are not the focus of this review, the drive to obtain enantiomerically pure material has fueled a flurry of activity in the area of supported Lewis acid catalyst development that is worth noting for its potential in HTPS (113). Many ligand systems have been adopted to the solid for the purpose of asymmetric catalysis, including Salen (114), TADDOL (115,116), MPO (117), BINOL (118), and BINAP (119), to name a few.

Seebach's group developed ($\alpha,\alpha,\alpha',\alpha'$-tetraaryl-1,3-dioxolane-4,5-dimethanol titanate (Ti-TADDOLate) catalysts to promote asymmetric additions of dialkylzinc reagents to aldehydes as well as Diels–Alder cycloaddition. The supported catalyst **289** was prepared by copolymerizing styrene with the homogeneous TADDOL ligand, which had been prepared with a *para*-vinylbenzene sub-

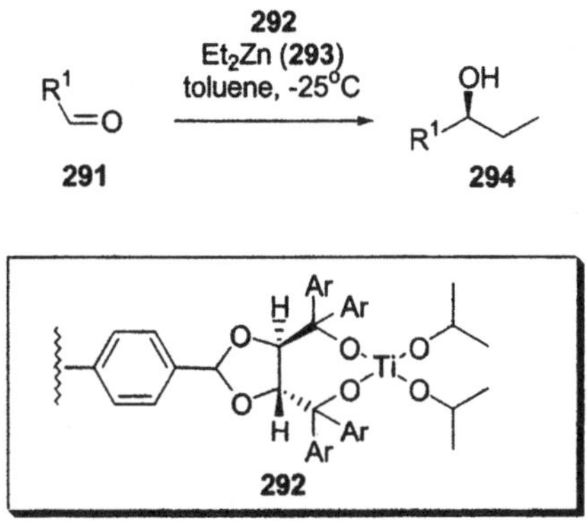

Scheme 54

stituent. In an interesting extension of this study, the catalyst was loaded into a polypropylene "tea bag" (120) to minimize physical degradation of the polymer from the strong agitation required for these reactions (Scheme 54) (121). In this case, the "tea bag" was reused 20 times without quenching or recovery of the polymer beads. Six different aldehydes **291** were used randomly in these 20 sequential runs, and cross contamination was less than 0.5%, indicating that this catalyst system could be amenable to a flow reactor system to produce more than just one product.

Kobayashi et al. (118) developed supported Zr-BINOL Lewis acid catalysts **297** to promote hetero Diels–Alder chemistry (Scheme 55). Their paper outlines the systematic optimization of both homogeneous and supported Zr-BINOL complexes. The yields and *ee* values obtained for the supported catalysts were not significantly different from those obtained in solution.

In contrast to the supported Zr-BINOL system of Kobayashi et al., Jorgensen and coworkers (122) developed a series of reliable homogeneous Cu catalysts to promote almost identical reactions. Although all ligands surveyed, including BINAP, provided excellent yields and good *ee* values, the supported Cu-BINAP catalyst **301** gave significantly reduced yield and essentially racemic product **302** in addition to the Mannich addition side product **303** (Scheme 56). Thus, not every Lewis acid system performs with equal efficacy when adopted to the solid.

2. Metal-Based Oxidation Catalysts

A number of transition-metal-based oxidation catalysts have been adapted to the solid support. Tetrapropylammonium perruthenate (TPAP) has emerged as one of the metal oxidants of choice for sensitive oxidations such as that of aromatic aldehydes. Ley and coworkers (123) successfully mounted the perruthenate on

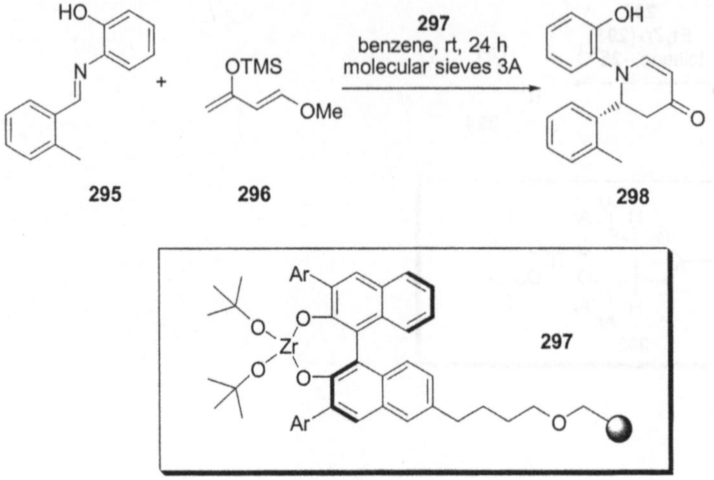

17 examples of different catalysts
yields 59-92%, ee 41-83%

Scheme 55

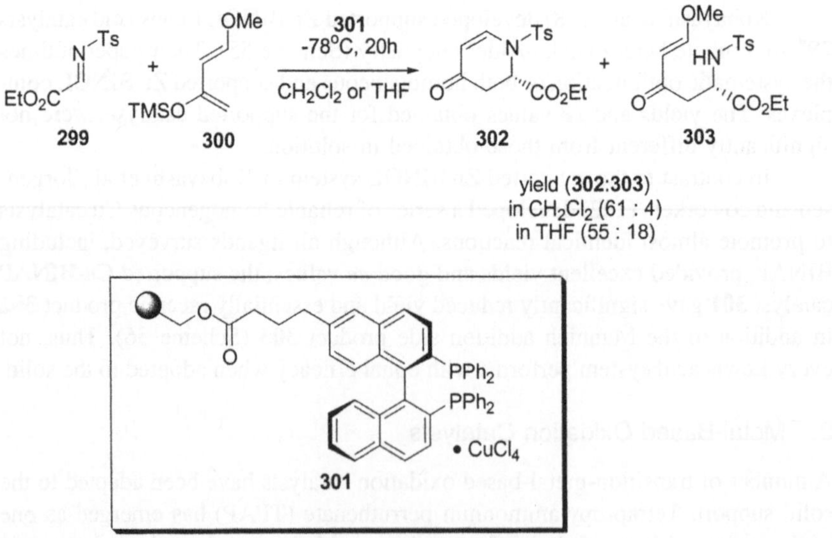

yield (**302**:**303**)
in CH$_2$Cl$_2$ (61 : 4)
in THF (55 : 18)

Scheme 56

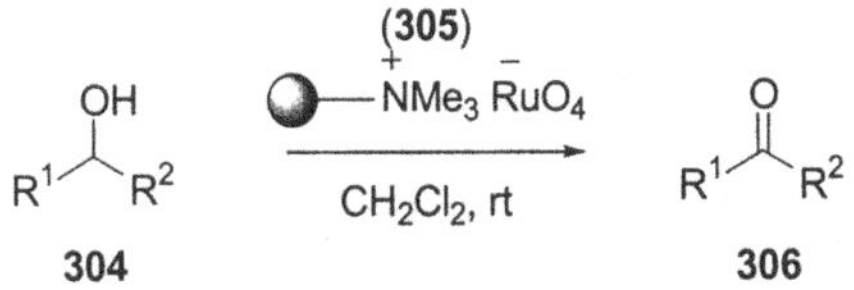

304 306

amberlyst anion
exchange resin (IR 27)

18 examples, yields 50-95%

Scheme 57

Amberlyst IR 27 resin to create a supported oxidant **305** that has shown excellent reactivity (Scheme 57). The metal is oxidized with either molecular oxygen or *N*-methylmorpholine *N*-oxide, the latter, of course, being potentially problematic for HTPS because it introduces a nitrogen-based by-product into the reaction mixture that may compromise purification. Ley's group reports that oxidations done under an atmosphere of oxygen are clean and purification generally involves simple filtration.

Alcohol oxidations were also carried out by Leadbeater and Scott (124) using Co(II) that was bound to polymer-supported triphenylphosphine **308** (Scheme 58). Comparison with the homogeneous catalyst **309** indicated that yields were not significantly different from the two-phase catalyst system. In addition to the usual advantages afforded by the filterable catalyst system, there were some noticeable differences in controlling overoxidation with **308** that could not be achieved with **309**. It is well known that controlling overoxidation of benzaldehyde to benzoic acid is a challenge. When **309** was used to oxidize benzyl alcohol, an equal amount of the acid was also formed. When **308** was used, overoxidation was reduced to only 14%, and if the reaction was halted before all the benzyl alcohol was fully consumed, no benzoic acid was formed, albeit with a diminished conversion of 80%.

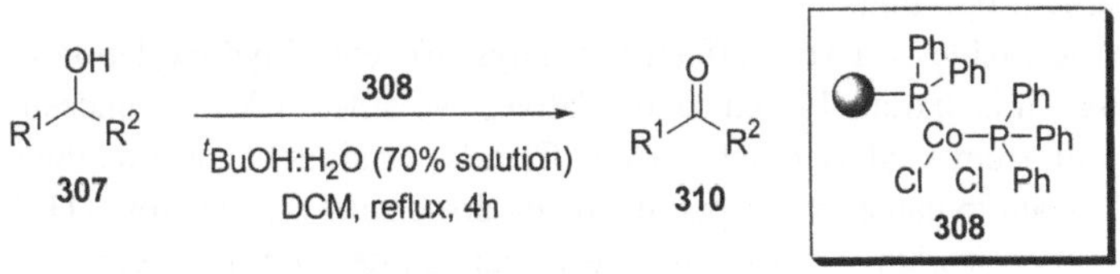

307 310 308

6 examples, yields 68-92%
purity by NMR and IR spectroscopy

$CoCl_2(PPh_3)_2$
309

Scheme 58

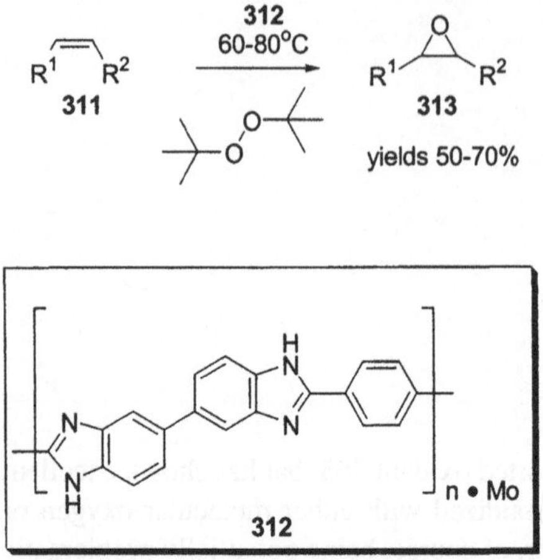

Scheme 59

A molybdenum-supported catalyst **312** was developed by Leinonin and Sherrington (125) for the epoxidation of olefins. Using t-butyl hydrogen peroxide as the co-oxidant, a variety of olefins **311** were oxidized in reasonable to good yield (Scheme 59).

Natarajan and Madalengoitia (126) have been working on the development of optimal epoxidation catalysts using supported amino acids as ligands for a variety of metals including Ru, Ni, Co, Mn(II), and Mn(III). Initially, they tried to epoxidize cis-stilbene **314** with the various catalysts using t-BHP as the co-oxidant. The only metal that displayed significant activity was Ru, and t-BHP was judged not to be a suitable co-oxidant because $trans$-stilbene oxide products were formed, suggesting the involvement of radicals. In subsequent experiments iodosobenzene was used as the co-oxidant, which provided a mixture of epoxide **317** and oxidative cleavage product **318** (Scheme 60). Despite trying two different ligands (i.e., **315** and **316**) and five different resins, **318** was always observed in roughly the same proportion.

Osmium tetroxide is a very effective catalyst for the dihydroxylation of olefins; however, it is extremely toxic and relatively volatile, making it undesirable to work with from a safety point of view. Clearly, osmium is not something that one would want to parcel out in parallel using dispensing stations for HTPS. Kobayashi has been working over the past few years on developing a supported Os catalyst that would address these issues and make it more practical to work with. In 1998, his group developed a microencapsulated OsO_4 catalyst that proved efficacious in the dihydroxylation of a variety of olefins (127). The rate of reactiv-

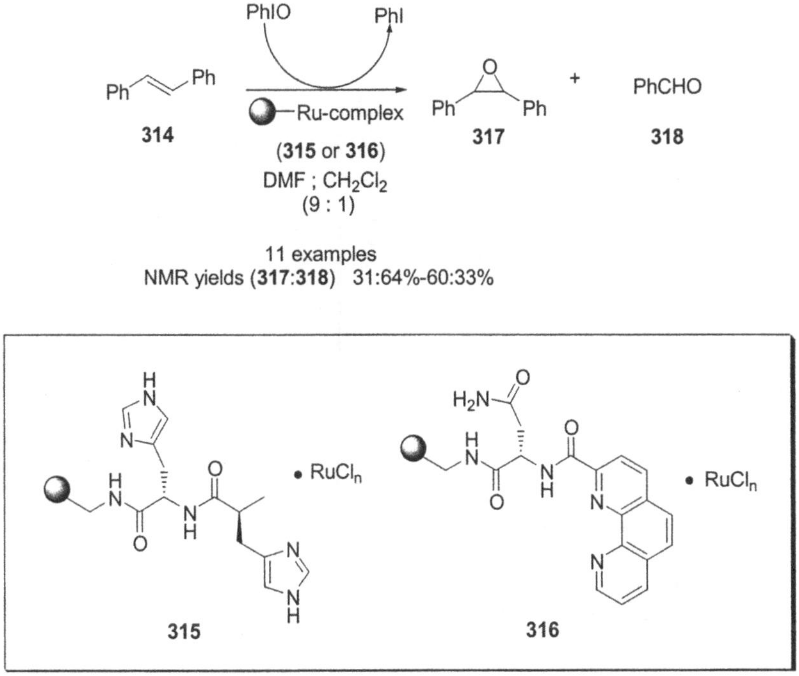

Scheme 60

ity appeared to mirror the homogeneous reaction, and catalyst activity did not noticeably diminish after the fifth recycling of the catalyst, which indicates that little of the Os was being leached out during the reactions, which is very important for the safety issues discussed above. Subsequent to that, Kobayashi et al. (128) pushed the envelope further by adopting their supported catalyst to perform asymmetric dihydroxylations as shown in Scheme 61. The enantioselection was good to excellent, which would support the close association of the alkaloid ligand with the metal on the support **320**.

Song et al. (129) also developed a supported osmium complex for the purpose of asymmetric dihydroxylation, but in this case it is the ligand that is supported on silica gel and the Os then binds to the ligand (**323**). They reported that the Os was more tightly bound to the supported ligand than to the corresponding homogeneous analog (DHQ)₂PHAL. Further, **323** could be reused many times with no loss of activity or enantioselectivity, and careful analysis of product mixtures revealed no trace of osmium. Song later teamed up with Sherrington and coworkers to apply this ligand system to asymmetric amino hydroxylation

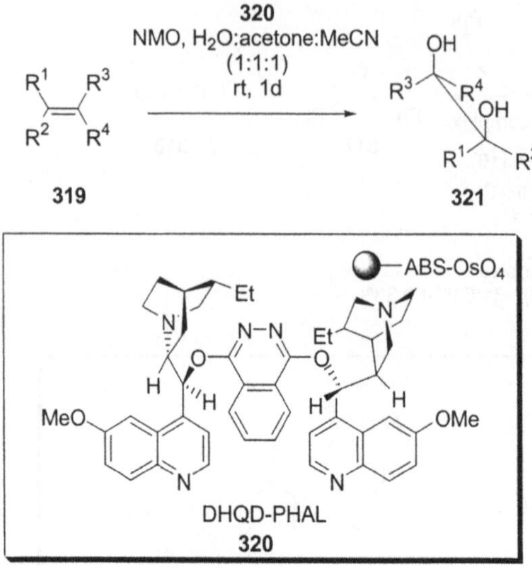

320
NMO, H₂O:acetone:MeCN
(1:1:1)
rt, 1d

6 examples, yields 36-98%
ee 60-92%, HPLC
(Daicel Chiracel OD)

Scheme 61

(Scheme 62) (130). With the examples tried, the yields are modest (none greater than 81%), but the enantioselectivity is excellent (most greater than 92%). The catalyst could be reused without a noticeable decrease in *ee*, but unlike the dihydroxylation studies Os was certainly being leached out into the supernatant. This must be addressed, not only for catalyst recycling but also for any potential use in HTPS and application of the products to biological systems.

3. Metal-Based Reduction Catalysts

Hydrogenation is one of the oldest and most widely used of the catalytic processes. Most hydrogenations are conducted in a protic solvent, such as ethanol, because catalysts such as Pd/C or Adam's catalyst (PtO₂) display little or no activity in ethereal or halogenated solvents. The problem in adapting such systems to the polymer support is that polymers, such as polystyrene, do not swell in high dielectric solvents and in fact contract in an attempt to expel the alcohol from the hydrocarbonaceous polymer.

A couple of reports have been made in the last few years outlining the use of highly cross-linked macroporous polymers to enable hydrogenation to be

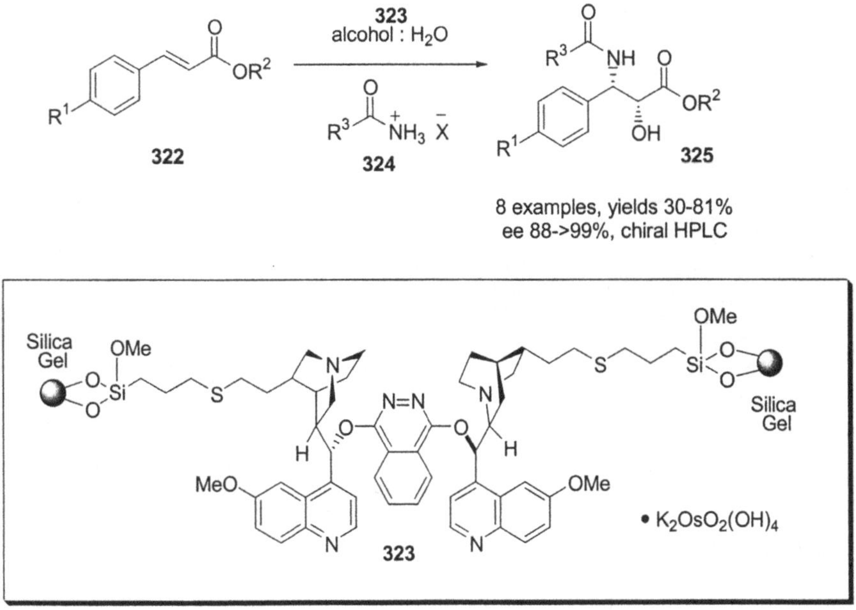

Scheme 62

conducted in alcoholic solvents. Although high cross-linking limits the amount a polymer can swell, it also prevents it from contracting, which allows the solvent carrying the starting materials to reach the active metal sites within the polymer. Corain et al. (131) reported a Pd-immobilized macroporous resin **327** that was prepared by supporting Pd crystallites on a styrene/2-methacryloxyethylsulfonic acid copolymer (Scheme 63). They varied the amount of methylene-bisacrylamide used to cross-link the polymer and tested each catalyst on cyclohexene. Not surprisingly, they found that small particles (i.e., <0.1 mm) were essentially insensitive to the amount of cross-linking whereas larger particles were essentially diffusion-limited.

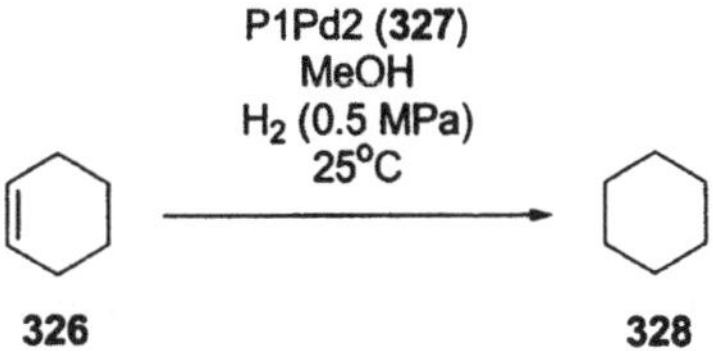

Scheme 63

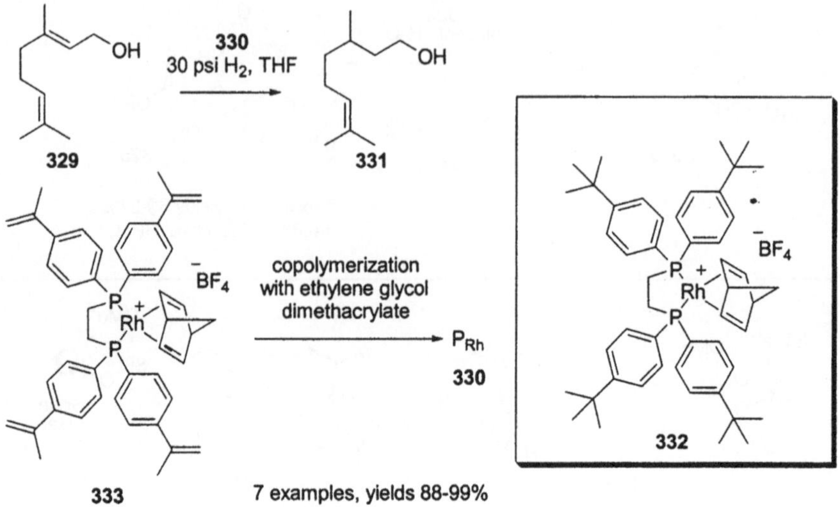

Scheme 64

Gagné and his group (132) developed an Rh-supported macroporous catalyst for the hydrogenation of olefins. Interestingly, the polymer-bound catalyst P_{Rh} (**330**) was formed by the copolymerization of the intact homogeneous Ru complexed monomer **333** (2%) with ethylene glycol dimethacrylate (98%) (Scheme 64). Heterogeneous catalyst **330** was tested with seven structurally and electronically diverse substrates, and each one proceeded to give 100% conversion and excellent yield. The catalyst was reused six times with no appreciable loss of activity. In one experiment, the polymer was removed halfway through the hydrogenation and the reaction ceased, strongly suggesting that the active metal species is indeed polymer-bound. Catalyst **330** was shown to react approximately one-fifth or one-sixth as fast as the corresponding homogeneous catalyst **332**. The authors reasoned that known mass transport problems associated with H_2 under homogeneous conditions are amplified with the added barrier of diffusion into the polymer.

B. Nonmetal Catalysts

1. Acid Catalysts

Supported nonmetallic acids can take a number of forms, including (a) Lewis acids incorporated coordinately in an inorganic or organic polymer matrix such as zeolites, clays, silica, and resins; (b) polymeric protonic (Brønsted) acids, such as proton-exchange zeolites, clays, and ion-exchange resins; and (c) organic cationic species attached to a resin. The use of such acids in a heterogeneous fashion has been practiced for years with great success (133). For example, the use of montmorillonite clay in the formation of sensitive acetals, such as the one derived from benzaldehyde, is well known. The reaction conditions are very gentle, and filtration, sometimes through celite if necessary, usually provides a clean crude product. With proper adaptation, any of the above-mentioned acids would be useful in an HTPS format.

Masaki et al. developed a neutral, supported π-acid **337** to promote an increasing number of transformations, including the monothioacetalization of acetals (133) and Mannich-type reactions of aldimines (formed by the condensation of **334** and **335**) and enol silyl ethers **336** (Scheme 65) (134). They showed that the imine need not be preformed in these addition reactions, thus increasing their utility for multicomponent transformation, which would make them more useful in HTPS applications. Further, in the absence of the catalyst, no product is formed, proving the involvement of the acid. The catalyst is readily recycled by washing with ethyl acetate and drying in vacuo for 4 h, and it shows no loss of activity.

Taylor's group used Nafion-H, a polymeric superacid **341**, to promote C—C bond formation via an epoxy-arene intramolecular cyclization of substrates **340** (Scheme 65) (135). At this stage, the yields reported are modest at best and

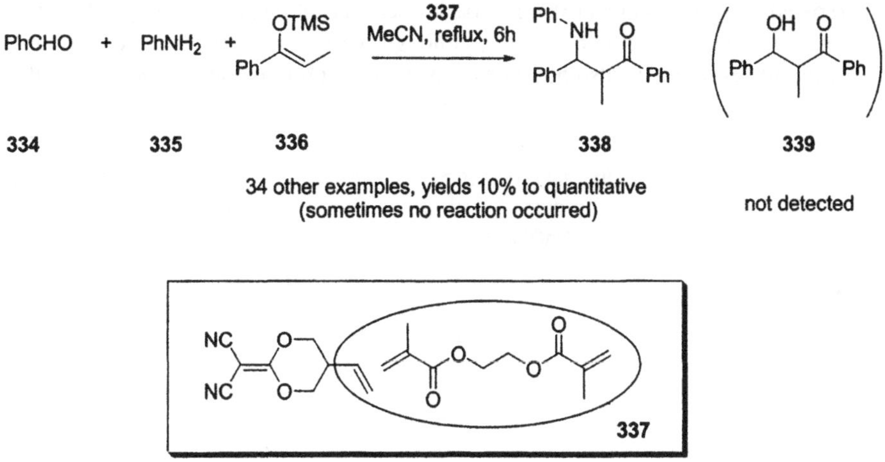

Scheme 65

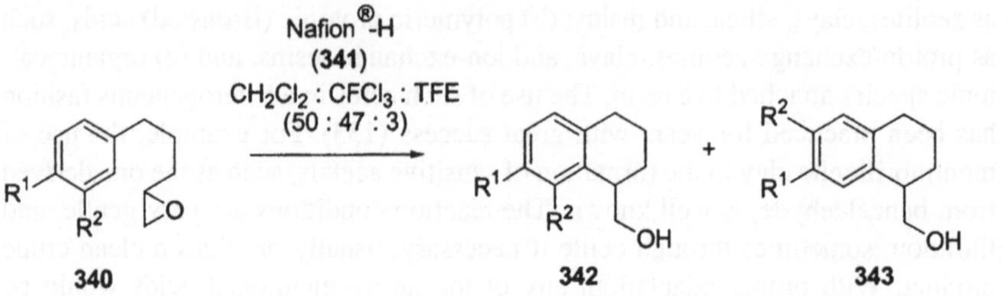
Nafion®-H
(341)
CH2Cl2 : CFCl3 : TFE
(50 : 47 : 3)
R1
R2
O
R2
OH
R1
OH
340
342
343

7 examples, yields 31-88%
purity >95%, capillary GC

Scheme 66

the procedure uses somewhat esoteric fluorinated solvents; thus it is not clear whether this method is suitable for HTPS without more work being done.

2. Base Catalysts

In truth, protonic bases are usually required stoichiometrically, and they are not regenerated during, or even after, the reaction; thus they are really "facilitators" of reactions, for lack of a better term. Strong kinetic bases (i.e., pK_a of the corresponding acid >40), such as Grignard reagents (136) and organolithium species (137), are certainly known on the solid phase. In fact, such species were first prepared many years ago (138). The basic species is often generated by halogen–metal exchange (Scheme 67) (137), although metal–metal exchange of lithium for tin, for example, is also possible, as is direct deprotonation with *n*-BuLi with

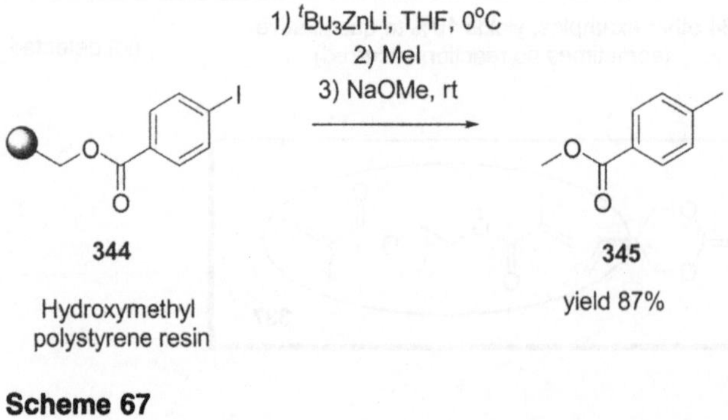
1) tBu3ZnLi, THF, 0°C
2) MeI
3) NaOMe, rt
I
O
O
O
O
344
345
Hydroxymethyl
polystyrene resin
yield 87%

Scheme 67

TMEDA (138). These supported organometallics have been used generally in substitution chemistry, i.e., as nucleophiles, not in classical acid–base chemistry. Although these strong bases could well be used as Brønsted bases, it is not very practical to consider their use in an HTPS format because they are so reactive and prone to quenching.

The less basic, less reactive family of amide bases (amine $pK_a = 35$), such as LDA or lithium tetramethylpiperidine, have also been produced on the solid. Patchornik and coworkers (139) illustrated the use of a supported LDA equivalent **346** to deprotonate the activated vinyl proton on **347**, which led to the formation of compound **351** in Scheme 68. One of the problems in working with such systems is polymer loading. For example, when deprotonating adjacent to a carbonyl, one needs to be sure that the amine is in excess of the *n*-butyllithium, or else carbonyl addition will certainly outcompete deprotonation. Exact loading and stoichiometry are always uncertain when working with polymer-supported compounds.

Supported ammonium hydroxide **352** is a comparatively mild anionic base that has been used to effect a variety of condensation reactions, including aldol, Henry, and Knoevenagel reactions. Kulkarni and Ganesan (140) reported the first solution-phase Dieckmann condensation reaction promoted by **352** (Scheme 69). The procedure worked in good yield and acceptable purity for HTPS purposes. The supported base offered the additional benefit of serving as a purification agent. The acidic product is readily deprotonated by the alkoxide, thus enabling

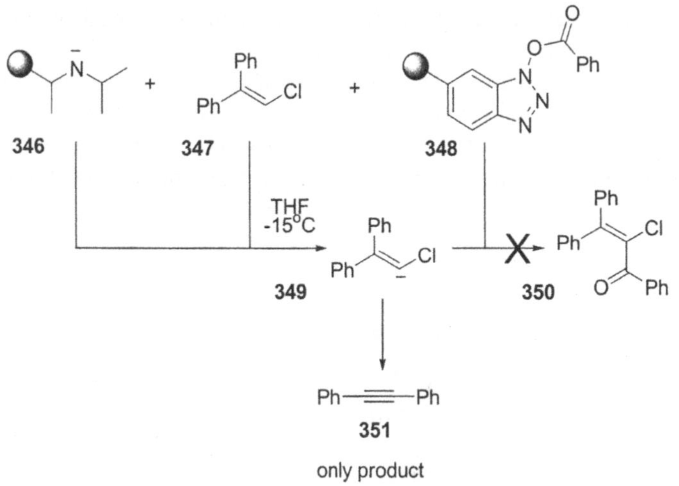

Scheme 68

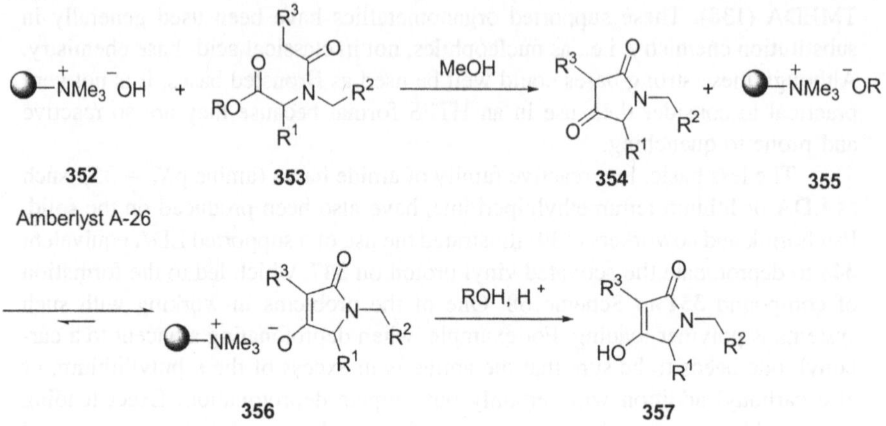

352   353   MeOH   354   355

Amberlyst A-26

356   ROH, H$^+$ 357

11 examples, yields 70-87%
purity 71-92%, HPLC (254 nm)

Scheme 69

it to be lifted out of solution onto the resin by the ammonium counterion to form the salt on resin. The uncyclized material **353** is not sufficiently acidic to allow it to undergo this catch and release.

A group of bases that should prove highly useful both in their adaptation to the solid and in synthetic applications, including HTPS, is the neutral bases, two of which are shown in Figure 3. The order of basicity increases as follows: tertiary amines, proton sponges (141), guanidines (e.g., DBU and DBN), urethanes, Verkade's phosphatrane **358** (142), and polyphosphazene **359** (143,144). One example discussed in Sec. III.B.1 (Scheme 15) was from the supported ure-

358   359

Figure 3 Examples of neutral homogeneous bases.

thane family of bases (i.e., PTBD resin), which was used as both a promoter and a base in nucleophilic substitution reactions (30).

Bertrand's group developed P-tris-(dimethylamino)-C-dimethylphosphonium ylide **367** as a strong neutral Brønsted base and used it to deprotonate a variety of acids including amides (N-alkylation) and carbonyls. They also adapted this base to the solid support, and its structure **360** is shown in Scheme 70 (145). The base proved useful in a variety of transformations, and purification was usually by filtration. They reported that the resin could be reused following washing with acetonitrile and that the yields obtained using **360** mirrored those obtained in solution using **367**. What remains unclear is the shelf stability of **367** and **360** for practical use of **360** in solution-phase HTPS. The detailed preparation of **367** is given in Ref. 145, including combustion analysis. It is stated that the method uses "conventional glassware," which would imply that **367** is stable to air, but this is not expressly written anywhere. Certainly **360** was always reported to be freshly prepared. This is not an insurmountable hurdle, and if **360** is stable and

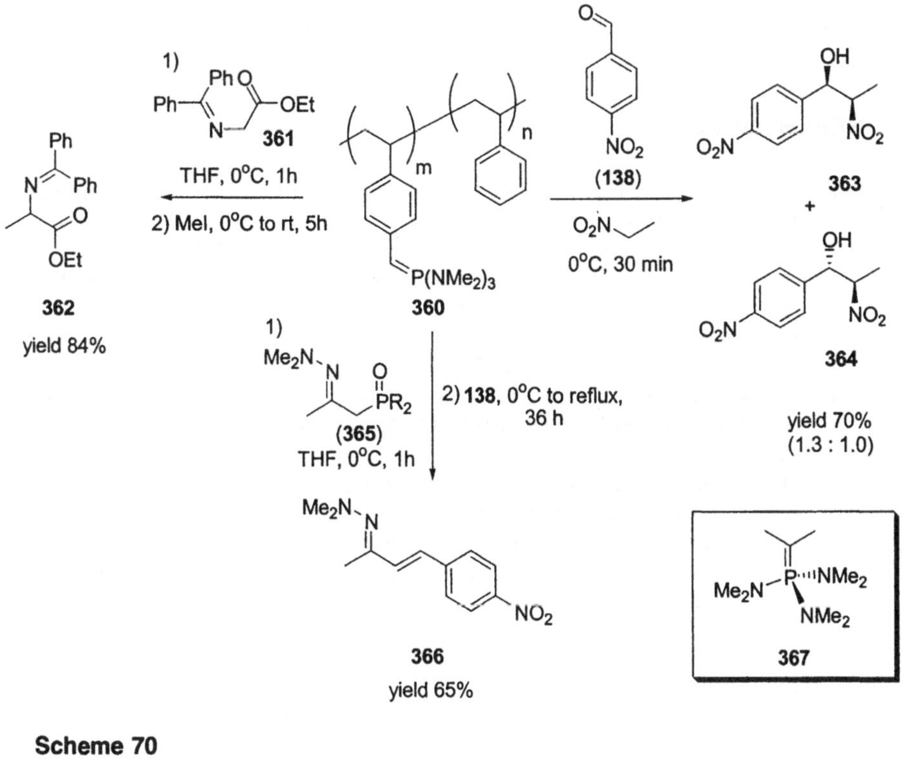

Scheme 70

can be prepared in bulk and dispensed in a parallel format, it will prove to be an enormous contribution to HTPS.

V. SUMMARY

One can see by reviewing the seminal references cited in this chapter that the development and use of solid-supported reagents and catalysts to expedite solution-phase chemistry is far from a new concept. As happens with a great deal of science, discoveries get revisited when the need arises, and HTPS has certainly created a renaissance of interest in both SPOS and its application to solution-phase synthesis. With advances in combinatorial chemistry, it is interesting to see that the catalysts used in HTPS are themselves targets for optimization by HTPS. The applications for supported reagents and catalysts appear to be limitless. For example, one could imagine using chemicals that can be affixed readily to the surface of devices, such as the "lab-on-a-chip," to perform functions that go well beyond HTPS, including, perhaps, biosensors and miniature medical devices.

REFERENCES

1. (a) CC Leznoff. Chem Soc Rev 3:65–85, 1974. (b) CC Leznoff. Acc Chem Res 11:327–333, 1978.
2. SD Brown, RD Armstrong. J Am Chem Soc 118:6331–6332, 1996.
3. (a) DL Flynn, RV Devraj, W Naing, JJ Parlow, JJ Weidner, S Yang. Med Chem Res 1998:219–243, 1998. (b) JJ Parlow, RV Devraj, MS South. Curr Opin Chem Biol 3:320–336, 1999. (c) JC Hodges. Synlett 1999:152–158, 1999. (d) SJ Shuttleworth, SM Allin, RD Wilson, D Nasturica. Synthesis 2000:1035–1074.
4. (a) NK Mathur, CK Narang, RE Williams. Polymers as Aids in Organic Chemistry. New York: Academic Press, 1980. (b) JMJ Fréchet. Tetrahedron 37:663–683, 1981.
5. M Fridkin, A Patchornik, E Katchalski. J Am Chem Soc 88:3164–3165, 1966.
6. A Link, S van Calenbergh, P Herdewijn. Tetrahedron Lett 39:5175–5178, 1998.
7. (a) M Adamczyk, JR Fishpaugh, PG Mattingly. Tetrahedron Lett 40:463–466, 1999. (b) M Adamczyk, JR Fishpaugh, PG Mattingly. Bioorg Med Chem Lett 9: 217–220, 1999.
8. Y Hamuro, MA Scialdone, WF DeGrado. J Am Chem Soc 121:1636–1644, 1999.
9. SV Ley, MH Bolli, B Hinzen, A-G Gervois, BJ Hall. J Chem Soc Perkin Trans I 1998:2239–2241.
10. M Tomoi, Y Akada, H Kakiuchi. Makromol Chem Rap Commun 3:537–542, 1982.
11. MC Desai, LMS Stramiello. Tetrahedron Lett 34:7685–7688, 1993.
12. JJ Parlow, DL Flynn. Tetrahedron 54:4013–4031, 1988.

13. M Fridkin, A Patchornik, E Katchalski. Pept Proc Eur Pept Symp 10th 1969, 1971, p 166.
14. CF Sturino, M Labelle. Tetrahedron Lett 39:5891–5894, 1998.
15. AD Wentworth, P Wentworth, Jr, UF Mansoor, KD Janda. Org Lett 2:477–480, 2000.
16. SL Regen, DP Lee. J Org Chem 40:1669–1670, 1975.
17. AR Tunoori, D Dutta, GI Georg. Tetrahedron Lett 39:8751–8754, 1998.
18. PH Toy, TS Reger, KD Janda. Org Lett 2:2205–2207, 2000.
19. (a) FEK Kroll, R Morphy, D Rees, D Gani. Tetrahedron Lett 38:8573–8576, 1997. (b) P Heinonen, H Lönnberg. Tetrahedron Lett 38:8569–8572, 1997.
20. KM Brummond, KD Gesenberg. Tetrahedron Lett 40:2231–2234, 1999.
21. J Haberman, SV Ley, R Smits. J Chem Soc Perkin Trans I 1999:2421–2423.
22. M Fréchet, MJ Farrel, LJ Nuyens. J Macromol Sci Chem 11:507–514, 1977.
23. (a) MG Finn, KB Sharpless. J Am Chem Soc 113:113–126, 1991. (b) W Zhang, JL Loebach, SR Wilson, EN Jacobsen. J Am Chem Soc 112:2801–2803, 1990. (c) R Irie, K Noda, Y Ito, N Matsumoto, T Katsuki. Tetrahedron Lett 31:7345–7348, 1990.
24. CAM Afonso, NML Vieira, WB Motherwell. Synlett 2000:382–384.
25. (a) A Kirschning, H Monenschein, C Schmeck. Angew Chem Int Ed Engl 38:2594–2596, 1999. (b) A Kirschning, M Jesberger, H Monenschein. Tetrahedron Lett 40:8999–9002, 1999.
26. (a) KC Nicolaou, J Pastor, S Barluenga, N Winssinger. J Chem Soc Chem Commun 1998:1947–1948. (b) K-i Fujita, K Watanabe, A Oishi, Y Ikeda, Y Taguchi. Synlett 1999: 1760–1762.
27. KC Nicolaou, G Cao, JA Pfefferkorn. Angew Chem Int Ed 39:734–739, 739–743, 2000.
28. A Patchornik, MA Kraus. J Am Chem Soc 92:7587–7589, 1970.
29. MG Organ, SW Kaldor, C Dixon, DJ Parks, U Singh, DJ Lavorato, PK Isbester, MG Siegel. Tetrahedron Lett 41:8407–8411, 2000.
30. W Xu, R Mohan, MM Morrissey. Tetrahedron Lett 38:7337–7340, 1997.
31. MJ Kurth, LA Ahlberg Randall, C Chen, C Melander, RB Miller. J Org Chem 59: 5862–5864, 1994.
32. LF Tietze, A Steinmetz. Synlett 1996:667–668.
33. M Rottlander, P Knochel. J Comb Chem 1:181–183, 1999.
34. P Garibay, PH Toy, T Hoeg-Jensen, KD Janda. Synlett 1999:1438–1440.
35. S Kobayashi, I Hachiya, S Suzuki, M Moriwaki. Tetrahedron Lett 37:2809–2812, 1996.
36. T Mukaiyama, K Banno, K Naraska. Chem Lett 1973: 1011–1014.
37. F Aznar, C Valdés, M-P Cabal. Tetrahedron Lett 41:5683–5687, 2000.
38. (a) SV McKinley, JW Rakshys Jr. J Chem Soc Chem Commun 1972:134–135. (b) W Heitz, R, Michels. Angew Chem Int Ed Engl 11:298–299, 1972. (c) I Hughes. Tetrahedron Lett 37:7595–7598, 1996. (d) JM Salvino, TJ Kiesow, S Darnbrough, R Labaudiniere. J Comb Chem 1:134–139, 1999.
39. AGM Barrett, SM Cramp, RS Roberts, FJ Zecri. Org Lett 1:579–582, 1999.
40. KC Nicolaou, AJ Roecker, JA Pfefferkorn, G-Q Cao. J Am Chem Soc 112:2966–2967, 2000.

41. S Berteina, S De Masmaeker. Tetrahedron Lett 39:5759–5762, 1998.
42. X Zhu, A Ganesan. J Comb Chem 1:157–162, 1999.
43. PJ Stang, VV Zhadankin. Chem Rev 96:1123–1178, 1996.
44. (a) Y Yamada, M Okawara. Makromol Chem 152:163–176, 1972. (b) M Zupan, A Pollak. J Chem Soc Chem Commun 1975:715–716.
45. H Togo, G Nogami, M Yokomaya. Synlett 1998:534–536.
46. G-P Wang, Z-C Chen. Synth Commun 29:2859–2866, 1999.
47. SV Ley, AW Thomas, H Finch. J Chem Soc Perkin Trans I 1999: 669–671.
48. K Geethakumari, K Sreekumar. Ind J Chem 37B:331–337, 1998.
49. S Abraham, PK Rajan, K Sreekumar. Polym Int 45:271–277, 1998.
50. JM Harris, Y Liu, Y Chai, MD Andrews, JC Vederas. J Org Chem 63:2407–2409, 1998.
51. C Bolm, T Fey. J Chem Soc Chem Commun 1999:1795–1796.
52. A Kessat, A Babadjamian, A Iraqi. Eur Polym J 34:323–325, 1998.
53. SS Mitra, K Sreekumar. Eur Polym J 34:561–565, 1998.
54. (a) AV Rama Rao, MK Gurjar, PA Sharma. Tetrahedron Lett 32:6613–6616, 1991. (b) A McKillop, L McLaren, RJK Taylor, RJ Watson. Synlett 1992: 201–203. (c) P Wipf, Y Kim. J Org Chem 58:7195–7203, 1993.
55. TR Boehlow, PC Buxton, EL Grocock, BA Marples, VL Waddington. Tetrahedron Lett 39:1839–1842, 1998.
56. (a) U Gerigk, M Gerlach, WP Neumann, R Vieler, V Weintritt. Synthesis 1990: 448–452. (b) G Ruel, NK The, G Dumartin, B Delmond, M Pereyre. J Organomet Chem 444:C18–C20, 1993.
57. G Dumartin, M Pourcel, B Delmond, O Donard, M Pereyre. Tetrahedron Lett 39: 4663–4666, 1998.
58. RO Hutchins, NR Natale, IM Taffer. J Chem Soc Chem Commun 1978: 1088–1089.
59. SV Ley, MH Bolli, B Hinzen, A-G Gervois, BJ Hall. J Chem Soc Perkin Trans I 1998: 2239–2241.
60. SV Ley, O Schucht, AW Thomas, PJ Murry. J Chem Soc Perkin Trans I 1999: 1251–1252.
61. MW Creswell, GL Bolton, JC Hodges, M Meppen. Tetrahedron 54:3983–3998, 1998.
62. H Bönneman, G Braun, W Brijoux, R Brinkmann, A Schulze-Tilling, K Seevogel, K Siepen. J Organomet Chem 520:143–162, 1996.
63. M Makosza, P Nieczpor, K Grela. Tetrahedron 54:10827–10836, 1998.
64. S-B Jang. Tetrahedron Lett 38:1793–1796, 1997.
65. (a) H Bruner, JC Bailar Jr. Inorg Chem 12:1465–1470, 1973. (b) M Terasawa, K Kaneda, T Imanaka, S Teranishi. J Organomet Chem 162:403–414, 1978.
66. I Fenger, C Le Drain. Tetrahedron Lett 39:4287–4290, 1998.
67. DS Ennis, J McManus, W Wood-Kaczmar, J Richardson, GE Smith, A Carstairs. Org Process Res Dev 3:248–252, 1999.
68. G Marck, A Villiger, R Buchecker. Tetrahedron Lett 35:3277–3280, 1994.
69. DE Bergbreiter, PL Osburn, Y-S Liu. J Am Chem Soc 121:9531–9538, 1999.
70. DE Bergbreiter, Y-S Liu, PL Osburn. J Am Chem Soc 120:4250–4521, 1998.
71. J Schwarz, VPW Böhm, MG Gardiner, M Grosche, WA Herrmann, W Hieringer, G Raudaschl-Sieber. Chem Eur 6:1773–1780, 2000.

72. MP Leese, JMJ Williams. Synlett 1999:1645–1647.
73. MS Anson, AR Mirza, L Tonks, JMJ Williams. Tetrahedron Lett 40:7147–7150, 1999.
74. TY Zhang, MJ Allen. Tetrahedron Lett 40:5813–5816, 1999.
75. Y Li, XM Hong, DM Collard, MA El-Sayed. Org Lett 2:2385–2388, 2000.
76. (a) NA Bumagin, VV Bykov, LI Sukhomlinova, TP Tolstaya, IP Beletskaya. J Organomet Chem 486:259–262, 1995. (b) S Lemaire-Audoire, M Savignac, C Dupuis, JP Genêt. Tetrahedron Lett 37:2003–2006, 1996.
77. Y Uozumi, H Danjo, T Hayashi. Tetrahedron Lett 38:3557–3560, 1997.
78. Y Uozumi, H Danjo, T Hayashi. J Org Chem 64:3384–3388, 1999.
79. H Danjo, D Tanaka, T Hayashi, Y Uozumi. Tetrahedron 55:14341–14352, 1999.
80. Y Uozumi, T Watanabe. J Org Chem 64:6921–6923, 1999.
81. S-K Kang, T-G Baik, S-Y Song. Synlett 1999:327–329.
82. (a) S Fritschel, J Ackerman, T Keyser, JK Stille. J Org Chem 44:3152–3157, 1979. (b) CU Pittman Jr, Y Kawabata, LI Flowers. J Chem Soc Chem Commun 1982: 473–474.
83. K Nozaki, N Sakai, T Nanno, T Higashijima, S Mano, T Horiuchi, H Takaya. J Am Chem Soc 119:4413–4423, 1997.
84. K Nozaki, F Shibahara, Y Itoi, E Shirakawa, T Ohta, H Takaya. Bull Chem Soc Jpn 72:1911–1918, 1999.
85. J-AM Andersen, N Karodia, DJ Miller, D Stones, D Gani. Tetrahedron Lett 39: 7815–7818, 1998.
86. RR Schrock, JS Murdzek, GC Bazan, J Robbins, M Dimare, M O'Regan. J Am Chem Soc 112:3875–3886, 1990.
87. (a) P Schwab, MB Marcia, JW Ziller, RH Grubbs. Angew Chem Int Ed 34:2039–2041, 1995. (b) RH Grubbs, S Chang. Tetrahedron 54:4413–4450, 1998.
88. D Meng, D-S Su, A Balog, P Bertinato, EJ Sorensen, SJ Danishefsky, Y-H Zheng, T-C Chou, L He, SB Horwitz. J Am Chem Soc 119:2733–2734, 1997.
89. DJ Wallace, CJ Cowden, DJ Kennedy, MS Ashwood, IF Cottrell, UH Dolling. Tetrahedron Lett 41:2027–2029, 2000.
90. KC Nicolaou, Y He, F Roschangar, NP King, D Vourloumis, T Li. Angew Chem Int Ed 37:84–87, 1998.
91. M Scholl, TM Trnka, JP Morgan, RH Grubbs. Tetrahedron Lett 40:2247–2250, 1999.
92. J Huang, ED Stevens, SP Nolan, GL Petersen. J Am Chem Soc 121:2674–2678, 1999.
93. M Ahmed, AGM Barrett, DC Braddock, SM Cramp, PA Procopiou. Tetrahedron Lett 40:8657–8662, 1999.
94. M Ahmed, T Arnauld, AGM Barrett, DC Braddock, PA Procopiou. Synlett 2000: 1007–1009.
95. AC Comely, SE Gibson, NJ Hales. J Chem Commun 2000:305–306.
96. WJ Kerr, DM Lindsay, SP Watson. J Chem Soc Chem Commun 1999:2551–2552.
97. WJ Kerr, DM Lindsay, M McLaughlin, PL Pauson, J Chem Soc Chem Commun 2000:1467–1468.
98. B Bhatia, MM Reddy, J Iqbal. J Chem Soc Chem Commun 1994:713–714.
99. EN Prabhakaran, J Iqbal. J Org Chem 64:3339–3341, 1999.

100. J Fossey, D Lefort, J Sorba. Free Radicals in Organic Synthesis. Chichester, UK: Wiley, 1995.

101. (a) X Du, RW Armstrong. J Org Chem 62:5678–5679, 1997. (b) X Du, RW Armstrong Tetrahedron Lett 39:2281–2284, 1998.

102. X Zhu, A Ganesan. J Comb Chem 1:157–162, 1999.

103. S Berteina, S De Masmaeker. Tetrahedron Lett 39:5759–5762, 1998.

104. H Nagashima, U-I Ara, H Wakamatsu, K Itoh. J Chem Soc Chem Commun 1985: 518–519.

105. GM Lee, M Parvez, SM Weinreb. Tetrahedron 44:4671–4678, 1988.

106. H Nagashima, N Ozaki, M Ishii, K Seki, M Washiyama, K Itoh. J Org Chem 58: 464–470, 1993.

107. AJ Clark, RP Filik, DM Haddleton, A Radigue, CJ Sanders, GH Thomas, ME Smith. J Org Chem 64:8954–8957, 1999.

108. E Lindner, T Schneller, F Auer, HA Mayer. Angew Chem Int Ed Engl 38:2154–2174, 1999.

109. S Kobayashi, S Nagayama. J Am Chem Soc 118:8977–8978, 1996.

110. S Kobayashi, S Nagayama. J Am Chem Soc 120:2985–2986, 1998.

111. S Nagayama, S Kobayashi. Angew Chem Int Ed 39:567–569, 2000.

112. S Saito, M Murase, H Yamamoto. Synlett 1999:57–58.

113. L Pu. Chem Rev 98:2405–2494, 1998.

114. DA Annis, EN Jacobsen. J Am Chem Soc 121:4147–4154, 1999.

115. D Seebach, RE Marti, T Hintermann. Helv Chim Acta 79:1710–1740, 1996.

116. B Altava, MI Burguette, JM Fraile, JI Garcia, SV Luis, JA Mayoral, MJ Vicent. Angew Chem Int Ed 39:1503–1506, 2000.

117. Y Uozumi, H Danjo, T Hayashi. Tetrahedron Lett 39:8303–8306, 1998.

118. S Kobayashi, K-i Kusakabe, H Ishitani. Org Lett 2:1225–1227, 2000.

119. A Fujii, M Sodeoka. Tetrahedron Lett 40:8011–8014, 1999.

120. RA Houghton. Proc Natl Acad Sci USA 82:5131–5135, 1985.

121. PJ Comina, AK Beck, D Seebach. Org Process Res Dev 2:18–26, 1998.

122. S Yao, S Sabby, RG Hazell, KA Jorgensen. Chem Eur J 6:2435–2448, 2000.

123. (a) B Hinzen, SV Ley. J Chem Soc Perkin Trans I 1997:1907–1908. (b) SV Ley, MH Bolli, B Hinzen, A-G Gervois, BJ Hall. J Chem Soc Perkin Trans I 1998: 2239–2240.

124. NE Leadbeater, KA Scott. J Org Chem 65:4770–4772, 2000.

125. SM Leinonen, DC Sherrington. J Chem Res (S) 1999:572–573.

126. (a) A Natarajan, JS Madalengoitia. Tetrahedron Lett 41:5783–5787, 2000. (b) A Natarajan, JS Madalengoitia. Tetrahedron Lett 41:5788–5793, 2000.

127. S Nagayama, E Massahiro, S Kobayashi. J Org Chem 63:6094–6095, 1998.

128. S Kobayashi, M Endo, S Nagayama. J Am Chem Soc 121:11229–11230, 1999.

129. CE Song, JW Yang, H-J Ha. Tetrahedron Asymm 8:841–844, 1997.

130. CE Song, CR Oh, SW Lee, S-g Lee, L Canali, DC Sherrington. J Chem Soc Chem Commun 1998:2435–2436.

131. B Corain, AA D'Archivio, L Galantini, K Jeràbek, M Kràlik, S Lora, G Palma, M Zecca. Roy Soc Chem 216 (Supported Reagents and Catalysts in Chemistry) 1998: 182.

132. RA Taylor, BP Santora, MR Gagné. Org Lett 2:1781–1783, 2000.

133. Y Masaki, N Tanaka, T Miura. Tetrahedron Lett 39:5799–5802, 1998 and references therein.
134. Y Masaki, N Tanaka, T Miura. Synlett 2000:406–408.
135. SK Taylor, MG Dickinson, SA May, DA Pickering, PC Sadek. Synthesis 1998: 1133–1136.
136. M Rottlander, P Knochel. J Comb Chem 1:181–183, 1999.
137. Y Kondo, T Komine, M Fujinami, M Uchiyama, T Sakamoto. J Comb Chem 1: 123–126, 1999.
138. RT Taylor, DB Crawshaw, JB Paperman, LA Flood, RA Cassel. Macromolecules 14:1134–1135, 1981.
139. BJ Cohen, MA Kraus, A Patchornik. J Am Chem Soc 103:7620–7629, 1981.
140. BA Kulkarni, A Ganesan. Angew Chem Int Ed Engl 36:2454–2455, 1997.
141. RW Alder, PS Bowman, WRS Steele, DR Winterman. J Chem Soc Chem Commun 1968:723–724.
142. C Lensink, SK Xi, LM Daniels, JG Verkade. J Am Chem Soc 111:3478–3479, 1989.
143. R Schwesinger, H Schlemper. Angew Chem Int Ed Engl 26:1167–1169, 1987.
144. S Goumri, O Guerret, H Gornitzka, JB Cazaux, D Bigg, F Palacios, G Betrand. J Org Chem 64:3741–3744, 1999.
145. F Palacios, D Aparicio, JM de los Santos, A Baceiredo, G Bertrand. Tetrahedron 56:663–669, 2000.

Index

T - #0029 - 111024 - C0 - 229/152/23 - PB - 9780367396534 - Gloss Lamination